Lecture Notes in Mathematics 2230

More information about this series at http://www.springer.com/series/304

Enno Keßler

Supergeometry, Super Riemann Surfaces and the Superconformal Action Functional

Springer

Enno Keßler
Max-Planck-Institut für
Mathematik in den Naturwissenschaften
Leipzig, Germany

ISSN 0075-8434 ISSN 1617-9692 (electronic)
Lecture Notes in Mathematics
ISBN 978-3-030-13757-1 ISBN 978-3-030-13758-8 (eBook)
https://doi.org/10.1007/978-3-030-13758-8

Mathematics Subject Classification (2010): Primary: 58A50; Secondary: 32C11, 17C70, 81T60, 83E30

This Springer imprint is published by the registered company Springer Nature Switzerland AG.
The registered company address is: Gewerbestrasse 11, 6330 Cham, Switzerland

Preface

This book grew out of my dissertation thesis accepted by the Universität Leipzig in early 2017. For the book at hand, I have rewritten and expanded several chapters. The leading question for the thesis was how the action functional of the two-dimensional non-linear supersymmetric sigma model, or spinning string, is related to the geometry of super Riemann surfaces. The necessary tools from supergeometry to answer that question were quite spread out in the literature or did not exist. It was necessary to gather results from quite diverse places in the literature and reformulate them in a common language and fill in the remaining gaps.

Hence the resulting book starts with a general introduction to superalgebra and supergeometry in the first part. The second part, building on the first part, develops the theory of super Riemann surfaces and the superconformal class of U(1)-structures on them. With this preparation at hand, it is possible to explain all symmetries of the two-dimensional non-linear supersymmetric sigma model as supergeometric properties of the superconformal action functional on super Riemann surfaces. With this outline of the text, I hope that the book can be useful not only for mathematicians to learn supergeometry and some of its applications inspired by high energy physics but also for physicists interested in learning supergeometry to apply it in supersymmetric field theories.

I wish to thank various people for their contribution to this project.

I am grateful to Jürgen Jost for proposing this fascinating and challenging topic to me. His advice, tremendous knowledge and mathematical vision guided me towards very interesting research. His patience and trust allowed me to complete this work.

I wish to express my gratitude to Jürgen Tolksdorf. He invested much time and effort for discussions, profound criticism and careful explanations. Without his encouragement and support I probably could not have tackled the problems surrounding the superconformal action functional.

I am indebted to Ron Donagi for accepting to write the second review for my thesis and for various helpful comments on my work. The anonymous referees for the book edition gave many detailed comments and valuable hints.

I greatly benefited from discussions and conversations with my colleagues and working group. It is hard to grasp what I have learned from them during the past years. In particular, I would like to thank Pierre-Yves Bourguignon, Jakob Bullerjahn, Michael Gransee, Julian Hofrichter, Andreas Kübel, Eric Noeth, Peter Schlicht and Ruijun Wu.

The Max Planck Institute for Mathematics in the Sciences offered me a friendly, open-minded and creative work environment. The library and staff are very kind and helpful; special thanks to Antje Vandenberg.

I want to thank the International Max Planck Research School Mathematics in the Sciences for financial support which made my work possible in the first place. The research leading to these results has also received funding from the European Research Council under the European Union's Seventh Framework Programme (FP7/2007–2013)/ERC grant agreement no 267087.

Finally, and just as important, I wish to thank my family and my friends. Without their love and care I could not have finished this work.

Leipzig, Germany Enno Keßler
December 2018

Contents

List of Symbols

(f, g)	For $f: L \to M$ and $g: L \to N$ the map $(f, g): L \to M \times N$ is given by the universal property of the cartesian product 45, 52
$\langle v, m, w \rangle$	Application of the (multi-) vectors v and w to the (multi-) linear map m from the left and right, respectively 16
Ad	Adjoint action of a Lie group on itself 85
ad	Adjoint representation of a Lie group on its Lie algebra 86
$\mathrm{Aut}_R(E)$	Group of R-linear, invertible endomorphisms of the R-module E 15
Ber E	Berezinian line bundle associated with the vector bundle E 128
Ber M	Berezinian of the matrix M, generalization of determinant 19
$b^\vee$	Bilinear form on $E^\vee$ dual to b on E 34
χ	Gravitino field, an even differential form with values in S 188
$\mathcal{D}$	Holomorphic distribution $\mathcal{D} \subset TM$ of complex rank $0\|1$ of a super Riemann surface M 140
$d(\cdot)$	Cohomological degree of an element of a tensor algebra, symmetric algebra or exterior algebra 29
δ_γ	Quantization map 198, 249, 285
$\mathrm{Der}_R(A, E)$	Set of R-linear derivations from the R-algebra A into the A-module E 31
df	Differential $df: TM \to f^*TN$ of a map $f: M \to N$ between supermanifolds 73
$\not{D}, \not{D}^{LC}$	Twisted Dirac operator, defined on twisted spinors, i.e. sections of $S^\vee \otimes \varphi^* TN$, with respect to an unspecified connection on $S^\vee$ or Levi-Civita connection respectively 261
$\mathrm{div}_b\, X$	Divergence of the vector field X with respect to the Berezinian form b 129

$\Omega_R(A)$	A-module of R-linear differentials, $\mathrm{Der}_R(A)^\vee$ 31
ω	Connection form with values in a Lie algebra 99
$\Omega^k(M, E)$	Differential forms of degree k on the supermanifold M with values in the vector bunlde E over M 75
$\Omega^k(M)$	Differential forms of degree k on the supermanifold M 73
$\Omega^k(P, V)_{hor}^{(G,\rho)}$	Sheaf of horizontal, G-equivariant differential k-forms on P with values in V 102
$\mathrm{O}_R(m\|2n)$	Group of orthogonal matrices 35
$\mathfrak{o}(m\|2n)$	Lie algebra of the orthogonal group $\mathrm{O}(m\|2n)$ 90
$p(\cdot)$	Parity of an element of a superring or supermodule. The value may be 0 (even) or 1 (odd) 14, 15
$\mathbb{P}_B^{m\|n}$	Relative projective superspace of dimension $m\|n$ 64
$P\chi$	The $\frac{1}{2}$-part of the gravitino χ 198, 249, 286
ΠE	Supermodule E with reversed parity 17
$\mathbb{P}^{m\|n}$	Projective superspace of dimension $m\|n$ 51
$\mathbb{P}_{\mathbb{C}}^{m\|n}$	Complex projective superspace of dimension $m\|n$ 121
$Q\chi$	The $\frac{3}{2}$-part of the gravitino χ 198, 249, 286
R^∇	Curvature tensor of the covariant derivative ∇ 78
T^∇	Torsion tensor of the covariant derivative ∇ 78
$\mathbb{R}^{m\|n}$	Real linear supermanifold of dimension $m\|n$ 43
$R^{m\|n}$	Free R-supermodule of rank $m\|n$ 17
S	Spinor bundle on $T\|M\|$, $S \otimes S = T\|M\|$ 186, 285
$S(E)$	Symmetric algebra of E 30
$\mathrm{Sec}(E)$	Sheaf of sections of the fiber bundle E 69
$S\mathbb{H}$	Upper half plane in $\mathbb{C}^{1\|1}$ 142
$\mathrm{Sp}_R(2m\|n)$	Group of supersymplectic matrices 35
$SR^N(\psi)$	Contraction of the curvature tensor of the target with ψ to the third order 224, 275
M^{sT}	Supertranspose of the matrix M 20
$\mathrm{sTr}\, M$	Supertrace of the matrix M 19
$\mathrm{supp}(s)$	support of the section s of a vector bundle 57
$T(E)$	Tensor algebra of E 29
Tf	Tangent map to the map $f\colon M \to N$, $Tf\colon TM \to TN$ 73
TM	Tangent bundle of the supermanifold M 73
$T^\vee M$	Cotangent bundle of the supermanifold M 73
$\mathrm{U}_S(m\|n)$	Group of unitary matrices 39
$\mathfrak{u}(m\|n)$	Lie algebra of the unitary group $\mathrm{U}(m\|n)$ 90
ζ^{MC}	Maurer–Cartan form, a differerential form on a Lie group with values in its Lie algebra 88

Chapter 1
Introduction

The motivating question for this work is how the superconformal action functional, a supersymmetric extension of the harmonic action functional on Riemann surfaces, is related to super Riemann surfaces and their moduli. The results lay the groundwork for a treatment of the moduli space of super Riemann surfaces via the superconformal action functional and show beautiful similarities to the theory of Riemann surfaces and harmonic maps.

The superconformal action functional appeared in the context of string theory already in the 1970s (see, for example, Deser and Zumino 1976; Brink et al. 1976). The mathematical theory of supergeometry developed around the same time, an early overview is given by Leites (1980). Super Riemann surfaces, a supergeometric analogue of Riemann surfaces, appeared only a little later and their significance for string theory was realized (Friedan 1986; D'Hoker and Phong 1988). It turned out that super Riemann surfaces behave quite similar to Riemann surfaces in their algebraic and differential geometry (LeBrun and Rothstein 1988; Crane and Rabin 1988). However, there seems to be a gap in the understanding of how the superconformal action functional is related to super Riemann surfaces, and for a lot of things that are known, explicit proofs are missing. It is the goal of this book to build a bridge between the mathematical theory of supergeometry and super Riemann surfaces with the superconformal action functional motivated by physics.

E. Keßler, *Supergeometry, Super Riemann Surfaces and the Superconformal Action Functional*, Lecture Notes in Mathematics 2230,
https://doi.org/10.1007/978-3-030-13758-8_1

1.1 The Supersymmetric Extension of the Harmonic Action Functional

Before turning to its supersymmetric extension let us first take a look at the harmonic action functional. For a compact closed surface $|M|$ with Riemannian metric g and a function $\varphi\colon |M| \to \mathbb{R}$ the harmonic action functional is given by

$$A(\varphi, g) = \int_{|M|} \|d\varphi\|^2 \, dvol_g. \tag{1.1.1}$$

The analytic and geometric properties of the harmonic action functional and its non-linear siblings are very well studied (see, for example, Jost 2001, 2006 and references therein). Its main properties are:

- Conformal invariance: That is, $A(\varphi, \lambda^2 g) = A(\varphi, g)$ for all non-vanishing functions $\lambda\colon |M| \to \mathbb{R}$. Hence the harmonic action functional is a functional on the space of conformal classes of metrics rather than on the space of metrics. But in the two dimensional case conformal classes of metrics coincide with almost complex and complex structures on the manifold $|M|$.
- The energy-momentum tensor T is the variation of $A(\varphi, g)$ with respect to g. Thus

$$T = \frac{\delta A}{\delta g}$$

  is a symmetric 2-tensor on $|M|$. The energy-momentum tensor T is trace free because of the conformal invariance of $A(\varphi, g)$. From the diffeomorphism invariance it follows that T is divergence free onshell (that is, using the Euler–Lagrange equations for φ). Consequently, the energy-momentum tensor T can be identified with a holomorphic quadratic differential and as a cotangent vector to the moduli space of Riemann surfaces.

It turns out that the harmonic action functional provides an efficient framework to describe the moduli space of Riemann surfaces, see Wolf (1989), Tromba (1992), and Jost and Yau (2010). For example, by help of this framework, it is possible to prove the Teichmüller theorem stating that the Teichmüller space—an infinite cover of the moduli space of Riemann surfaces of genus p—is a ball of real dimension $6p - 6$.

In superstring theory and supergravity one studies a supersymmetric extension of the harmonic action functional where respectively the field φ and the Riemannian metric g on $|M|$ get as superpartners the spinor field ψ and the gravitino χ. See, for example, Deser and Zumino (1976) and Brink et al. (1976). Choose a spin structure on $|M|$ and a spinor bundle S on $|M|$. Let ψ be a section of $S^\vee$ and χ a section of

$T^\vee|M| \otimes_{\mathbb{R}} S$. Then the superconformal action functional is given by

$$A(\varphi, \psi, g, \chi) = \int_{|M|} \Big(\| d\varphi\|_g^2 + g_S^\vee\left(\psi, D\!\!\!/\psi\right) + 2\left\langle \psi, \gamma^a\gamma^b\chi_a\right\rangle \partial_{x^b}\varphi + \frac{1}{2} g_S\left(\chi_a, \gamma^b\gamma^a\chi_b\right) g_S^\vee\left(\psi, \psi\right)\Big)\, dvol_g. \tag{1.1.2}$$

Here, g_S denotes the induced metric on S. For the non-vanishing of the Dirac-term it is however necessary that (the components of) the fields ψ and χ are odd, that is for any commutation of those fields in a product an additional factor of -1 appears. In physics, the additional sign factor is motivated by the fact that the odd fields represent fermions and consequently follow a Fermi statistic and anti-commutation relation.

The superconformal action functional does not only come with more fields and terms but also with more symmetries:

- Conformal invariance: $A(\varphi, \psi, \lambda^2 g, \chi) = A(\varphi, \psi, g, \chi)$
- Super Weyl invariance: $A(\varphi, \psi, g, \chi + \gamma s) = A(\varphi, \psi, g, \chi)$, for any spinor $s \in \Gamma(S)$
- Supersymmetry:

$$\begin{aligned}
\mathrm{susy}_q\, \varphi &= \langle \psi, q\rangle\\
\mathrm{susy}_q\, \psi &= -\left(\partial_{x^k}\varphi + \langle \psi, \chi_k\rangle\right)\gamma^k \vee q\\
\mathrm{susy}_q\, f_a &= 2g_S\left(\gamma^b\chi(f_a), q\right) f_b\\
\mathrm{susy}_q\, \chi(f_a) &= -\nabla^{LC}_{\partial_{x^a}} q - g_S\left(\gamma^b\chi_b, \chi(\mathrm{I}\, f_a)\right)\mathrm{I}q
\end{aligned}$$

Here f_a is a g-orthonormal frame, q a section of S and $\vee q$ its metric dual.

Again, to show the invariance of the action functional under these symmetries it is crucial that the fields ψ and χ as well as the super Weyl parameter s and the supersymmetry parameter q are odd in the way mentioned above.

Natural questions related to the functional (1.1.2) are:

- What is the geometrical significance of the odd quantities and the additional sign rule?
- Do the functional (1.1.2) and its symmetries have a similar geometric meaning as the harmonic functional (1.1.1) has for Riemann surfaces?
- What is the geometric interpretation of the supersymmetry transformation?

At a first glance, these questions seem to be answered already. The additional signs can be incorporated in geometry using the language of supergeometry. The appropriate geometrical setup for the superconformal action functional (1.1.2) are so called "super Riemann surfaces". However, delving deeper into these theories, it becomes clear that the details are hard to find and that proofs for longstanding claims are missing. For example, why does a pair of a Riemannian metric g and gravitino χ on a surface $|M|$ describe a super Riemann surface? And how is it possible that the sections ψ, χ and q are odd, even though they live on a purely even object $|M|$? It is the goal of this book to formulate the existing results in a common language, refine the existing results, and fill in the gaps between the world of supergeometry, super Riemann surfaces, their moduli and the action functional (1.1.2).

It should be mentioned that it has been tried to study the action functional (1.1.2) and special cases thereof in the realm of "classical", that is non-super, differential geometry. A prominent example in this direction are Dirac-harmonic maps that have been introduced in Chen et al. (2006). In the framework of Dirac-harmonic maps, anti-commuting variables can be avoided at the expense of using a different Clifford algebra. We have later introduced Dirac harmonic maps with gravitinos in Jost et al. (2018a) and studied its symmetries in Jost et al. (2018b) and further analytic properties have been derived in Jost et al. (2017b). However, in order to obtain a geometric interpretation of supersymmetry, it is necessary to use supergeometry, as was argued in Keßler and Tolksdorf (2016).

1.2 Super Differential Geometry

The basic idea of all supermathematics is to replace commutativity by graded commutativity or supercommutativity. The best known example for graded commutativity is the Grassmann algebra $\bigwedge_n$, an algebra with generators η^α for $\alpha = 1, \ldots, n$ such that

$$\eta^\alpha \eta^\beta = -\eta^\beta \eta^\alpha .$$

The degree induces a decomposition $\bigwedge_n = \bigwedge_n^0 \oplus \bigwedge_n^1$ of the Grassmann algebra into an even part and an odd part, such that elements of the odd part anti-commute with each other.

Even though graded commutative algebras have been studied even earlier, the pioneer of supermathematics is Felix Berezin, who was the first to study systematically the application of anti-commutative variables in the context of the second quantization (see Berezin 1987). Different constructions in supergeometry carry his name. In the 1970s the use of anti-commuting variables was adopted in high-energy physics for supersymmetric field theories, partly due to Wess and Zumino (1974). The interaction with the physics community led to a rise of supermathematics around 1980, see, for example, Leites (1980), Kostant (1977) or Manin (1988). At that time it was shown that large parts of algebra and geometry allow for a supercommutative generalization.

In this work we will mainly be concerned with super differential geometry, where smooth manifolds are extended by anti-commuting variables. Surprisingly, different definitions of supermanifolds exist. Even though the different approaches should coincide in principle, it is not always easy to transfer results from one approach to another. Different choices of sign conventions may add to the confusion. In this book we use the approach by Berezin (1987), Kostant (1977), and Leites (1980) and the sign conventions given in Deligne and Freed (1999a). In Chap. 3 we will give several references to other approaches to supergeometry.

In the Berezin–Kostant–Leites approach to supermanifolds, the structure sheaf of an ordinary manifold is extended by nilpotent, anti-commuting functions. Even though the sheaf theoretic approach to manifolds might look unfamiliar, it generalizes easily to supermanifolds. The model space $\mathbb{R}^{m|n}$ with m even and n odd dimensions is given by the topological space $\mathbb{R}^m$ together with the structure sheaf

$$\mathcal{O}_{\mathbb{R}^{m|n}} = C^\infty(\mathbb{R}^m, \mathbb{R}) \otimes_{\mathbb{R}} \bigwedge\nolimits_n .$$

Supermanifolds are ringed spaces which are locally of the form $\mathbb{R}^{m|n}$. Many geometric concepts have been carried over successfully to supermanifolds.

In the case that the odd dimension n of the supermanifold is zero, supermanifolds are nothing but ordinary manifolds. However, this is not desired for the applications we have in mind. Remember that for the superconformal action functional odd fields on the smooth manifold $|M|$ are needed. In order to incorporate odd fields into the Berezin–Kostant–Leites approach to supermanifolds, it is necessary use families of supermanifolds over an arbitrary base supermanifold together with a concept of base change. As the base may possess odd directions, a family of smooth manifolds without odd dimensions has the necessary odd functions in the structure sheaf. The idea of families of supermanifolds can be found already in Leites (1980) and more prominently in Deligne and Morgan (1999).

Part I of this book is devoted to present the super differential geometry of families of supermanifolds with base change. The selection of topics is determined by the applications which are developed in Part II and includes most notably principal bundles on supermanifolds, complex supermanifolds and the Berezin integral. The theory of families of supermanifolds contains the special case of trivial families. Consequently, very little knowledge on supergeometry has to be assumed and Part I is mostly self-contained.

The theory and results of super differential geometry of families of supermanifolds presented here look quite similar to their classical differential geometric counterpart. The main reason is that in many cases commutativity can be replaced by graded commutativity with the sole expense of additional sign prefactors. Additionally, for the applications we have in mind it is sufficient to consider even generalizations of, for example, almost complex structures and covariant derivatives. From a purely supergeometric perspective it would be consistent to consider, for example, almost complex structures of any parity, yet this is beyond the scope of this work.

A phenomenon unique to families of supermanifolds are different embeddings of underlying even manifolds. The reduced space of a single supermanifold is an ordinary manifold that embeds into the supermanifold. An appropriate generalization of the reduced space to families of supermanifolds is the underlying even manifold, defined in Sect. 3.3. An underlying even manifold is a family of supermanifolds of dimension $m|0$ that embeds into a supermanifold of dimension $m|n$ with the same topology. The non-uniqueness of the embedding of such an underlying even manifold to a supermanifold plays an important role in Part II.

1.3 Super Riemann Surfaces

Super Riemann surfaces are particular complex supermanifolds of dimension $1|1$ with additional structure in their tangent bundle. One of the first appearances of the definition accepted nowadays is in Friedan (1986) and in Baranov et al. (1987). It was quickly recognized that super Riemann surfaces share a lot of beautiful properties with Riemann surfaces. Their moduli spaces have been studied from different perspectives, see, for example, Giddings and Nelson (1988), LeBrun and Rothstein (1988), Natanzon (2004), Sachse (2009), and Donagi and Witten (2015).

The concept of a super Riemann surface was originally motivated by its application to physics, notably in superstring theory and supergravity. In the early 1980s superspaces were used to study the superconformal action functional or "spinning string", see, for example, Howe (1979) and D'Hoker and Phong (1988). The geometric properties of the superspaces in question are given by a supervielbein and a supercovariant derivative with certain "torsion constraints". Metric and gravitino fields appeared as certain coefficients of the vielbein with respect to odd coordinates. It was shown later on in Giddings and Nelson (1988) that some of the torsion constraints can be interpreted as integrability conditions for a super Riemann surface. However, the relation between super Riemann surfaces, metrics and gravitinos and the superconformal action functional was never fully clarified. In Part II of this work, we will explain their relation in the language of families of supermanifolds.

Our starting point will be the theorem of Giddings and Nelson that interprets super Riemann surfaces as an integrable reduction of the structure group of a $2|2$-dimensional real supermanifold. The concept of the reduction of the structure group allows for a formulation of supermetrics that are compatible with the super Riemann surface as $\mathrm{U}(1)$-structures. As in classical differential geometry a $\mathrm{U}(1)$-connection on a super Riemann surface is completely determined by its torsion tensor. Furthermore, the integrability conditions from the theorem of Giddings and Nelson are encoded in the torsion tensor of such a $\mathrm{U}(1)$-connection. However, the supergravity torsion constraints can in general not be realized globally on an arbitrary super Riemann surface.

We propose a global definition of metric and gravitino on an underlying even manifold of a super Riemann surface that depends on the choice of $\mathrm{U}(1)$-structure. Variations of the $\mathrm{U}(1)$-structure lead to conformal and super Weyl transformations of the metric and gravitino, whereas an infinitesimal change of the embedding of the underlying even manifold leads to supersymmetry of metric and gravitino. As suggested by the supergravity approach, a super Riemann surface is completely determined by metric and gravitino fields.

For any map $\Phi\colon M \to N$ from a super Riemann surface and U(1)-metric m on M the Berezin integral

$$A(m, \Phi) = \int_M \| d\Phi|_{\mathcal{D}} \|_m^2 [dvol_m]$$

defines an action functional. Similar to the case of harmonic maps on Riemann surfaces, it can be shown that $A(m, \Phi)$ depends only on the super Riemann surface structure and not on the choice of U(1)-metric m. The map Φ determines fields φ and ψ on an underlying even manifold and it can be shown that $A(\varphi, g, \psi, \chi)$ coincides with $A(m, \Phi)$. Again, an infinitesimal change of the embedding of the underlying even manifold yields supersymmetry of φ and ψ. Hence, all symmetries of $A(\varphi, g, \psi, \chi)$ obtain a supergeometric interpretation.

The description of super Riemann surfaces in terms of metrics and gravitinos should lead to a new description of the moduli space of super Riemann surfaces. It is known that the moduli space of super Riemann surfaces is a complex supermanifold of dimension $3p - 3|2p - 2$ for genus $p \geq 2$. Its reduced space is the spin moduli space, which itself is a finite cover of the moduli space of super Riemann surfaces. Consequently, the odd directions of the supermoduli space must be encoded in the gravitino. Indeed, a family of super Riemann surface for which the gravitino cannot be gauged to zero must be a non-trivial family over a base with odd dimensions and correspond to higher points of the supermoduli space. We show that infinitesimal deformations of the super Riemann surface correspond to certain infinitesimal deformations of the metric and gravitino. By its symmetries, the superconformal action functional can be interpreted as a functional on the moduli space of super Riemann surfaces. The fact that it can be consistently formulated for a non-linear target space and the analogy to the case of Riemann surfaces leads to the hope that the superconformal action functional might be a useful tool in the further study of the moduli space of super Riemann surfaces.

1.4 Main Results

Certainly, large parts of what is presented in this book is known to mathematicians and physicists in some form. In addition to a detailed and organized presentation, the following main results are rather new:

- A super Riemann surface is determined by a metric, spinor bundle, and gravitino field on an underlying even manifold. Conversely, a super Riemann surface determines a metric, spinor bundle, and gravitino on an underlying even manifold which are unique up to conformal and super Weyl transformations. Supersymmetry of metric and gravitino is given by a change of the embedding of the underlying even manifold. This demonstrates the usefulness of the concept of underlying even manifolds as introduced here.

- On a super Riemann surface with given U(1)-structure the compatible covariant derivatives are classified by a certain "small" part of the torsion tensor. The "larger" part of the torsion tensor is given by integrability conditions. In order to formulate the results, we develop the theory of connections on principal bundles over supermanifolds in the ringed-space approach.
- We present a detailed proof that the superconformal action functional given in Eq. (1.1.2) is actually a Berezin integral on a super Riemann surface. The different fields in the superconformal action functional are interpreted on the super Riemann surfaces. The symmetries of the superconformal action functional allow the latter to be interpreted as an action functional on the moduli space of super Riemann surfaces. This leads to a reformulation of insights on the infinitesimal structure of the moduli space of super Riemann surfaces, and opens the possibility for a Teichmüller theorem in the setting of super Riemann surfaces.

Material drawn from this book has been used in the research articles (Jost et al. 2017a; Keßler 2016; Keßler and Tolksdorf 2016).

1.5 Organization

This book consists of two parts. In the first part we describe the differential geometry of families of supermanifolds, and the second part is devoted to the theory of super Riemann surfaces and the superconformal action functional.

In the first part we have put emphasis on consistent notation and conventions. This will be indispensable for the second part, where a lot of calculations have to be performed, also in local frames and coordinates. Furthermore a major theme is to generalize the geometry on supermanifolds to families of supermanifolds and to assure functoriality under base change.

The first part consists of Chaps. 2–8. In Chap. 2 we treat the linear algebra of free and finitely generated supermodules over superrings. In this chapter, the results are often comparable to the well-known linear algebra over commutative rings. However, the additional signs and notation have to be chosen with care in order to consistently prepare for the tensor calculus on supermanifolds.

Chapter 3 introduces supermanifolds and families of supermanifolds together with justification for this generalization. Furthermore, the concept of underlying even manifold is introduced which is crucial for the second part.

Vector bundles as well-understood examples of fiber bundles on supermanifolds are studied in Chap. 4. As a first example of connections in a fiber bundle over a supermanifold, linear connections in vector bundles are described as covariant derivatives.

Super Lie groups are treated in Chap. 5.

Connections in principal bundles over families of supermanifolds are the main subject of Chap. 6. Different descriptions of connections in principal fiber bundles are given and the link to reductions of the structure group and covariant derivatives on associated vector bundles is explained.

An example of supermanifolds with reduced structure group is given by complex supermanifolds (Chap. 7). A generalization of the famous Newlander–Nirenberg theorem shows that smooth families of complex supermanifolds correspond to an integrable reduction of the structure group of the tangent bundle to matrices that commute with an almost complex structure.

Integrals on families of supermanifolds take values in the base of the family. In Chap. 8, we show that any integral on a supermanifold can be reduced to an integral on the underlying even manifold.

The second part studies super Riemann surfaces as smooth families of complex supermanifolds building on the detailed knowledge of super differential geometry obtained in the first part. The description of super Riemann surfaces in terms of metrics and gravitinos is given here as well as the relation to the superconformal action functional.

Chapter 9 lays groundwork, as it describes super Riemann surfaces as certain integrable reductions of the structure group of the frame bundle of the supermanifold. This leads also to a class of $U(1)$-structures on super Riemann surfaces that we will call superconformal class of metrics on the super Riemann surfaces, and can be compared to the conformal class of metrics on a Riemann surface.

Connections on the reduced frame bundles of a super Riemann surface are studied in Chap. 10. Early descriptions of super Riemann surfaces are given in terms of torsion constraints of a supercovariant derivative. In order to connect to this literature, we show that the torsion of a connection on the tangent bundle of a super Riemann surface is related to the integrability conditions.

In Chap. 11, it is shown that the underlying even manifold of a super Riemann surface is a $2|0$-dimensional manifold with a Riemannian metric, a spinor bundle and a gravitino field up to conformal and super Weyl transformations. Supersymmetry of metric and gravitino is identified to be an infinitesimal change of the embedding of the underlying even manifold in the super Riemann surface.

Finally, in Chap. 12 we come to the superconformal action functional. We show that the action functional (1.1.2) can be obtained as the reduction of a Berezin integral on a super Riemann surface to an underlying even manifold. The conformal invariance, super Weyl invariance and supersymmetry of the action functional are explained in terms of the structure of the super Riemann surface. Consequently, the action functional allows for a study of the moduli space of super Riemann surfaces, at least infinitesimally.

Some proofs in Part II rely on long and complicated calculations. In order not to overburden the presentation, those calculations are regrouped in Chap. 13.

The annex consists of two chapters. In Appendix A, known facts on spinors on Riemann surfaces are summarized. Appendix B gives a direct proof of the supersymmetry of the superconformal action functional.

For the convenience of the reader, a List of Symbols can be found after the table of contents and an index is placed at the end of the book.

Part I
Super Differential Geometry

Chapter 2
Linear Superalgebra

Principle 2.0.1 *The guiding principle in all supermathematics is that every object has an additional $\mathbb{Z}_2$-grading or parity. Whenever an odd object in any operation is passed over another odd object, it acquires an additional factor* -1.

The goal of this chapter is to describe the necessary pieces of linear superalgebra. A good understanding of linear superalgebra is necessary to understand the geometry to be treated in later chapters.

To realize the Principle 2.0.1 in linear algebra, one has to grade all objects, that is, declare which part is even and which part is odd. The sign rule of supercommutativity then forces us to keep track of the order of certain objects. Consequently, there will be left- and right modules, left- and right-coordinates, different types of matrices etc. which are not different in classical, non-super, linear algebra. After the introduction of the general principles and notation for modules, bases and matrices, we turn to more specific topics, like bilinear forms, metrics and almost complex structures.

Of course many of the concepts have been studied before. Important references are Leites (1980), Tuynman (2004), and Manin (1988, Chapter 3). Many proofs and concepts from "classical" linear algebra apply directly to the super case. As a consequence, proofs will be brief or omitted altogether. However, care is taken to introduce notation and concepts that will be helpful later on. The super summation convention formulated in Principle 2.4.5 will be particularly helpful for the tensor calculus in super differential geometry in later chapters.

We are following here the sign conventions given in Deligne and Freed (1999a).

E. Keßler, *Supergeometry, Super Riemann Surfaces and the Superconformal Action Functional*, Lecture Notes in Mathematics 2230,
https://doi.org/10.1007/978-3-030-13758-8_2

2.1 Superrings and Algebras

Definition 2.1.1 A ring R is called $\mathbb{Z}_2$-graded if it possesses a direct sum decomposition $R = R_0 \oplus R_1$ in the category of groups such that

$$R_i \cdot R_j \subseteq R_{i+j}.$$

Elements $r \in R_0$ are called even, elements $r \in R_1$ are called odd. Even and odd elements are also called homogeneous, and we define the parity function p for homogeneous elements such that $p(r) = i$ for $r \in R_i$.

A supercommutative ring, or short a superring, is a $\mathbb{Z}_2$-graded, unital ring $R = R_0 \oplus R_1$ such that for all homogeneous elements

$$a \cdot b = (-1)^{p(a)p(b)} b \cdot a.$$

Every ring homomorphism between supercommutative rings preserves the $\mathbb{Z}_2$-grading.

The elements of the odd part R_1 are nilpotent. Let us denote by $\mathcal{I}^{nil}$ the ideal of nilpotent elements in R, and by

$$R_{red} = {}^{R}\!/_{\mathcal{I}^{nil}}$$

the reduced ring.

A superalgebra A over R is a superring homomorphism $R \to A$.

Example 2.1.2 Any real Grassmann algebra $\bigwedge_n$, is a superalgebra over $\mathbb{R}$. It is generated by n linearly independent generators $\eta^1, \ldots, \eta^n$ such that for any $\alpha, \beta \in \{1, \ldots, n\}$

$$\eta^\alpha \eta^\beta = -\eta^\beta \eta^\alpha.$$

The even, respectively odd, part of $\bigwedge_n$ are the linear span of monomials with even/odd number of the generators. The reduced ring of $\bigwedge_n$ is $\mathbb{R}$. Similarly, the complex Grassmann algebra $\bigwedge_n^{\mathbb{C}}$ is a superalgebra over $\mathbb{C}$.

Example 2.1.3 Any commutative ring is a purely even superring. Conversely, for every superring $R = R_0 \oplus R_1$, the even part R_0 is a commutative subring, with nilpotent elements whenever $R_1 \neq \{\}$.

2.2 Modules

Definition 2.2.1 A left-supermodule (respectively right-module) E over a superring R is a left-module (respectively right-module) over R with subgroups E_0 and E_1 with respect to the additive group structure, such that

- $E = E_0 \oplus E_1$ and
- $R_a E_b \subseteq E_{a+b}$ (resp. $E_b R_a \subseteq E_{a+b}$).

We say $e \in E$ is pure of parity $p(e)$ if $e \in E_{p(e)}$.

The module E is called a superbimodule if it is a right- and a left-supermodule and in addition the left- and right-supermodule structures are compatible. That is the respective subgroups of even and odd elements coincide and the multiplication fulfils

$$r \cdot e = (-1)^{p(r)p(e)} e \cdot r.$$

In the following, any module will tacitly be assumed to be equipped with a superbimodule structure.

Definition 2.2.2 Let E and F be R-supermodules and $l \colon E \to F$ a map. l is called linear, if for all $e, e' \in M$ and $r \in R$

$$l(e + e') = l(e) + l(e'),$$
$$l(e \cdot r) = l(e) \cdot r.$$

We say that l is pure of parity $p(l)$ if for all pure $e \in E$ it holds that $l(e)$ is pure of parity $p(l(e)) = p(l) + p(e)$.

Remark 2.2.3 In Definition 2.2.2 we have actually used right-linearity. With this convention it holds that

$$l(r \cdot e) = (-1)^{p(e)p(r)} l(e \cdot r) = (-1)^{p(e)p(r)} l(e) \cdot r = (-1)^{p(l)p(r)} r \cdot l(e).$$

We prefer not to introduce left-linear maps to avoid confusion and to adhere strictly to the Principle 2.0.1. Thus scalars will be moved out to the left of a linear map by the above formula.

Definition 2.2.4 The set of linear maps $l \colon E \to F$ can be endowed in an obvious way with the structure of an R-module. This module is called $\mathrm{Hom}_R(E, F)$. The module $\mathrm{Hom}_R(E, R)$ is also denoted by $E^\vee$ and called dual space of E. The module $\mathrm{Hom}_R(E, E)$ is also denoted by $\mathrm{End}_R(E)$ and the group of invertible elements in $\mathrm{End}_R(E)$ is denoted by $\mathrm{Aut}_R(E)$.

Definition 2.2.5 Let $E_1, \ldots, E_n$ and F be R-supermodules and $m\colon E_1 \times \cdots \times E_n \to F$ be a map. We call m multilinear if for all $e_i, e_i' \in E_i$ and $r \in R$ we have

$$\begin{aligned} &m(e_1, \ldots, e_{i-1}, e_i + e_i', e_{i+1}, \ldots, e_n) \\ &\quad = m(e_1, \ldots, e_{i-1}, e_i, e_{i+1}, \ldots, e_n) + m(e_1, \ldots, e_{i-1}, e_i', e_{i+1}, \ldots, e_n), \end{aligned}$$

and

$$m(\ldots, e_{i-1}, e_i \cdot r, e_{i+1}, \ldots) = m(\ldots, e_{i-1}, e_i, r \cdot e_{i+1}, e_{i+2}, \ldots),$$
$$m(\ldots, e_n \cdot r) = m(\ldots, e_n) \cdot r.$$

We will introduce the following additional notation in order to reduce sign prefactors in later formulas:

$$\langle e_1, \ldots, e_{i-1}, m, e_i, \ldots, e_n \rangle = (-1)^{p(m)(p(e_1)+\cdots+p(e_{i-1}))} m(e_1, \ldots, e_n). \tag{2.2.6}$$

The set of multilinear maps $m\colon E_1 \times \cdots E_n \to F$ can again be endowed with the structure of an R-module denoted by $\mathrm{Mult}_R(E_1, \ldots, E_n; F)$.

Remark 2.2.7 For linear maps $m\colon E \to F$ and $n\colon F \to G$ the systematic application of the sign rule gives the following formula:

$$\begin{aligned} \langle e, n \circ m \rangle &= (-1)^{p(e)(p(m)+p(n))} (n \circ m)(e) = (-1)^{p(e)(p(m)+p(n))} n(m(e)) \\ &= (-1)^{p(e)(p(m)+p(n))+(p(e)+p(m))p(n)} \langle m(e), n \rangle \\ &= (-1)^{p(m)p(n)} \langle \langle e, m \rangle, n \rangle \end{aligned}$$

Thus one might be tempted to introduce an extra symbol for left-composition.

Definition 2.2.8 Let $E = E_0 \oplus E_1$ be R-modules. Any submodule in the ordinary sense $F \subseteq E$ acquires a grading $F = F_0 \oplus F_1$, turning it into a super subbimodule. Similarly, the quotient $Q = {}^{E}\!/_{F}$ has the structure of a superbimodule.

Definition 2.2.9 Let $E = E_0 \oplus E_1$ and $F = F_0 \oplus F_1$ be R-modules. The direct sum $E \oplus F$ has a canonical grading given by

$$(E \oplus F)_0 = E_0 \oplus F_0, \qquad (E \oplus F)_1 = E_1 \oplus F_1.$$

Similarly, the tensor product $E \otimes F$ is given the structure of a R-superbimodule via

$$(E \otimes F)_0 = (E_0 \otimes F_0) \oplus (E_1 \oplus F_1), \quad (E \otimes F)_1 = (E_0 \otimes F_1) \oplus (E_1 \otimes F_0).$$

Direct sum and tensor product of R-superbimodules fulfill the usual universal properties. Furthermore the usual laws of associativity and distributivity of tensor products hold. However, commutativity of the tensor product is given by:

$$E \otimes F \to F \otimes E$$

$$e \otimes f \mapsto (-1)^{p(e)p(f)} f \otimes e$$

Definition 2.2.10 Let E be a module over R. We denote by ΠE the module with reversed parity. That is for every element $e \in E$ the same element in ΠE is denoted Πe but it holds $p(\Pi e) = p(e) + 1$. If $m \colon E \to F$ is a linear map we define $\Pi m \colon \Pi E \to \Pi F$ by $\Pi m(\Pi e) = \Pi(m(e))$.

Example 2.2.11 Trivially, the ring R is an R-module. Consequently, also ΠR is an R-module. ΠR coincides with R as a set, but the $\mathbb{Z}_2$-grading differs:

$$(\Pi R)_0 = R_1, \qquad (\Pi R)_1 = R_0.$$

Notice, however, that ΠR is not a superring, as the supercommutativity rule would be violated.

A very important example in the following is the free module of rank $m|n$:

$$R^{m|n} = R^m \oplus \Pi R^n.$$

2.3 Example: The Algebra of Supermatrices

Definition 2.3.1 Let $\mathrm{Mat}_R(m|n \times p|q)$ denote the set of matrices with entries in R of size $(m+n) \times (p+q)$. A matrix M of size $(m+n) \times (p+q)$ with entries in R and a decomposition in blocks

$$M = \begin{pmatrix} A & B \\ C & D \end{pmatrix} \tag{2.3.2}$$

is called a matrix of size $m|n \times p|q$. Here A is a block of size $m \times p$, B a block of size $m \times q$, C a block of size $n \times p$ and D a block of size $n \times q$. We say that M is of even parity, if the blocks A and D consist of even entries and B and C of odd entries. We say that M is odd, if it is the other way around. This gives the set $\mathrm{Mat}_R(m|n \times p|q)$ of matrices of size $m|n \times p|q$ in the obvious way the structure of a free R-module of rank $mp + nq|mq + np$.

As usual, matrices of size $m|n \times p|q$ can be left multiplied with matrices of size $p|q \times r|s$ giving a matrix of dimension $m|n \times r|s$. This multiplication respects the parity, that is, $p(MN) = p(M) + p(N)$.

Let M be an $m|n \times p|q$-dimensional matrix and $v \in R^{p|q}$ a vector understood as a column of elements in R. Then the multiplication from the left with M

$$v \mapsto M \cdot v$$

is a linear map $R^{p|q} \to R^{m|n}$ of the same parity as M. Analogously if $w \in R^{m|n}$ is a vector understood as row of elements in R. Then the multiplication with M from the right

$$w \mapsto w \cdot M$$

is a linear map $R^{m|n} \to R^{p|q}$ of the same parity as M.

Definition 2.3.3 Denote by $\mathrm{Mat}_R(m|n)$ the free R-module $\mathrm{Mat}_R(m|n \times m|n)$ of quadratic matrices. Together with the usual matrix multiplication $\mathrm{Mat}_R(m|n)$ is a R-superalgebra with identity $\mathrm{id}_{m|n}$. Denote the group of even invertible matrices by $\mathrm{GL}_R(m|n)$.

Lemma 2.3.4 *A square matrix $M \in \mathrm{Mat}_R(m|n \times m|n)$ is invertible if and only if its reduction $M_{red} \in \mathrm{Mat}_{R_{red}}(m|n \times m|n)$ is.*

Proof If M is invertible, so is M_{red}. To show the converse, let us first consider the special case $M_{red} = \mathrm{id}_{m|n}$. In that case, $M = \mathrm{id}_{m|n} - N$ for some matrix $N \in \mathrm{Mat}_R(m|n \times m|n)$ such that all entries of N are nilpotent. Consequently, also the matrix N is nilpotent and

$$\left(\mathrm{id}_{m|n} - N\right)^{-1} = \mathrm{id}_{m|n} + \sum_{i>0} N^i.$$

For the general case, M_{red} arbitrary, let $\tilde{M}$ be such that $\tilde{M}_{red} = (M_{red})^{-1}$. Then $M\tilde{M} = \mathrm{id}_{m|n} - N$, for some matrix N with nilpotent entries. Consequently,

$$M^{-1} = \tilde{M}\left(\mathrm{id}_{m|n} + \sum_{i>0} N^i\right).$$

□

Corollary 2.3.5 *An even square matrix is M is invertible if and only if its upper left block A and lower right block D are.*

Proof The reduction of an even matrix M in block form as in Eq. (2.3.2) is given by

$$M_{red} = \begin{pmatrix} A_{red} & 0 \\ 0 & D_{red} \end{pmatrix}.$$

Hence, M is invertible precisely if A_{red} and D_{red} are invertible. Again by Lemma 2.3.4, A_{red} and D_{red} are invertible if and only if A and D are invertible. □

Remark 2.3.6 Notice that Lemma 2.3.4 also holds for odd matrices. An odd matrix is invertible if the off-diagonal blocks B and C are invertible. This implies in particular that the matrix is of dimension $m|m$. Hence in general, odd invertible matrices do not exist.

Definition 2.3.7 Let M be an even invertible square matrix. Define the Berezinian of

$$M = \begin{pmatrix} A & B \\ C & D \end{pmatrix}$$

as

$$\operatorname{Ber} M = \det(A - BD^{-1}C)\det(D)^{-1}.$$

Remark that D is invertible and $\det D$ non-zero.

Lemma 2.3.8 *Let M and N be two invertible matrices. Then it holds that*

$$\operatorname{Ber}(MN) = \operatorname{Ber} M \cdot \operatorname{Ber} N.$$

Proof Let M and N in block form be given by

$$M = \begin{pmatrix} A & B \\ C & D \end{pmatrix} = \begin{pmatrix} 1 & BD^{-1} \\ 0 & 1 \end{pmatrix} \begin{pmatrix} A - BD^{-1}C & 0 \\ C & D \end{pmatrix},$$

$$N = \begin{pmatrix} E & F \\ G & H \end{pmatrix} = \begin{pmatrix} E & 0 \\ G & H - GE^{-1}F \end{pmatrix} \begin{pmatrix} 1 & E^{-1}F \\ 0 & 1 \end{pmatrix}.$$

Hence, it suffices to show the following three cases, all of which follow from elementary calculation and the multiplicativity of the determinant:

- $M = \begin{pmatrix} 1 & B \\ 0 & 1 \end{pmatrix}$, N arbitrary,
- $M = \begin{pmatrix} A & 0 \\ C & D \end{pmatrix}$, $N = \begin{pmatrix} E & 0 \\ G & H \end{pmatrix}$,
- $M = \begin{pmatrix} A & 0 \\ C & D \end{pmatrix}$, $N = \begin{pmatrix} 1 & F \\ 0 & 1 \end{pmatrix}$.

□

Definition 2.3.9 The supertrace is the linear functional $\operatorname{sTr}\colon \operatorname{Mat}_R(m|n) \to R$ given by

$$\operatorname{sTr} M = \operatorname{Tr} A - (-1)^{p(M)} \operatorname{Tr} D.$$

Lemma 2.3.10 *Let M and N be two square matrices, then it holds that*

$$\operatorname{sTr} MN = \operatorname{sTr} NM.$$

Proposition 2.3.11 *For any matrix-valued function* $M : \mathbb{R} \to \mathrm{GL}(m|n)$ *we have*

$$\left.\frac{d}{dt}\right|_{t=0} \operatorname{Ber} M = \operatorname{Ber} M_0 \operatorname{sTr}\left(M_0^{-1} \left.\frac{d}{dt} M\right|_{t=0} \right).$$

Proof By replacing M by $M_0^{-1}M$, we may assume that $M_0 = \mathrm{id}$. Let us write M in block form, see Eq. (2.3.2). Then,

$$\begin{aligned}\left.\frac{d}{dt}\right|_{t=0} \operatorname{Ber} M &= \left.\frac{d}{dt}\right|_{t=0} \left(\det\left(A - BD^{-1}C\right)(\det D)^{-1}\right) \\ &= \left(\operatorname{Tr} \left.\frac{d}{dt}\right|_{t=0} A\right) - \left(\operatorname{Tr} \left.\frac{d}{dt}\right|_{t=0} D\right) = \operatorname{sTr} \left.\frac{d}{dt}\right|_{t=0} M.\end{aligned}$$

Here, we have used the formula for the derivative of the determinant. □

Definition 2.3.12 The supertransponse of a block matrix is defined by

$$\begin{pmatrix} A & B \\ C & D \end{pmatrix}^{sT} = \begin{cases} \begin{pmatrix} A^T & C^T \\ -B^T & D^T \end{pmatrix} & \text{if the matrix is of even parity.} \\ \begin{pmatrix} A^T & -C^T \\ B^T & D^T \end{pmatrix} & \text{if the matrix is of odd parity.} \end{cases}$$

2.4 Bases and Coordinates

Definition 2.4.1 We call a supermodule free if it has a linearly independent generating set consisting of homogeneous elements. An ordered, linearly independent generating set is called basis.

If the basis consists of m even and n odd elements, the module is isomorphic to $R^{m|n} = R^m \oplus \Pi R^n$. We say that $R^{m|n}$ has dimension $m|n$. In this case we denote the even vectors of the basis by e_a, $a = 1, \ldots, m$ and the odd basis vectors by e_α, $\alpha = 1, \ldots, n$. The even and the odd basis vectors together are denoted by $e_A = {}_Ae$ where A is the index running over even and odd indices.

Principle 2.4.2 (Indices) *The lower case Latin alphabet is used for indices running over only even entries, whereas the lower case Greek alphabet is used for odd indices. Capital Latin letters are used for indices running over even and odd entries. We denote by* $p(A)$ *the parity of the index, that is,* $p(a) = 0$*, whereas* $p(\alpha) = 1$.

Until the end of this Chap. 2, we will assume that any R-module is free and finitely generated.

Definition 2.4.3 Let $v \in E$ be a vector and $e_A = {}_Ae$ a basis for E. Then there are $v^A \in R$ such that

$$v = \sum_A v^A \, {}_Ae.$$

The coefficients v^A are called left-coordinates of v. We define furthermore ${}^Av = (-1)^{p(v)p(A)} v^A$ and call them right-coordinates of v. With this convention we have that:

$$v = \sum_A v^A \, {}_Ae = \sum_A (-1)^{p(A)(p(v)+p(A))} \, {}_Ae v^A = \sum_A (-1)^{p(A)} e_A \, {}^Av \qquad (2.4.4)$$

Principle 2.4.5 (Summation Convention) *If an index appears twice in an expression, first upper right to an object and then lower left of an object it is summed over. If an capital Latin index A appears twice in an expression, first lower right to an object and then upper left of an object it is summed over with an additional factor of* $(-1)^{p(A)}$. *As an example the above Eq. (2.4.4) can now be written shorter as*

$$v = v^A \, {}_Ae = e_A \, {}^Av.$$

Proposition 2.4.6 *The choice of a basis* e^A *of a free module E corresponds to linear isomorphisms*

$$\begin{aligned} {}_\bullet e \colon E &\to R^{m|n} & \qquad e_\bullet \colon E &\to R^{m|n} \\ v &\mapsto (v^A) & v &\mapsto ({}^Av) \end{aligned}$$

Proposition 2.4.7 *Let E be a free supermodule with basis* e_A. *The dual module* $E^\vee$ *has a basis* ${}^Be = e^B$ *such that*

$$\left\langle {}_Ae, e^B \right\rangle = \delta_A{}^B.$$

This basis is called right dual basis. Here we have used the notation to apply the vector ${}_Ae$ *to the linear form* e^B *from the left, as introduced in Eq. (2.2.6).*

Lemma 2.4.8 *Let* $v \in E$ *be a vector and, respectively,* v^A *and* ${}^Av$ *be its left- resp. right-coordinates with respect to the basis* e_A *of E. It holds for the left dual basis* ${}_Ae$ *that*

$$v^A = \left\langle v, e^A \right\rangle, \qquad {}^Av = {}^Ae(v),$$

that is,

$$v = \left\langle v, e^A \right\rangle {}_Ae = e_A \, {}^Ae(v).$$

Let now $l \in E^\vee$ *be a linear form and define its left- and right-coordinates by* $l = e^A\,{}_Al = l_A\,{}^Ae$. *Then*

$${}_Al = \langle {}_Ae, l\rangle, \qquad\qquad l_A = l(e_A),$$

that is,

$$l = e^A\langle {}_Ae, l\rangle = l(e_A)\,{}^Ae.$$

From the above, we can conclude that

$$\langle v, l\rangle = v^A\,{}_Al, \qquad\qquad l(v) = l_A\,{}^Av.$$

Proof To prove the identities for the coordinates of v we just use the linearity of the right-dual basis. That is

$$\left\langle v, e^A\right\rangle = \left\langle v^B\,{}_Be, e^A\right\rangle = v^B\left\langle {}_Be, e^A\right\rangle = v^B\delta_B{}^A = v_A,$$
$${}^Ae(v) = {}^Ae(e_B\,{}^Bv) = {}^Ae(e_B)\,{}^Bv = (-1)^{p(A)p(B)}\delta^A{}_B\,{}^Bv = {}^Av.$$

The equalities for the coordinates of l follow analogously. Linearity also implies that

$$\langle v, l\rangle = v^A\langle {}_Ae, l\rangle = v^A\,{}_Al,$$
$$l(v) = l(e_A)\,{}^Av = l_A\,{}^Av. \qquad \square$$

2.5 Matrices of Linear Maps

Definition 2.5.1 Let E and F be R-modules with bases e_A and f_B and right dual bases ${}^Ae$ and ${}^Bf$ respectively. Let $l\colon E \to F$ be a linear map and define

$${}^BL_A = {}^Bf(l(e_A)), \qquad\qquad L_A{}^B = \left\langle l(e_A), f^B\right\rangle,$$
$${}^B{}_AL = {}^Bf(\langle {}_Ae, l\rangle), \qquad\qquad {}_AL^B = \left\langle \langle {}_Ae, l\rangle, f^B\right\rangle.$$

By definition the following identities are fulfilled:

$${}^BL_A = (-1)^{p(B)(p(l)+p(A))}L_A{}^B = (-1)^{p(l)p(A)}\,{}^B{}_AL$$
$$= (-1)^{p(l)(p(A)+p(B))+p(A)p(B)}\,{}_AL^B$$

Lemma 2.5.2 *Let $v \in E$. It holds that*

$$l(v) = f_B{}^B L_A{}^A v = L_A{}^B{}_B f^A v,$$
$$\langle v, l\rangle = v^A f_B{}^B{}_A L = v^A{}_A L^B{}_B f.$$

Proof

$$l(v) = l(e_A)^A v = f_B{}^B f(l(e_A))^A v = f_B{}^B L_A{}^A v$$
$$l(v) = l(e_A)^A v = \left\langle l(e_A), f^B\right\rangle{}_B f^A v = L_A{}^B{}_B f^A v$$
$$\langle v, l\rangle = v^A \left\langle {}_A e, l\right\rangle = v^A f_B{}^B f(\langle {}_A e, l\rangle) = v^A f_B{}^B{}_A L$$
$$\langle v, l\rangle = v^A \left\langle {}_A e, l\right\rangle = v^A \left\langle \left\langle {}_A e, l\right\rangle, f^B\right\rangle{}_B f = v^A{}_A L^B{}_B f$$

□

Proposition 2.5.3 *The matrices ${}^\bullet L_\bullet$ and ${}_\bullet L^\bullet$ fit into the following commutative diagrams*

$$\begin{array}{ccc} E & \xrightarrow{\quad l \quad} & F \\ \downarrow {\scriptstyle {}_\bullet e} & & \downarrow {\scriptstyle {}_\bullet f} \\ R^{m|n} & \xrightarrow{\cdot {}_\bullet L^\bullet} & R^{p|q} \end{array} \qquad \begin{array}{ccc} E & \xrightarrow{\quad l \quad} & F \\ \downarrow {\scriptstyle e_\bullet} & & \downarrow {\scriptstyle f_\bullet} \\ R^{m|n} & \xrightarrow{{}^\bullet L_\bullet \cdot} & R^{p|q} \end{array}$$

where $\cdot {}_\bullet L^\bullet$ denotes right multiplication with the appropriate matrix and ${}^\bullet L_\bullet \cdot$ denotes left multiplication with the appropriate matrix including the additional factor $(-1)^\bullet$ from the summation convention. In the first case the elements of $R^{m|n}$ have to be written as row vectors, in the second case as column vectors.

Proposition 2.5.4 *Let E, F, G be modules with bases e_A, f_B, g_C and $l\colon E \to F$ and $m\colon F \to G$ linear maps with matrices L and M with respect to the given bases. Then the matrix N of $m \circ l$ is given by*

$$ {}^C N_A = {}^C M_B{}^B L_A, \qquad {}_A N^C = (-1)^{p(l)p(m)}{}_A L^B{}_B M^C,$$
$$ N_A{}^C = (-1)^{p(m)(p(l)+p(A))} L_A{}^B{}_B M^C, \qquad {}^C{}_A N = (-1)^{p(m)p(A)}{}^C M_B{}^B{}_A L.$$

Proof

$$\begin{aligned} {}^C N_A &= {}^C g(m \circ l(e_A)) = {}^C g(m(l(e_A))) = {}^C g(m(f_B))^B f(l(e_A)) \\ &= {}^C M_B{}^B L_A \end{aligned}$$

$$\begin{aligned} {}_A N^C &= \left\langle \left\langle {}_A e, m \circ l\right\rangle, g^C\right\rangle = (-1)^{p(l)p(m)} \left\langle \left\langle \left\langle {}_A e, l\right\rangle, m\right\rangle, g^C\right\rangle \\ &= (-1)^{p(l)p(m)} \left\langle \left\langle {}_A e, l\right\rangle f^B\right\rangle \left\langle \left\langle {}_B f, m\right\rangle, g^C\right\rangle = (-1)^{p(l)p(m)}{}_A L^B{}_B M^C \end{aligned}$$

$$
\begin{aligned}
N_A{}^C &= \left\langle m \circ l(e_A), g^C \right\rangle = (-1)^{p(m)(p(l)+p(A))} \left\langle \langle l(e_A), m\rangle , g^C \right\rangle \\
&= (-1)^{p(m)(p(l)+p(A))} \left\langle l(e_A), f^B \right\rangle \left\langle \langle_B f, m\rangle , g^C \right\rangle \\
&= (-1)^{p(m)(p(l)+p(A))} L_A{}^B{}_B M^C \\
{}^C{}_A N &= {}^C g(\langle_A e, m \circ l\rangle) = (-1)^{p(m)p(A)}\, {}^C g(m(\langle_A e, l\rangle)) \\
&= (-1)^{p(m)p(A)}\, {}^C g(m(f_B))\, {}^B f(\langle_A e, l\rangle) = (-1)^{p(m)p(A)}\, {}^C M_B{}^B{}_A L \qquad \square
\end{aligned}
$$

Proposition 2.5.5 *The set* End(E) *of endomorphisms of E has the structure of an R-algebra, where the product of the endomorphisms m and l is given by $m \circ l$. Let e_A be a basis of the module E of dimension $m|n$. Mapping an endomorphism to the particular type of matrix*

$$
\begin{aligned}
\mathrm{End}(E) &\to \mathrm{Mat}_R(m|n) \\
l &\mapsto {}^\bullet L_\bullet
\end{aligned}
$$

is an isomorphisms of R-algebras. The map restricts to an isomorphism of groups $(\mathrm{Aut}(E))_0 \to \mathrm{GL}_R(m|n)$. *Recall that the block matrix ${}^\bullet L_\bullet$ is defined to include the sign from the summation convention.*

To understand how the analogous statement for ${}_\bullet L^\bullet$ needs to be formulated see Remark 2.2.7.

2.6 Change of Coordinates

Definition 2.6.1 Let E be an R-module, and ${}_A e$ and ${}_B\tilde{e}$ two bases of E. There are even matrices $b(\tilde{e}, e)$ and $b(e, \tilde{e})$ such that

$$
{}_B\tilde{e} = {}_B b(\tilde{e}, e)^A{}_A e, \qquad {}_A e = {}_A b(e, \tilde{e})^B{}_B\tilde{e}.
$$

These matrices are called matrices of base change. They are mutually inverse to each other. The basis e_A transforms as

$$
\tilde{e}_B = e_A{}^A b(\tilde{e}, e)_B, \qquad e_A = \tilde{e}_B{}^B b(e, \tilde{e})_A.
$$

Lemma 2.6.2 *The left-, respectively right-coordinates transform then by the inverse of the base change. Let v_A, ${}_A v$ be the coordinates with respect to e*

and $\tilde{v}_B$, ${}_B\tilde{v}$ *the coordinates with respect to* $\tilde{e}$. *Then*

$$\tilde{v}^B = v^A{}_A b(e,\tilde{e})^B, \qquad v^A = \tilde{v}^B{}_B b(\tilde{e},e)^A,$$
$$^B\tilde{v} = {}^B b(e,\tilde{e})_A{}^A v, \qquad {}^A v = {}^A b(\tilde{e},e)_B{}^B\tilde{v}.$$

Proof We have

$$v = v^A{}_A e = v^A{}_A b(e,\tilde{e})^B{}_B\tilde{e}.$$

This shows the first equation. The others follow analogously. □

Proposition 2.6.3 *The matrices* $b(e,\tilde{e})$ *and* $b(\tilde{e},e)$ *fit into the following commutative diagrams*

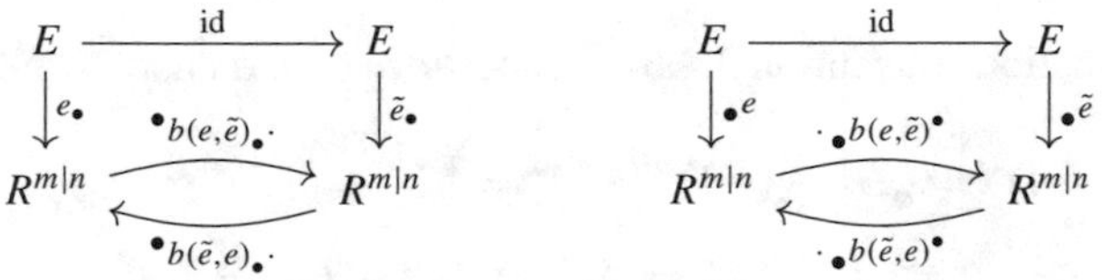

Lemma 2.6.4 *The right dual bases* e^C *and* $\tilde{e}^D$ *to* e_A *and* $\tilde{e}_B$ *respectively are related by the following formulas:*

$$\tilde{e}^D = e^C{}_C b(e,\tilde{e})^D \qquad e^C = \tilde{e}^B{}_B b(\tilde{e},e)^C$$

Proof We are showing that $\tilde{e}^D = e^C{}_C b(e,\tilde{e})^D$ is the right dual base to ${}_B\tilde{e}$:

$$\left\langle {}_B\tilde{e}, \tilde{e}^D \right\rangle = \left\langle {}_B b(\tilde{e},e)^A{}_A e, e^C{}_C b(e,\tilde{e})^D \right\rangle = {}_B b(\tilde{e},e)^A{}_A b(e,\tilde{e})^D = \delta_B{}^D$$ □

Lemma 2.6.5 *Let* E *and* F *be* R*-modules,* $l\colon E \to F$ *a linear map and* e_A, $\tilde{e}_B$ *and* f_C, $\tilde{f}_D$ *two bases for each. Let* L *be the matrix of* l *with respect to the bases* e_A, f_C *and* $\tilde{L}$ *the matrix of* l *with respect to* $\tilde{e}_B$, $\tilde{f}_D$. *Then it holds that*

$$
\begin{aligned}
{}_B\tilde{L}^D &= {}_B b(\tilde{e},e)^A{}_A L^C{}_C b(f,\tilde{f})^D,\\
{}^D\tilde{L}_B &= {}^D b(f,\tilde{f})_C{}^C L_A{}^A b(\tilde{e},e)_B,\\
\tilde{L}_B{}^D &= (-1)^{p(D)(p(A)+p(B))} L_A{}^C{}_C b(f,\tilde{f})^D{}^A b(\tilde{e},e)_B,\\
{}_B{}^D\tilde{L} &= (-1)^{p(B)(p(C)+p(D))}{}^D b(f,\tilde{f})_C{}_B b(\tilde{e},e)^A{}_A{}^C L.
\end{aligned}
$$

Proof We only prove the first equation, using the definition of the particular matrix type:

$$
\begin{aligned}
{}_B\tilde{L}^D &= \left\langle \left\langle {}_B\tilde{e}, l \right\rangle, \tilde{f}^D \right\rangle = {}_B b(\tilde{e}, e)^A \left\langle \left\langle {}_A e, l \right\rangle, f^C \right\rangle {}_C b(f, \tilde{f})^D \\
&= {}_B b(\tilde{e}, e)^A \; {}_A L^C \; {}_C b(f, \tilde{f})^D
\end{aligned}
$$

□

Definition 2.6.6 The supertrace of an endomorphism $l \colon E \to E$ is defined as

$$
\operatorname{sTr} l = \sum_a {}_a L^a - (-1)^{p(l)} \sum_\alpha {}_\alpha L^\alpha ,
$$

where ${}_A L^A$ is the matrix of l with respect to any basis of E. By Lemmas 2.3.10 and 2.6.5 the right-hand side is independent of the chosen basis.

With the summation convention 2.4.5 the supertrace of and endomorphism reads

$$
\begin{aligned}
\operatorname{sTr} l &= {}_a L^a - (-1)^{p(l)} \, {}_\alpha L^\alpha = {}^a L_a + (-1)^{p(l)} \, {}^\alpha L_\alpha \\
&= {}^a{}_a L + {}^\alpha{}_\alpha L = {}^A{}_A L = L_a{}^a - L_\alpha{}^\alpha = L_A{}^A .
\end{aligned}
$$

If furthermore l is written in a particular basis as left multiplication with a block matrix

$$
l \colon \begin{pmatrix} {}^a v \\ {}^\alpha v \end{pmatrix} \mapsto \begin{pmatrix} {}^a L_b & -{}^a L_\beta \\ {}^\alpha L_b & -{}^\alpha L_\beta \end{pmatrix} \begin{pmatrix} {}^b v \\ {}^\beta v \end{pmatrix} \tag{2.6.7}
$$

the supertrace of the block matrix coincides with the supertrace of l.

Definition 2.6.8 The Berezinian of an even endomorphism $l \colon E \to E$ is defined as

$$
\operatorname{Ber} l = \det \left({}_a L^b - {}_a L^\gamma \, {}_\gamma L^\delta \, {}_\delta L^b \right) \left(\det {}_\alpha L^\beta \right)^{-1} = \operatorname{Ber} {}_\bullet L^\bullet .
$$

This is independent of the choice of basis by Lemmas 2.3.8 and 2.6.5.

Notice that the formula for the Berezinian is not independent of the matrix type, that is, position and order of the indices. However, if l is written as left-multiplication with a block matrix as, in Eq. (2.6.7), the Berezinian of the block matrix coincides with the Berezinian of l.

2.7 The Dual Map

Definition 2.7.1 Let E and F be R-modules and let $l: E \to F$ be a linear map. Define the dual map $l^\vee: F^\vee \to E^\vee$ by the requirement

$$\left\langle e, l^\vee(f)\right\rangle = \langle\langle e, l\rangle , f\rangle ,$$

for all $f \in F^\vee$ and $e \in E$.

Lemma 2.7.2 *Let E and F be R-modules with bases e_A and f_B and right dual bases ${}^A e$ and ${}^B f$ respectively. Let $l: E \to F$ be a linear map with matrices ${}^B L_A$, $L_A{}^B$, ${}^B{}_A L$ and ${}_A L^B$. Then the dual map $l^\vee: F^\vee \to E^\vee$ has the same matrix. More explicitly:*

$$l^\vee(f^B) = e^A{}_A L^B,$$
$$\left\langle {}^B f, l^\vee \right\rangle = {}^B L_A{}^A e.$$

Proof We calculate

$$\left\langle e_A, l^\vee(f^B)\right\rangle = \left\langle \langle e_A, l\rangle , f^B\right\rangle = {}_A L^B.$$

The second equation follows analogously. □

Definition 2.7.3 Furthermore, if one introduces the matrices

$$L^B{}_A = (-1)^{p(A)p(B)} L_A{}^B, \qquad {}_A{}^B L = (-1)^{p(A)p(B)}\, {}^B{}_A L.$$

one gets the following

$$l^\vee(f^B) = e^A{}_A L^B = L^B{}_A{}^A e,$$
$$\left\langle {}^B f, l^\vee \right\rangle = e^A{}_A{}^B L = {}^B L_A{}^A e.$$

Remark 2.7.4 (Supertranspose and Dual Map) According to Proposition 2.5.5 the equality $l(e_A{}^A v) = e_A{}^A l_B{}^B v$ can be written as the left multiplication of the column vector ${}^A v$ with the matrix L:

$$l: \begin{pmatrix} {}^a v \\ {}^\alpha v \end{pmatrix} \mapsto \begin{pmatrix} {}^a L_b & -{}^a L_\beta \\ {}^\alpha L_b & -{}^\alpha L_\beta \end{pmatrix} \begin{pmatrix} {}^b v \\ {}^\beta v \end{pmatrix}. \tag{2.7.5}$$

Recall that the sign from the summation convention is embraced in the block matrix. By Lemma 2.7.2, the dual map $l^\vee$ of l has the same matrix. Consequently, for any

$w = f^A{}_A w \in F$, we have in block form:

$$l^\vee : \begin{pmatrix} {}_a w \\ {}_\alpha w \end{pmatrix} \mapsto \begin{pmatrix} {}_a L^b & {}_a L^\beta \\ {}_\alpha L^b & {}_\alpha L^\beta \end{pmatrix} \begin{pmatrix} {}_b w \\ {}_\beta w \end{pmatrix} \tag{2.7.6}$$

Notice that in Eq. (2.7.6) there are no signs due to the summation convention. Using the usual rule

$${}^A L_B = (-1)^{p(A)p(B)+p(L)(p(A)+p(B))} {}_B L^A,$$

one can see that the block matrix in Eq. (2.7.6) is the supertranspose of the block matrix in Eq. (2.7.5).

2.8 Tensor Algebra

Let $A, B, C, \ldots$ be free supermodules with bases $a_A, b_B, c_C, \ldots$. The tensor product $A \otimes B \otimes C \otimes \ldots$ has the basis $a_A \otimes b_B \otimes c_C \ldots$. An element $t \in A \otimes B \otimes C \otimes \ldots$ can have different coordinate expressions, to the left, between two consecutive factors and to the right. We denote them as follows:

$$\begin{aligned} t &= t_{\ldots}{}^{CBA}{}_A a \otimes {}_B b \otimes {}_C c \otimes \cdots = a_A {}^A t^{\ldots CB} \otimes {}_B b \otimes {}_C c \otimes \cdots \\ &= a_A \otimes b_B {}^{BA} t^{\ldots C} \otimes {}_C c \otimes \cdots = a_A \otimes b_B \otimes c_C \otimes \cdots {}^{\ldots CBA} t \end{aligned}$$

Of course the coordinates $t^{\ldots CBA}$, ${}^A t^{\ldots CB}$, ... differ by signs which can be calculated:

$$t^{\ldots CBA} = (-1)^{p(A)(p(t)+p(\ldots)+p(C)+p(B))} {}^A t^{\ldots CB}$$

This generalizes to the other positions of the indices as well as to dual bases (lower indices) and is summarized in the following principle:

Principle 2.8.1 (Order of Indices) *The summation convention for objects with several indices needs to respect the order of the indices and their position to the left or to the right of the object. An index to the left of an object must always be paired with an index to the right of an object. Between a pair of indices that is summed over, there may be no other free indices. That is, one sums from the innermost pair to the outermost pair.*

Remark that we have followed this principle already for all constructs we have used up to now: bases, coordinates and matrices of linear maps.

Lemma 2.8.2 *Let $t \in \cdots \otimes C^\vee \otimes B^\vee \otimes A^\vee \otimes Z$ be a tensor. t has coordinates $t_{ABC\ldots}{}^Z$, ${}_A t_{BC\ldots}{}^Z$, ..., ${}_{ABC\ldots} t^Z$ with respect to a basis $a^A \otimes b^B \otimes c^C \otimes \cdots \otimes z_Z$.*

We associate to it a multilinear map $m\colon A \times B \times C \cdots \to Z$ *of the same parity via*

$$m(a_A, b_B, c_C, \dots) = t_{ABC\dots}{}^{Z}{}_{Z}z.$$

It holds

$$\langle a_A, m, b_B, c_C, \dots\rangle = {}_At_{BC\dots}{}^{Z}{}_{Z}z,$$

$$\vdots$$

$$\langle a_A, b_B, c_C, \dots, m\rangle = {}_{ABC\dots}t_{Z}{}_{Z}z.$$

Furthermore, when considered as an element of $Z \otimes \cdots \otimes C^\vee \otimes B^\vee \otimes A^\vee$, *the tensor* t *can have the coordinates* ${}^{Z}t_{ABC\dots}$, ${}^{Z}{}_{A}t_{BC\dots}$, $\dots$, ${}^{Z}{}_{ABC\dots}t$. *For those it holds that:*

$$m(a_A, b_B, c_C, \dots) = z_Z{}^{Z}t_{ABC\dots}$$

$$\langle a_A, m, b_B, c_C, \dots\rangle = z_Z{}^{Z}{}_{A}t_{BC\dots}$$

$$\vdots$$

$$\langle a_A, b_B, c_C, \dots, m\rangle = z_Z{}^{Z}{}_{ABC\dots}t$$

Proof Verify that both right and left hand side acquire the same sign when passing from one equality to the next. □

Definition 2.8.3 (Tensor Algebra) For any supermodule E over R we define the tensor algebra

$$T(E) = \bigoplus_{n \geq 0} E^{\otimes n},$$

where $E^{\otimes n}$ is the n-th tensor power of E, and $E^0 = R$. The multiplication in the tensor algebra is given by the tensor product. That is for $t \in E^{\otimes n}$ and $t' \in E^{\otimes m}$ their product is given by

$$t \otimes t' \in E^{\otimes m+n}.$$

In addition to the super, $\mathbb{Z}_2$-grading, the algebra $T(E)$ is a $\mathbb{N}_0$-graded algebra with the degree n-part giving by $T^n(E) = E^{\otimes n}$. For elements $t \in T(E)$ that are homogeneous with respect to the $\mathbb{N}_0$-grading, we define $d(t) = n$ if $t \in T^n(E)$. Following Deligne and Freed (1999a), we call $d(t)$ the cohomological degree of t.

Definition 2.8.4 Denote by $\mathcal{I}_\pm$ the two sided ideal of $T(E)$ generated by expressions of the form

$$e \otimes e' \pm (-1)^{p(e)p(e')} e' \otimes e$$

for homogeneous elements e and e' of E.

The symmetric algebra

$$S(E) = {}^{T(E)}\!/_{\mathcal{I}_-}$$

is a supercommutative, $\mathbb{N}_0$-graded algebra. For the induced product $\odot$ it holds that

$$t \odot t' = (-1)^{p(t)p(t')} t' \odot t.$$

The degree n part of $S(E)$ is denoted by $S^n(E)$.

Similarly, the exterior algebra

$$\bigwedge E = {}^{T(E)}\!/_{\mathcal{I}_+}$$

is a $\mathbb{N}_0$-graded R-algebra. The inherited product $\wedge$ on $\bigwedge E$ fulfills

$$t \wedge t' = (-1)^{p(t)p(t')+d(t)d(t')} t' \wedge t$$

for elements t and t' that are homogeneous with respect to parity and cohomological degree. Notice that $\bigwedge E$ is bounded if and only if the odd part E_1 of E vanishes.

Remark 2.8.5 Notice that the parity of $t \odot t'$ and $t \wedge t'$ is given by

$$p(t \odot t') = p(t \wedge t') = p(t) + p(t').$$

Thus for any (parity) homogeneous $r \in R$ it holds that

$$rt \odot t' = (-1)^{p(r)(p(t)+p(t'))} t \odot t'r, \qquad rt \wedge t' = (-1)^{p(r)(p(t)+p(t'))} t \wedge t'r.$$

In particular, for the definition of parity and commutativity of the exterior product there are different choices in the literature. The sign conventions here follow Deligne and Freed (1999a), but differ from Leites (1980), for example. Notice that in Deligne and Morgan (1999), Appendix to §1; Deligne and Freed (1999a) and Manin (1988), Chapter 3,§4, there are comparisons and rationale for the different choices of signs. Essentially it is a matter of taste, and we prefer to follow Deligne and Freed (1999a) in applying the Principle 2.0.1 "relentlessly".

2.9 Derivations

In this section we give the definition of derivations useful for the definition of tangent spaces and tangent maps in Sect. 4.3.

Definition 2.9.1 Let A be an R-algebra and E an A-module. An R-linear map $X\colon A \to E$ which fulfils the Leibniz rule

$$X(a \cdot a') = X(a) \cdot a' + (-1)^{p(a)p(X)} a X(a')$$

is called a derivation from A with values in E. The module of derivations on A with values in E is denoted $\mathrm{Der}_R(A, E)$. If $E = A$ we write $\mathrm{Der}_R(A)$.

Example 2.9.2 The example that gives the name to derivations are first order differential operators acting on functions.

Example 2.9.3 (Canonical Derivation) There is a canonical derivation d on A with values in $\Omega_R(A) = \mathrm{Der}_R(A)^\vee$. It is given by

$$\begin{aligned} d\colon A &\to \mathrm{Der}_R(A)^\vee \\ f &\mapsto \begin{cases} df\colon \mathrm{Der}_R(A) \to A \\ \qquad\qquad X \mapsto \langle X, df\rangle = Xf \end{cases} \end{aligned}$$

In fact every derivation $X \in \mathrm{Der}_R(A, E)$ can be factorized as $X = l \circ d$ where l is the linear map given by

$$\begin{aligned} \operatorname{im} d &\to E \\ df &\mapsto Xf \end{aligned}$$

The map l is indeed linear, as both X and d are derivations.

Notice the difference between the module $\Omega_R(A)$ and the module of Kähler differentials $\Omega_{A/R}$ as defined for example in Matsumura (1989, §25). The module of Kähler differentials does not coincide with the dual space of derivations in all cases, in particular in the case of $A = C^\infty(\mathbb{R})$ to which we will return in Sect. 4.3. A discussion of this fact can be found online, see Speyer (2009) and Kähler differential (2015).

Example 2.9.4 Let A be an R-algebra and B an S-algebra, $\phi\colon R \to S$ a ring homomorphism and $\varphi\colon A \to B$ an algebra homomorphism over ϕ. Denote the canonical derivations $d_A\colon A \to \Omega_R(A)$ and $d_B\colon B \to \Omega_S(B)$. Then $d_B \circ \varphi$ is a derivation on

A with values in $\Omega_S(B)$. Thus there is a A-linear map $d\varphi\colon \Omega_R(A) \to \Omega_S(B)$ such that $d_B \circ \varphi = d\varphi \circ d_A$.

$$\begin{array}{ccc}
\Omega_R(A) & \xrightarrow{d\varphi} & \Omega_S(B) \\
{\scriptstyle d_A}\uparrow & & \uparrow{\scriptstyle d_B} \\
A & \xrightarrow{\varphi} & B \\
\uparrow & & \uparrow \\
R & \xrightarrow{\phi} & S
\end{array}$$

The map $d\varphi$ extends to a B-linear map $d\varphi\colon B \otimes_A \Omega_R(A) \to \Omega_S(B)$.

Example 2.9.5 The derivation $d\colon A \to \Omega_R(A)$ can be extended to its exterior algebra $\bigwedge \Omega_R(A)$ by imposing the following rules

$$d^2 = 0,$$
$$d\,(\alpha \wedge \beta) = (d\alpha) \wedge \beta + (-1)^{d(\alpha)} \alpha \wedge d\beta.$$

By similarity to the definition of derivations, one might call the extended d a "graded derivation". Notice that d raises the cohomological degree by one.

Example 2.9.6 There is another class of "graded derivations" on $\bigwedge \Omega_R(A)$ that reduce the cohomological degree by one. For any derivation X define the contraction operator ι_X of parity $p(X)$ by $\iota_X a = 0$ for any $a \in A$, $\iota_X \alpha = \langle X, \alpha\rangle$ for $\alpha \in \Omega_R(A)$ and

$$\iota_X\,(\alpha \wedge \beta) = (\iota_X \alpha) \wedge \beta + (-1)^{d(\alpha)+p(\alpha)p(X)} \alpha \wedge \iota_X \beta$$

for higher order terms. The operators d and ι_X are not independent. For example for any $\alpha \in \Omega_R(A)$ and $X, Y \in \mathrm{Der}_R(A)$ it holds that

$$\iota_X \iota_Y\, d\alpha = X\,(\iota_Y \alpha) - (-1)^{p(X)p(Y)} Y\,(\iota_X \alpha) - \iota_{[X,Y]}\alpha, \tag{2.9.7}$$

where $[X, Y]$ is the commutator of X and Y explained in Example 2.11.4. Equation (2.9.7) can be derived easily using the fact that one can write $\alpha = \alpha_A\, d f^A$ for suitable $\alpha_A, f^A \in A$. Similar expressions can be deduced for α with higher cohomological degree, however they will not be needed in this work.

2.10 Bilinear Forms

Definition 2.10.1 The bilinear form $b\colon E \times E \to R$ on the R-module E is called symmetric, if for any two $v, v' \in E$ it holds that

$$b(v, w) = (-1)^{p(v)p(w)} b(w, v).$$

The bilinear form b is called non-degenerate if the map

$$\vee_b\colon E \to E^\vee$$

$$\vee_b(v) = \langle \cdot, b, v\rangle = \begin{cases} E \to R \\ \langle w, \vee_b(v)\rangle = \langle w, b, v\rangle \end{cases}$$

is an isomorphism.

One can check easily that the map ${}_b\vee\colon E \to E^\vee$ given by

$$\left\langle v, {}_b\vee\right\rangle(w) = \langle v, b, w\rangle$$

coincides with $\vee_b$. Thus we will drop the b from the notation occasionally, if it is clear with respect to which bilinear form we are dualizing.

Example 2.10.2 (Standard Symmetric and Antisymmetric Bilinear Forms) Let E be a free module over R of rank $m|2n$ and e_A a basis of E. Then the bilinear form given by

$$b(e_a, e_b) = \delta_{ab}, \qquad b(e_\alpha, e_\beta) = \varepsilon_{\alpha\beta}, \qquad b(e_a, e_\beta) = b(e_a, e_\beta) = 0,$$

is symmetric and non-degenerate. Here ε is the completely anti-symmetric tensor given by

$$\varepsilon_{\alpha\beta} = \begin{cases} 1 & \text{if } \alpha = 2l-1,\ \beta = 2l, \text{ for } l = 1, \ldots, n \\ -1 & \text{if } \alpha = 2l,\ \beta = 2l-1, \text{ for } l = 1, \ldots, n \\ 0 & \text{else} \end{cases}$$

This bilinear form is called the standard symmetric bilinear form on $R^{m|2n}$.

Similarly, for E of rank $2m|n$ the bilinear form

$$b(e_a, e_b) = \varepsilon_{ab}, \qquad b(e_\alpha, e_\beta) = \delta_{\alpha\beta}, \qquad b(e_a, e_\beta) = b(e_a, e_\beta) = 0,$$

is anti-symmetric and non-degenerate. It is called standard anti-symmetric bilinear form or standard symplectic form.

Definition 2.10.3 Let b be a non-degenerate, symmetric bilinear form on E. There is a non-degenerate, symmetric bilinear form $b^\vee$ on $E^\vee$ such that

$$\left\langle \langle v, \vee \rangle, b^\vee, \vee(w) \right\rangle = \langle v, b, w \rangle . \tag{2.10.4}$$

Remark 2.10.5 The notation $\langle \cdot, b, \cdot \rangle$ helps again to reduce signs in the formulas. At first sight the defining equality

$$b^\vee(\vee(v), \vee(w)) = b(v, w) \tag{2.10.6}$$

might look more natural than Eq. (2.10.4). However, the definition given in Eq. (2.10.6) implies that dualizing with respect to $b^\vee$ is inverse to dualizing with respect to b only up to the sign $(-1)^{p(b)}$. With the definition of $b^\vee$ given in (2.10.4), dualizing with respect to $b^\vee$ is inverse to dualizing with respect to b.

Lemma 2.10.7 *Let E be a module with basis e_A, and e^A the corresponding right-dual basis of $E^\vee$, b a symmetric bilinear form on E and $v = v^A{}_A e = e_A{}^A v$. Then the map $\vee = \vee_b$ is given by*

$$\langle v, \vee \rangle = v^A{}_A b_B{}^B e, \qquad \vee(v) = e^A{}_A b_B{}^B v.$$

In particular, b is non-degenerate if and only if the coordinate matrix ${}_A b_B$ of b is invertible. The dual metric $b^\vee$ is then given by the inverse matrix of ${}_A b_B$.

Proof Denote the matrix of $b^\vee$ by ${}^C b^D$. Then the condition (2.10.4) implies:

$${}_A b_B = \left\langle {}_A e, b, e_B \right\rangle = \left\langle \left\langle {}_A e, \vee \right\rangle, b^\vee, \vee(e_B) \right\rangle = {}_A b_C{}^C b^D{}_D b_B$$

We see that this is possible, if and only if ${}_A b_C{}^C b^D = \delta_A^D$, that is ${}^C b^D$ is the inverse of $\vee$ which exists, as b is non-degenerate. □

Lemma 2.10.8 *Let b be an even, symmetric, non-degenerate bilinear form on the free R-module E. Then there exists a basis f_A of E such that b is given by*

$$b(f_a, f_b) = r_a \delta_{ab}, \qquad b(f_\alpha, f_\beta) = r_\alpha \varepsilon_{\alpha\beta}, \qquad b(f_a, f_\beta) = 0,$$

for some $r_A \in R$. Such a basis is called an orthogonal basis.

If all the r_A possess a square root in R, it is possible to define $e_A = \frac{1}{\sqrt{r_A}} f_A$. In this case, the bilinear form b coincides with the standard bilinear form from Example 2.10.2. The basis e_A is then called an orthonormal basis and b is called positive.

Proof This is an adaption of the Gram-Schmidt orthogonalization procedure. See also Hanisch (2009, Proposition 3.39). □

Definition 2.10.9 Let E be a module of dimension $m|2n$ over the ring R and e^A a basis of E. A linear automorphism m of E preserving the bilinear form, that is,

$$b(m(v), m(w)) = b(v, w)$$

for all $v, w \in E$ is called an orthogonal automorphism. The group of even orthogonal automorphisms, denoted by $\mathrm{O}_R(E, b)$, is a subgroup of $\mathrm{GL}(E)$.

If $E = R^{m|2n}$ and b the standard symmetric bilinear form, the subgroup of orthogonal matrices of $\mathrm{GL}_R(m|2n)$ is denoted by $\mathrm{O}_R(m|2n)$. If b is positive, $\mathrm{O}_R(E)$ can be identified with $\mathrm{O}_R(m|2n)$ after a choice of orthonormal basis.

If $E = R^{2m|n}$ and b the standard anti-symmetric form, the subgroup of matrices of $\mathrm{GL}_R(2m|n)$ preserving b is denoted by $\mathrm{Sp}_R(2m|n)$ and called supersymplectic matrices.

Remark 2.10.10 Of course, for the case $R = \mathbb{R}$ and $n = 0$ the definitions above reduce to the usual definition of orthogonal automorphisms and matrices. On the other hand, the group $\mathrm{O}_{\mathbb{R}}(0|2n)$ is isomorphic to the group of symplectic matrices $\mathrm{Sp}(2n)$. For this reason, in some texts $\mathrm{O}(m|2n)$ is called $\mathrm{OSp}(m|2n)$.

Example 2.10.11 For later reference, we will give an explicit description of $\mathrm{Sp}_R(2|1)$ following Manin (1991, Chapter 2.1). Let ${}_AL^B \in \mathrm{Sp}_R(2|1)$ be given by the block matrix

$$\begin{pmatrix} a & c & \gamma \\ b & d & \delta \\ \alpha & \beta & e \end{pmatrix}.$$

The condition that L preserves the standard symplectic form b is ${}_Ab_D = {}_AL^B\, {}_Bb_C\, {}^CL_D$, or in block form

$$\begin{pmatrix} 0 & 1 & 0 \\ -1 & 0 & 0 \\ 0 & 0 & -1 \end{pmatrix} = \begin{pmatrix} a & c & \gamma \\ b & d & \delta \\ \alpha & \beta & e \end{pmatrix} \begin{pmatrix} 0 & 1 & 0 \\ -1 & 0 & 0 \\ 0 & 0 & -1 \end{pmatrix} \begin{pmatrix} a & b & -\alpha \\ c & d & -\beta \\ \gamma & \delta & e \end{pmatrix}$$

$$= \begin{pmatrix} 0 & ad - bc - \gamma\delta & c\alpha - a\beta - e\gamma \\ -ad + bc + \gamma\delta & 0 & d\alpha - b\beta - e\delta \\ -a\beta + c\alpha - e\gamma & -b\beta + d\alpha - e\delta & -e^2 - 2\alpha\beta \end{pmatrix}$$

Recall that we include the additional signs from the summation convention to the left matrix. Hence the constraints for $L \in \mathrm{Sp}_R(2|1)$ are

$$\begin{aligned} ad - bc - \gamma\delta &= 1, \quad a\beta - c\alpha + e\gamma = 0, \\ e^2 + 2\alpha\beta &= 1, \quad b\beta - d\alpha + e\delta = 0. \end{aligned} \tag{2.10.12}$$

Notice that those equations imply

$$e^2\gamma\delta = (c\alpha - a\beta)\,(d\alpha - b\beta) = (ad - bc)\,\alpha\beta,$$

hence $\alpha\beta\gamma\delta = 0$ which implies $\left(e^2 + 2\alpha\beta\right)\gamma\delta = (ad - bc - \gamma\delta)\,\alpha\beta$ and consequently $\alpha\beta = \gamma\delta$. Furthermore, up to the choice of a sign the constraints (2.10.12) can be solved for α, β and e as functions of a, b, c, d, γ and β such that $ad - bc + \gamma\delta = 1$ by

$$e = \pm\,(1 - \gamma\delta)\,, \qquad \alpha = \pm\,(b\gamma - a\delta)\,, \qquad \beta = \pm\,(d\gamma - c\delta)\,.$$

2.11 Lie Algebras

Definition 2.11.1 A module E over R, equipped with a symmetric bilinear mapping

$$[\cdot,\cdot]\colon E \times E \to E$$

is called a Lie superalgebra provided it fulfils the Jacobi identity

$$[e,[f,g]] + (-1)^{p(e)(p(f)+p(g))}[f,[g,e]] + (-1)^{p(g)(p(f)+p(e))}[g,[e,f]] = 0$$

for all $e, f, g \in E$. An R-linear map $l\colon E \to F$ between Lie superalgebras is called a homomorphism of Lie superalgebras, if for all $e, e' \in E$ it holds that

$$[l(e), l(e')] = l([e,e']).$$

Example 2.11.2 The square matrices in $\mathrm{Mat}(m|n)$ form a Lie algebra with the commutator

$$[M,N] = MN - (-1)^{p(N)p(M)}NM.$$

Only the Jacobi identity is to verify:

$$\begin{aligned}
&[M,[N,P]] + (-1)^{p(M)(p(N)+p(P))}[N,[P,M]] \\
&\quad + (-1)^{p(P)(p(N)+p(M))}[P,[M,N]] \\
&= [M, NP - (-1)^{p(N)p(P)}PN] \\
&\quad + (-1)^{p(M)(p(N)+p(P))}[N, PM - (-1)^{p(P)p(M)}MP]M \\
&\quad (-1)^{p(P)(p(N)+p(M))}[P, MN - (-1)^{p(N)p(M)}NM]
\end{aligned}$$

$$
\begin{aligned}
&= MNP - (-1)^{p(M)(p(N)+p(P))} NPM \\
&\quad - (-1)^{p(N)p(P)} MPN - (-1)^{p(N)p(P)+p(M)(p(N)+p(P))} PNM \\
&\quad + (-1)^{p(M)(p(N)+p(P))} NPM - (-1)^{p(P)(p(M)+p(N))} PMN \\
&\quad - (-1)^{p(M)p(N)} NMP + (-1)^{p(N)p(P)} MPN \\
&\quad + (-1)^{p(P)(p(N)+p(M))} PMN - MNP \\
&\quad - (-1)^{p(P)(p(N)+p(M))+p(N)p(M)} PNM + (-1)^{p(M)p(N)} NMP \\
&= 0
\end{aligned}
$$

Example 2.11.3 Let $F \subseteq E$ be a sub supermodule of the Lie algebra E that is closed under the Lie-bracket, that is, for any two $f, f' \in F$ the Lie bracket $[f, f']$ is in F as well. Then F is a Lie algebra, called sub Lie algebra of E. In Sect. 5.3 we will encounter several sub Lie algebras of $\mathrm{Mat}_R(m|n)$.

Example 2.11.4 The derivations on A with values in A form a Lie algebra with the Lie bracket of $X, Y \in \mathrm{Der}_R(A)$ given by

$$[X, Y]a = XYa - (-1)^{p(X)p(Y)} YXa.$$

To demonstrate that $[X, Y]$ is indeed a derivation of parity $p(X) + p(Y)$, let $a, a' \in A$:

$$
\begin{aligned}
[X, Y](a \cdot a') &= X\left(Y(a) \cdot a' + (-1)^{p(Y)p(a)} a \cdot Y(a')\right) \\
&\quad - (-1)^{p(X)p(Y)} Y\left(X(a) \cdot a' + (-1)^{p(X)p(a)} a \cdot X(a')\right) \\
&= XY(a) \cdot a' + (-1)^{p(X)(p(Y)+p(a))} Y(a) \cdot X(a') \\
&\quad + (-1)^{p(Y)p(a)} X(a) \cdot Y(a') + (-1)^{(p(Y)+p(X))p(a)} a \cdot XY(a') \\
&\quad - (-1)^{p(X)p(Y)} XY(a) \cdot a' - (-1)^{p(Y)p(a)} X(a) \cdot Y(a') \\
&\quad - (-1)^{p(X)(p(Y)+p(a))} Y(a) \cdot X(a') \\
&\quad + (-1)^{p(X)p(Y)+(p(X)+p(Y))p(a)} a \cdot YX(a') \\
&= [X, Y](a) \cdot a' + (-1)^{(p(X)+p(Y))p(a)} a \cdot [X, Y](a')
\end{aligned}
$$

The Jacobi identity can be checked similar to the case of matrices in $\mathrm{Mat}_R(m|n)$.

2.12 Almost Complex Structures

Let S be a supercommutative R-algebra. Any S-module possesses the structure of an R-module. Conversely, for any R-module E the R-module $S \otimes_R E$ is also an S-module. As an R-module the modules E and $S \otimes_R E$ may be quite different.

A classical situation where this difference is well understood is the case $R = \mathbb{R}$ and $S = \mathbb{C}$. In this section we list generalizations of some of the classical results to the case where R is a supercommutative $\mathbb{R}$ algebra and $S = \mathbb{C} \otimes_{\mathbb{R}} R$ a supercommutative $\mathbb{C}$-algebra. The classical proofs, to be found, for example, in Huybrechts (2005, Chapter 1.2), extend to the super setting.

Definition 2.12.1 Let E be an R module. An almost complex structure on E is an even automorphism I of E such that $\mathrm{I}^2 = -\,\mathrm{id}_E$.

Example 2.12.2 The standard almost complex structure on $R^{2m|2n}$ is given by

$$\mathrm{I}\, e_{2k-1} = e_{2k}, \qquad\qquad \mathrm{I}\, e_{2k} = -e_{2k-1}.$$

Proposition 2.12.3 *Let E be an R-module and $S = \mathbb{C} \otimes R$. The following are equivalent:*

i) an S-module structure for E that reduces to the given R-module structure
ii) an almost complex structure on E

Proposition 2.12.3 has interesting consequences.

- If E is free as an S module of dimension $m|n$ with basis e_A, then the vectors e_A, Ie_A form an R-basis of dimension $2m|2n$. After reordering, the almost complex structure is in standard form in this R-basis.
- If E is free as an R-module of dimension $2m|2n$ and has a basis, such that the almost complex structure is in the standard form, then E is also free as an S-module.
- Any R-basis of E such that the almost complex structure I is in standard form gives rise to a real structure on E. Indeed, the vector space generated by all e_{2k-1} spans an R-module F of dimension $m|n$ such that $E = S \otimes_R F$.
- The S-linear endomorphisms of E are the subset of the R-linear endomorphisms of E that commute with the almost complex structure. In particular for the even automorphisms, we have

$$\mathrm{GL}_S(m|n) = \{A \in \mathrm{GL}_R(2m|2n) \mid A\,\mathrm{I} = \mathrm{I}\,A\}.$$

- The complex linear continuation of I to the S-module $E \otimes \mathbb{C}$ gives a decomposition

$$E \otimes \mathbb{C} = E^{1,0} \oplus E^{0,1}$$

where

$$E^{1,0} = \{e \in E \otimes \mathbb{C} \mid \mathrm{I}\, e = \mathrm{i}e\}, \qquad E^{0,1} = \{e \in E \otimes \mathbb{C} \mid \mathrm{I}\, e = -\mathrm{i}e\}.$$

Consequently, $E^{1,0}$ is isomorphic to E as an S-module. Furthermore the complex conjugation on $E \otimes \mathbb{C}$ gives an R-linear isomorphism on $E^{1,0} \simeq E^{0,1}$.

We now turn to the relation between bilinear forms and hermitian forms.

Definition 2.12.4 A bilinear form b on an R-module E is called compatible with the almost complex structure I, if the following holds for all $e, e' \in E$:

$$b(\mathrm{I}\, e, \mathrm{I}\, e') = b(e, e')$$

Definition 2.12.5 We denote the complex conjugate $\overline{c} \otimes r$ of $s = c \otimes r \in S = \mathbb{C} \otimes R$ by $\overline{s}$. Let E be an S-module. Then a sesquilinear form on E is an R-bilinear form on E such that in addition for all $s, s' \in S$

$$h(e, e's) = h(e, e')s, \qquad h(e, se') = h(e\overline{s}, e').$$

The parity of h is given by the formula

$$p(h(e, e')) = p(h) + p(e) + p(e').$$

The sesquilinear form is called hermitian, if it is non-degenerate and

$$h(e, e') = (-1)^{p(e)p(e')}\overline{h(e, e')}$$

for e, e' of pure parity.

Example 2.12.6 The standard hermitian form h on $S^{m|n}$ is given in the standard basis by

$$h(e_a, e_a) = 1, \qquad h(e_\alpha, e_\alpha) = \mathrm{i}.$$

Similar to Lemma 2.10.8 and under certain conditions on S and h, one can find on every free S-module a hermitian basis e_A such that h is in standard form. In that case we call h positive.

Definition 2.12.7 The subgroup of the even automorphisms of E that preserve the hermitian form h is called the group of unitary matrices $\mathrm{U}(E, h)$. For $E = S^{m|n}$ and h the standard hermitian form, we also write $\mathrm{U}_S(m|n)$.

Proposition 2.12.8 *Let (E, I) be an R-module with almost complex structure. The following are equivalent:*

i) a symmetric, non-degenerate R-bilinear form b on E that is compatible with the almost complex structure
ii) a hermitian form on $E \otimes \mathbb{C}$
iii) a hermitian form on (E, I)

The hermitian form is positive if and only if the R-bilinear form is positive.

Proof Any symmetric, non-degenerate, positive R-bilinear form b extends to a hermitian form $b^{\mathbb{C}}$ on $E \otimes \mathbb{C}$ by setting

$$b^{\mathbb{C}}\left(e \otimes c, e' \otimes c'\right) = \bar{c}c' b(e, e').$$

The restriction of $\frac{1}{2}b^{\mathbb{C}}$ to $(E, \mathrm{I}) = E^{(1,0)}$ gives a hermitian form h on (E, I). The prefactor $\frac{1}{2}$ is chosen such that the identification of (E, I) and $E^{(1,0)}$ given by

$$v \mapsto \frac{1}{2}(v - \mathrm{i}\,\mathrm{I}\,v)$$

is hermitian. The bilinear form on E associated to a hermitian form h on (E, I) is given by

$$b(e, e') = \frac{1}{2}\left(h(e, e') + \overline{h(e, e')}\right). \qquad \square$$

Chapter 3
Supermanifolds

There are a number of different approaches to supermanifolds that can roughly be divided in three classes: the Rogers–DeWitt approach, the approach via ringed spaces (sometimes called the Berezin–Kostant–Leites approach) and the approach via functors of points.

In the Rogers–DeWitt approach the super vector space $\mathbb{R}^{m|n}$ is given a topology and supermanifolds are constructed by glueing those topological spaces. Depending on the applications, different topologies on the odd directions and different regularity classes for functions on the resulting supermanifolds are used. References for this approach are for example the textbooks (DeWitt 1992; Rogers 2007; Bartocci et al. 1991).

In contrast, in the ringed space approach, no odd topological points are introduced. Rather the sheaf of functions on the supermanifold is $\mathbb{Z}_2$-graded and contains nilpotent functions. An early overview article for this approach is Leites (1980), see also Berezin (1987) and Kostant (1977).

The functor of points approach is a categorical reformulation of the ringed space approach which allows to generalize it, for example, to infinitely many dimensions or supermanifolds with singularities. Building on Molotkov (2010), the thesis Sachse (2009) works out an example of the functor of points approach in all details.

The different approaches to supermanifolds have developed in parallel and, to a certain extent, produced similar results. However, despite results showing the equivalence of the approaches in certain cases (most notably Batchelor 1980), it is often difficult to transfer results from the Rogers–DeWitt approach and vice versa. In this work we will exclusively use the ringed space approach. The more algebraic language seems to be well suited for the ultimate goal to study moduli spaces of super Riemann surfaces. In particular non-trivial families of supermanifolds will play an important role in Part II. Families of supermanifolds have been used more or less implicitly for a long time. In Deligne and Morgan (1999), it is argued that one should always consider families of supermanifolds and allow arbitrary base change.

E. Keßler, *Supergeometry, Super Riemann Surfaces and the Superconformal Action Functional*, Lecture Notes in Mathematics 2230,
https://doi.org/10.1007/978-3-030-13758-8_3

As Deligne and Morgan (1999) is rather brief and in order to make this work self-contained, we will repeat the argument below when introducing the precise notions.

In the first section, we define the building block of supergeometry: the superdomain $\mathbb{R}^{m|n}$. By examples, we will motivate families of supermanifolds.

In the second section, families of supermanifolds and base change are discussed. In particular the category of supermanifolds that we will work with in this book is defined.

The concept of underlying even manifold is discussed in Sect. 3.3. Roughly speaking an underlying even manifold of a family of supermanifolds is an embedded family of supermanifolds where all the fibers have only even dimensions. Thus the underlying even manifold is a generalization of the reduced space of a supermanifold that behaves well under change of basis. Even though, as we will see in Part II, the underlying even manifolds are intimately connected to supersymmetry, this concept does not seem to be studied before. We will prove the existence of underlying even manifolds for all supermanifolds (Theorem 3.3.7) and show that a superdiffeomorphism induces a diffeomorphism on the underlying even manifold and a change of the embedding (Corollary 3.3.14).

Section 3.4 gives a brief comparison to the theory of functor of points.

3.1 The Supermanifold $\mathbb{R}^{m|n}$

We formulate the theory of supermanifolds in terms of ringed spaces. We thus assume basic knowledge of sheaf theory, see, for example, Hartshorne (1977) or Grothendieck and Dieudonné (1960, Chapitre 0.4). Recall that a locally ringed space M is a pair $(\|M\|, \mathcal{O}_M)$, consisting of a topological space $\|M\|$ and a sheaf of rings $\mathcal{O}_M$ such that for every point $p \in \|M\|$ the stalk $\mathcal{O}_{M,p}$ is a local ring. A morphism $f: M \to N$ between the locally ringed spaces $M = (\|M\|, \mathcal{O}_M)$ and $N = (\|N\|, \mathcal{O}_N)$ is a pair $(\|f\|, f^\#)$ where $\|f\|: \|M\| \to \|N\|$ is a continuous map and $f^\#: \mathcal{O}_N \to f_*\mathcal{O}_M$ a map of sheaves preserving the maximal ideals of the stalks.

Every ringed space $M = (\|M\|, \mathcal{O}_M)$ has an associated reduced space $M_{red} = (\|M\|, {\mathcal{O}_M}/{\mathcal{I}_{nil}})$ where $\mathcal{I}_{nil} \subset \mathcal{O}_M$ is the ideal sheaf of nilpotent elements.The projection $i_{red}^\#: \mathcal{O}_M \to {\mathcal{O}_M}/{\mathcal{I}_{nil}}$ gives rise to an inclusion

$$i_{red}: M_{red} \hookrightarrow M$$

over the identity on $\|M\|$. Actually, taking the reduced space is a functor. In fact, for any map of ringed spaces $\Phi : M \to N$ there is a corresponding reduced map

$$\Phi_{red}: M_{red} \to N_{red},$$

such that the following diagram is commutative:

$$\begin{array}{ccc} M & \xrightarrow{\Phi} & N \\ \uparrow & & \uparrow \\ M_{red} & \xrightarrow{\Phi_{red}} & N_{red} \end{array}$$

The reducing functor is determined by its universal property that any map from a reduced space $M = M_{red}$ to N factors over N_{red}.

We now turn to smooth supermanifolds. The model space $\mathbb{R}^{m|n}$ is a generalization of the smooth manifold $\mathbb{R}^m$ and plays the same fundamental role for the geometry of supermanifolds as $\mathbb{R}^m$ plays for ordinary manifolds. In particular all supermanifolds are obtained through glueing of open subsets of $\mathbb{R}^{m|n}$.

Definition 3.1.1 The supermanifold $\mathbb{R}^{m|n}$ is the locally ringed space $(\mathbb{R}^m, \mathcal{O}_{\mathbb{R}^{m|n}})$, where $\mathbb{R}^m$ is the usual euclidean topological vector space and $\mathcal{O}_{\mathbb{R}^{m|n}}$ the sheaf of supercommutative $\mathbb{R}$-algebras

$$\mathcal{O}_{\mathbb{R}^{m|n}} = C^\infty(\mathbb{R}^m, \mathbb{R}) \otimes_{\mathbb{R}} \bigwedge\nolimits_n,$$

that is, the tensor product of smooth functions from $\mathbb{R}^m$ to $\mathbb{R}$ and a real Grassmann algebra in n generators. The sheaf $\mathcal{O}_{\mathbb{R}^{m|n}}$ inherits the $\mathbb{Z}_2$-grading from the grading of the Grassmann algebra.

The restriction of $\mathbb{R}^{m|n}$ to an open subset $U \subseteq \mathbb{R}^m$ is called a superdomain. A map between superdomains is a map of ringed spaces. The tuple consisting of the m coordinate functions $x^1, \dots, x^m$ on $\mathbb{R}^m$ together with the n generators $\eta^1, \dots, \eta^m$ of the Grassmann algebra are called standard coordinates on $\mathbb{R}^{m|n}$, often written as $X^A = (x^a, \eta^\alpha)$.

Example 3.1.2 $\mathbb{R}^{m|n}$ is a generalization of $\mathbb{R}^m$, as the manifold $\mathbb{R}^m$ coincides with $\mathbb{R}^{m|0}$ as ringed space. Any smooth map

$$\begin{aligned} f\colon \mathbb{R}^m &\to \mathbb{R}^p \\ x = (x^1, \dots, x^m) &\mapsto y = f(x) = (f^1(x), \dots, f^p(x)) \end{aligned}$$

induces a map of sheaves $f^\#\colon \mathcal{O}_{\mathbb{R}^p} \to \mathcal{O}_{\mathbb{R}^m}$ by $f^\# y^a = f^a(x)$. Furthermore $\mathbb{R}^{m|0}$ is the reduced space of $\mathbb{R}^{m|n}$.

Every function $f \in \mathcal{O}_{\mathbb{R}^{m|n}}$ can be expanded in terms of coordinates $X^A = (x^a, \eta^\alpha)$

$$f = \sum_{\underline{\alpha}} \eta^{\underline{\alpha}}\, {}_{\underline{\alpha}}f(x) = {}_0 f + \eta^\alpha\, {}_\alpha f(x) + \dots .$$

Here $\underline{\alpha}$ is a $\mathbb{Z}_2$-multiindex, that is, a multiindex such that every entry is either zero or one. The functions ${}_{\underline{\alpha}}f(x)$ are usual smooth functions on $\mathbb{R}^m$. The reduced function of f, that is, the image of f under $i^\#_{red}$ is given by $f_{red} = i^\#_{red} f = {}_0 f$. The following theorem explains how to express maps of superdomains in terms of coordinates, thus generalizing Example 3.1.2.

Theorem 3.1.3 (Charts Theorem, see also Leites 1980) *Let $U \subset \mathbb{R}^{m|n}$ and $V \subset \mathbb{R}^{p|q}$ be superdomains with coordinates $X^A = (x^a, \eta^\alpha)$ and $Y^B = (y^b, \theta^\beta)$ respectively.*

i) Any morphism of superdomains $\phi\colon U \to V$ yield the functions $f^B = \phi^\# Y^B$, that is, p even and q odd elements of $\mathcal{O}_U(\|U\|)$. The functions f^b satisfy

$$\left(f^1{}_{red}(x), \ldots, f^p{}_{red}(x)\right) \in \|V\| \text{ for all } x \in \|U\|. \tag{3.1.4}$$

ii) Conversely, for any tuple $f^B = (f^b, f^\beta)$ of p even and q odd elements of $\mathcal{O}_U(\|U\|)$ satisfying (3.1.4), there is a unique morphism of superdomains $\phi\colon U \to V$ such that $\phi^\# Y^B = f^B$.

Proof The claim i) follows since $f_{red}\colon U_{red} \to V_{red}$ is a smooth map from an open domain in $\mathbb{R}^m$ to an open domain in $\mathbb{R}^p$. The condition (3.1.4) reflects that the image of U_{red} under f_{red} should lie in V_{red}.

In order to show claim ii), one has to define a map $\|\phi\|\colon \|U\| \to \|V\|$ and the pullback $\phi^\# g$ for arbitrary sections g of $\mathcal{O}_V$. Since g is at most polynomial in the odd coordinates and $\phi^\#$ is an algebra homomorphism, we can restrict our attention to the case that $g = g(y)$ depends only on the even coordinates y^b. We know that the functions $f^b{}_{red}$ determine a smooth map $\tilde{\phi}\colon U_{red} \to V_{red}$ such that $\tilde{\phi}^\# y^b = f^b{}_{red}$ and $\left(\tilde{\phi}^\# g_{red}\right)(x) = g_{red}(f_{red}(x))$. Since $\|\phi\| = \phi_{red} = \tilde{\phi}$, we only have to find the nilpotent corrections for $\phi^\#$. To this end, we use Hadamard's Lemma: There are smooth functions $g_{\underline{b}}$ on V such that in a neighbourhood of $v \in V$

$$g(y) = g(v) + \sum_{|\underline{b}| \leq q} \frac{\partial g}{\partial y^{\underline{b}}}(y - v)^{\underline{b}} + \sum_{|\underline{b}| = q+1} (y - v)^{\underline{b}} g_{\underline{b}}(y).$$

Here $\underline{b}$ is a multiindex of order $|\underline{b}|$. Now write $f^b = f^b{}_{red} + f^b_{nil}$ where f^b_{nil} is the nilpotent part with $\left(f^b_{nil}\right)^{q+1} = 0$. Then by Hadamard's Lemma and since $\phi^\#$ is an algebra homomorphism, we obtain

$$\left(\phi^\# g\right)(x, \eta) = g(f_{red}(x)) + \sum_{|\underline{b}| \leq q} \left(\frac{\partial g}{\partial y^{\underline{b}}}\right)(x, \eta) f^{\underline{b}}_{nil}(x, \eta).$$

Hence, the functions f^b determine $\phi^\# g$ uniquely for all g. By restricting to suitable open subsets of U and V we obtain a morphism of sheaves $\mathcal{O}_V \to \mathcal{O}_U$ over $\|\phi\|$.

□

Example 3.1.5 (Topological Points of Superdomains) Let us consider a map ϕ from $\mathbb{R}^{0|0} = (pt, \mathbb{R})$ to $\mathbb{R}^{m|n}$ with coordinates $X^A = (x^a, \eta^\alpha)$. By the Theorem 3.1.3 we know that we have to give the image of all coordinate functions under $\phi^\#$. As the image of $\phi^\#$ lies in $\mathbb{R}$ we know that we have to give m real values $p^1, \ldots, p^m$ such that $\phi^\#(x^a) = p^a$. But as $\mathbb{R}$ has no odd functions the image of the odd coordinates has to be zero, $\phi^\#(\eta^\alpha) = 0$.

This demonstrates an important difference between the $\mathbb{R}$-supermodule $\mathbb{R}^{m|n}$ and the superdomain $\mathbb{R}^{m|n}$. For the odd vectors of the supermodule $\mathbb{R}^{m|n}$ there are no corresponding points in the supermanifold $\mathbb{R}^{m|n}$. The geometric reason is that we did not add any odd points since the topological space of the manifold still is $\mathbb{R}^m$. Furthermore, the underlying scalars $\mathbb{R}$ are not sufficient as they do not have any odd part.

There is always a unique map $b\colon \mathbb{R}^{m|n} \to \mathbb{R}^{0|0}$. The map $\|b\|$ sends everything to a point and the map $b^\#\colon \mathcal{O}_{\mathbb{R}^{0|0}} = \mathbb{R} \to \mathcal{O}_{\mathbb{R}^{m|n}}$ is given by the inclusion.

Example 3.1.6 Let us consider a map $\phi\colon \mathbb{R}^{2|2} \to \mathbb{R}^{2|2}$ with coordinates $X^A = (x^a, \eta^\alpha)$ and $Y^B = (y^b, \theta^\beta)$ respectively. By Theorem 3.1.3 the map ϕ is completely determined by $\phi^\# Y^B$ as functions of X^A. However, the ring homomorphism $\phi^\#$ preserves parity and hence the functions $\phi^\# Y^B$ are of the form

$$\left(\phi^\# y^b\right)(x, \eta) = {}_0f^b(x) + \eta^1\eta^2\,{}_{12}f^b(x),$$

$$\left(\phi^\# \theta^\beta\right)(x, \eta) = \eta^\mu\,{}_\mu f^\beta(x).$$

Here the functions ${}_0f^b$, ${}_{12}f^b$ and ${}_\mu f^b$ are arbitrary functions in the variables x^a.

Intuitively one might have expected terms of any order in the coordinate expansion of $\phi^\# Y^A$. The restricted form of $\phi^\# Y^A$ is again due to the fact that the underlying scalars in $\mathbb{R}$ do not have any odd elements.

Examples 3.1.5 and 3.1.6 present clearly that morphisms of supermanifolds are restricted due to the fact that $\mathbb{R}$ is purely even. However, for the applications we have in mind, for example, Eq. (9.1.5), these restrictions need to be lifted. Consequently, we will generalize superdomains to families of superdomains over a superdomain B.

As a preparation we need to recall the definition of the cartesian product, which we give here in the special case of superdomains:

Definition 3.1.7 The cartesian product of the superdomains U and V is a superdomain $U \times V$ together with projections $p_U\colon U \times V \to U$ and $p_V\colon U \times V \to V$ with the following universal property: For any maps $f\colon W \to U$ and $g\colon W \to V$

there exists a unique map $(f, g)\colon W \to U \times V$ such that the following diagram commutes:

$$\begin{array}{ccccc} & & W & & \\ & \swarrow^{f} & \downarrow{\scriptstyle (f,g)} & \searrow^{g} & \\ U & \xleftarrow{p_U} & U \times V & \xrightarrow{p_V} & V \end{array}$$

Let now $h\colon U' \to U$ and $k\colon V' \to V$ and $p_{U'}\colon U' \times V' \to U'$ and $p_{V'}\colon U' \times V' \to V'$. We denote by $h \times k$ the unique map

$$h \times k = (h \circ p_{U'}, k \circ p_{V'})\colon U' \times V' \to U \times V$$

given by the universal property of the cartesian product $U \times V$.

Theorem 3.1.3 implies the following:

Lemma 3.1.8 *Let $X^A = (x^a, \eta^\alpha)$ be coordinates on $\mathbb{R}^{m|n}$, $L^B = (l^b, \lambda^\beta)$ coordinates on $\mathbb{R}^{p|q}$ and $Y^C = (y^c, \theta^\gamma)$ coordinates on $\mathbb{R}^{m+p|n+q}$. The product of superdomains $\mathbb{R}^{m|n} \times \mathbb{R}^{p|q}$ is the superdomain $\mathbb{R}^{m+p|n+q}$ together with the canonical projections $p_1\colon \mathbb{R}^{m+p|n+q} \to \mathbb{R}^{m|n}$ and $p_2\colon \mathbb{R}^{m+p|n+q} \to \mathbb{R}^{p|q}$ given in coordinates by*

$$p_1^\#(x^a) = y^a, \qquad p_1^\#(\eta^\alpha) = \theta^\alpha,$$
$$p_2^\#(l^c) = y^{m+c}, \qquad p_2^\#(\lambda^\gamma) = \theta^{n+\gamma}.$$

The analogous statement holds for arbitrary superdomains.

Definition 3.1.9 Let $B \subseteq \mathbb{R}^{p|q}$ be a superdomain. An $m|n$-dimensional superdomain over B is a superdomain $U \times B$ where $U \subseteq \mathbb{R}^{m|n}$ together with the projection $p_2\colon U \times B \to B$. A superdomain $U \subseteq \mathbb{R}^{m|n}$ is a superdomain over $\mathbb{R}^{0|0}$.

Let $b\colon B \to B'$ be a map of superdomains, $U \times B$ a superdomain over B and $V \times B'$ a superdomain over B'. A map of superdomains over b from $U \times B$ to $V \times B'$ is a map of superdomains $\phi\colon U \times B \to V \times B'$ such that the following diagram commutes:

$$\begin{array}{ccc} U \times B & \xrightarrow{\phi} & V \times B' \\ \downarrow & & \downarrow \\ B & \xrightarrow{b} & B' \end{array}$$

For $B' = B$ and $b = \mathrm{id}$ we say that the map ϕ is a map of superdomains over B.

From the definition of the product we know that every map of superdomains $\phi\colon U \times B \to V \times B'$ over $b\colon B \to B'$ can be decomposed in

$$\begin{array}{ccccc} U \times B & \xrightarrow{\overline{\phi}} & V \times B & \xrightarrow{\mathrm{id}_V \times b} & V \times B' \\ \downarrow & & \downarrow & & \downarrow \\ B & \xrightarrow{\mathrm{id}_B} & B & \xrightarrow{b} & B' \end{array}$$

The second square is referred to as "change of base". To understand the map $\overline{\phi}$ one can generalize the Theorem 3.1.3 as follows:

Theorem 3.1.10 (Relative Chart Theorem) *Let $U \subset \mathbb{R}^{m|n}$, $V \subset \mathbb{R}^{p|q}$ and $B \subset \mathbb{R}^{r|s}$ be superdomains with coordinates $X^A = (x^a, \eta^\alpha)$, $Y^B = (y^b, \theta^\beta)$ and $L^C = (l^c, \lambda^\gamma)$ respectively.*

i) Any morphism of superdomains $\phi\colon U \times B \to V \times B$ over B yields the functions $f^B = \phi^\# y^B$, that is, p even and q odd functions of $\mathcal{O}_{U\times B}(\|U \times B\|)$. The functions f^b satisfy

$$\left(f^1{}_{red}(x,l), \ldots, f^p{}_{red}(x,l)\right) \in \|V\| \textit{ for all } (x,l) \in \|U \times B\|. \qquad (3.1.11)$$

ii) Conversely, for any tuple $f^B = (f^b, f^\beta)$ of p even and q odd elements of $\mathcal{O}_{U\times B}(\|U \times B\|)$ satisfying (3.1.11), there is a unique morphism of superdomains $\phi\colon U \times B \to V \times B$ over B such that $\phi^\# Y^B = f^B$.

Proof Since $\phi^\# L^C = L^C$ for maps of superdomains over B, this theorem is a consequence of Theorem 3.1.3. □

Example 3.1.12 (B-Points of $\mathbb{R}^{m|n}$) Let us consider a map ϕ over B from $\mathbb{R}^{0|0} \times B$ to $\mathbb{R}^{m|n} \times B$. By Theorem 3.1.10 we know that we need to give the image $\phi^\# Y^A \in \mathcal{O}_{\mathbb{R}^{0|0}\times B} = \mathcal{O}_B$ of the standard coordinates $Y^A = (y^a, \theta^\alpha)$ on $\mathbb{R}^{m|n}$. The image of y^a under ϕ has to be even and the image of θ^α has to be odd. So B-maps from $\mathbb{R}^{0|0} \times B$ to $\mathbb{R}^{m|n} \times B$ are in one-to-one correspondence to even elements of $\mathcal{O}_B^m \oplus \Pi\mathcal{O}_B^n$. Note that, in contrast to Example 3.1.5, if B contains odd directions the B-points of $\mathbb{R}^{m|n}$ capture the odd directions.

Example 3.1.13 Let $l \in \mathrm{Hom}(\mathcal{O}_B^{m|n}, \mathcal{O}_B^{p|q})$ be an even linear map and ${}_B L^A$ its matrix with respect to the standard basis e_A of $\mathcal{O}_B^{m|n}$ and f_B of $\mathcal{O}_B^{p|q}$, that is, $\langle e_A, l\rangle = {}_A L^B f_B$. There is a map $l\colon \mathbb{R}^{m|n} \times B \to \mathbb{R}^{p|q} \times B$ of superdomains over B which is given in the standard coordinates X^A of $\mathbb{R}^{m|n} \times B$ and Y^B of $\mathbb{R}^{p|q} \times B$ by

$$l^\# Y^B = X^A {}_A L^B.$$

We will call those maps linear maps between superdomains.

Example 3.1.14 An arbitrary map $\varphi\colon \mathbb{R}^{2|2}\times B \to \mathbb{R}^{2|2}\times B$ of superdomains over B can be expressed in coordinates $X^A = (x^a, \eta^\alpha)$ and $Y^B = (y^b, \theta^\beta)$ respectively by

$$\left(\varphi^\# y^b\right)(x,\eta) = {}_0f^b(x) + \eta^\mu\ {}_\mu f^b(x) + \eta^1\eta^2\ {}_{12}f^b(x),$$
$$\left(\varphi^\# \theta^\beta\right)(x,\eta) = {}_0f^\beta(x) + \eta^\mu\ {}_\mu f^\beta(x) + \eta^1\eta^2\ {}_{12}f^\beta(x).$$

Notice that the full Taylor expansion in η appears here, in contrast to Example 3.1.6. The functions ${}_0f^B$, ${}_\mu f^B$ and ${}_{12}f^B$ depend on the coordinates x^a, but also on the functions on B. The functions ${}_0f^b$, ${}_\mu f^\beta$ and ${}_{12}f^b$ are even, whereas the functions ${}_0f^\beta$, ${}_\mu f^b$ and ${}_{12}f^\beta$ are odd.

In the following we will frequently need the following characterization of invertible maps, submersions and immersions in terms of the Jacobi matrix. We start with the generalization of the inverse function theorem for families of superdomains:

Theorem 3.1.15 (Inverse Function Theorem for Families of Superdomains) *Let $U \subset \mathbb{R}^{m|n}$ and $B \subset \mathbb{R}^{p|q}$ be superdomains with coordinates $X^A = (x^a, \eta^\alpha)$ and $L^C = (l^c, \lambda^\gamma)$ respectively. Let $f\colon U\times B \to U\times B$ be a morphism of superdomains over B and $u \in \|U \times B\|$ a point. The following are equivalent:*

i) There is an open neighbourhood $U' \subset U \times B$ of u and a neighbourhood $V' \subset U \times B$ of $f_{red}(u)$ such that the restriction $f|_U'\colon U' \to V'$ of f to U' is invertible.
ii) The matrix $\frac{\partial f^\# X^A}{\partial X^B}$ with entries in $\mathcal{O}_U(U')$ is invertible.
iii) The matrix $\left(\frac{\partial f^\# X^A}{\partial X^B}\right)_{red}(u)$ with entries in $\mathbb{R}$ is invertible.

The following proof is adapted from the proof of Theorem 2.3.1 in Leites (1980) to families of superdomains.

Proof The implication from i) to ii) follows from the chain rule. Indeed, if g is an inverse to f on the open set U', it follows

$$\delta_A^B = \partial_{X^A}\left(f^\# g^\# X^B\right) = \frac{\partial f^\# X^C}{\partial X^A} f^\# \frac{\partial g^\# X^B}{\partial X^C}.$$

The implication ii) to iii) follows from Lemma 2.3.4. It remains to show that iii) implies i) Let us write

$$f^\# x^a = {}_0f^a(x,l) + \sum_{|\underline{\beta}|+|\underline{\gamma}|>1} \eta^{\underline{\beta}}\lambda^{\underline{\gamma}}\ {}_{\underline{\gamma}\underline{\alpha}}f^a(x,l),$$

$$f^\# \eta^\alpha = \lambda^\gamma\ {}_{\gamma 0}f^\alpha(x,l) + \eta^\beta\ {}_{0\beta}f^\alpha(x,l) + \sum_{|\underline{\beta}|+|\underline{\gamma}|>2} \eta^{\underline{\beta}}\lambda^{\underline{\gamma}}\ {}_{\underline{\gamma}\underline{\alpha}}f^\alpha(x,l).$$

The condition that $\left(\frac{\partial f^{\#} X^A}{\partial X^B}\right)_{red}(u)$ is invertible then implies that the matrices $\left(\frac{\partial f^{\#} x^a}{\partial x^b}\right)_{red} = \frac{\partial\, {}_0 f^a}{\partial x^b}$ and $\left(\frac{\partial f^{\#} \eta^\alpha}{\partial \eta^\beta}\right)_{red} = {}_{0\beta} f^\alpha$ are invertible at the point u. Applying the standard implicit function theorem to the map $f_{red}\colon (U \times B)_{red} \to (U \times B)_{red}$, we obtain a diffeomorphism $g_{red}\colon (U \times B)_{red} \supset V'_{red} \to U'_{red} \subset (U \times B)_{red}$, that is $f^{\#}_{red} g^{\#}_{red} x^a = x^a$. Up to restricting U' and V' further, we may assume that ${}_{0\beta} f^\alpha(x, l)$ has an inverse ${}_{0\beta} G^\alpha(x, l)$ in a small neighbourhood of $f(u)$.

Define a homomorphisms $g_1\colon V' \to U'$ by setting

$$g_1^{\#} x^a = g_{red}^{\#} x^a, \qquad g_1^{\#} \eta^\alpha = -\lambda^\gamma\, {}_{\gamma 0} f^\alpha(g_{red}^{\#} x, l) + \eta^\beta\, {}_{0\beta} G^\alpha(g_{red}^{\#} x, l).$$

We obtain that $f^{\#} g_1^{\#} x^a = x^a + \left(f^{\#} g_1^{\#} x^a - x^a\right)$ and $f^{\#} g_1^{\#} \eta^\alpha = \eta^\alpha + \left(f^{\#} g_1^{\#} \eta^\alpha - \eta^\alpha\right)$ for some functions $f^{\#} g_1^{\#} x^a - x^a \in \mathcal{I}_{nil}(U')$ and $f^{\#} g_1^{\#} \eta^\alpha - \eta^\alpha \in \mathcal{I}^2_{nil}(U')$. Consequently for any $r \in \mathcal{I}^k_{nil}(U')$ we have $r - f^{\#} g_1^{\#} r \in \mathcal{I}^{k+1}_{nil}(U')$.

For $k \geq 1$, we define iteratively

$$g_{k+1}^{\#} X^A = g_k^{\#} X^A - g_1^{\#}\left(f^{\#} g_k^{\#} X^A - X^A\right)$$

and obtain

$$\begin{aligned} f^{\#} g_{k+1}^{\#} X^A &= f^{\#} g_k^{\#} X^A - f^{\#} g_1^{\#}\left(f^{\#} g_k^{\#} X^A - X^A\right) \\ &= X^A + \left(f^{\#} g_k^{\#} X^A - X^A\right) - f^{\#} g_1^{\#}\left(f^{\#} g_k^{\#} X^A - X^A\right). \end{aligned}$$

That is,

$$x^a - f^{\#} g_{k+1}^{\#} x^a \in \mathcal{I}^{k+1}_{nil}, \qquad \eta^\alpha - f^{\#} g_{k+1}^{\#} \eta^\alpha \in \mathcal{I}^{k+2}_{nil}.$$

Hence, for $k > n + q$, the morphism g_k is inverse to f. □

Proposition 3.1.16 (Submersions) *Let $U \subset \mathbb{R}^{m+k|n+l}$, $V \subset \mathbb{R}^{m|n}$ and $B \subset \mathbb{R}^{p|q}$ be superdomains with coordinates $X^A = (x^a, \eta^\alpha)$, $Y^B = (y^b, \theta^\beta)$ and $L^C = (l^c, l^\gamma)$ respectively. For a map of superdomains $f\colon U \times B \to V \times B$ over B and $u \in \|U \times B\|$ a point, the following are equivalent:*

i) There exists an open neighbourhood $V' \subset V \times B$ of $f_{red}(u)$ and a superdomain $U' \subset \mathbb{R}^{k|l}$ and an open neighbourhood of u which is isomorphic to $V' \times U'$ such that $f|_{V' \times U'}\colon V' \times U' \to V'$ is given by projection on the first factor.

ii) The matrix $\frac{\partial f^{\#} y^b}{\partial x^a}$ has rank k and the matrix $\frac{\partial f^{\#} \theta^\beta}{\partial \eta^\alpha}$ has rank l.

iii) The matrix $\left(\frac{\partial f^{\#} y^b}{\partial x^a}\right)_{red}(u)$ has rank k and the matrix $\left(\frac{\partial f^{\#} \theta^\beta}{\partial \eta^\alpha}\right)_{red}(u)$ has rank l.

In the case that any of the conditions hold, the map f is called a submersion at u. If f is a submersion for all $u \in \|U \times B\|$, it is called submersion.

Proof The implications i) via ii) to iii) are obvious. For the implication iii) to i) we use the Theorem 3.1.15. Up to reordering the coordinates we may assume that the matrices $\left(\frac{\partial f^\# y^b}{\partial x^a}\right)_{red}(u)$ for $a, b = 1, \ldots, k$ and $\left(\frac{\partial f^\# \theta^\beta}{\partial \eta^\alpha}\right)_{red}(u)$ for $\alpha, \beta = 1, \ldots, l$ have full rank. Then for the superdomain $\mathbb{R}^{k|l}$ with coordinates $Z^D = (z^d, \zeta^\delta)$ we define the map $F: U \times B \to V \times \mathbb{R}^{k|l} \times B$ over B by

$$F^\# Y^B = f^\# Y^B, \qquad F^\# z^d = x^{d+k}, \qquad F^\# \zeta^\delta = \eta^{\delta+l},$$

If we denote by $p_1: V \times \mathbb{R}^{k|l} \times B \to V \times B$ the projection on the first factor, we have that $f = p_1 \circ F$. The differential of F is invertible and hence by Theorem 3.1.15 the map F possesses an inverse G locally around $F_{red}(p)$. Using G as a change of coordinates on $U \times B$ we obtain $f = p_1$ as claimed. □

Proposition 3.1.17 (Immersions) *Let $U \subset \mathbb{R}^{m|n}$, $V \subset \mathbb{R}^{m+k|n+l}$ and $B \subset \mathbb{R}^{p|q}$ be superdomains with coordinates $X^A = (x^a, \eta^\alpha)$, $Y^B = (y^b, \theta^\beta)$ and $L^C = (l^c, l^\gamma)$ respectively. For a map of superdomains $f: U \times B \to V \times B$ over B and $u \in \|U \times B\|$ a point, the following are equivalent:*

i) There exists an open neighbourhood $U' \subset U \times B$ of u and a superdomain $V' \subset \mathbb{R}^{k|l}$ and an open neighbourhood of $f_{red}(u)$ which is isomorphic to $U' \times V'$ such that $f|_{U'}: U' \to U' \times V'$ is given by inclusion in the first factor.

ii) The matrix $\frac{\partial f^\# y^b}{\partial x^a}$ has rank m and the matrix $\frac{\partial f^\# \theta^\beta}{\partial \eta^\alpha}$ has rank n.

iii) The matrix $\left(\frac{\partial f^\# y^b}{\partial x^a}\right)_{red}(u)$ has rank m and the matrix $\left(\frac{\partial f^\# \theta^\beta}{\partial \eta^\alpha}\right)_{red}(u)$ has rank n.

In the case that any of the conditions hold, the map f is called an immersion at u. If f is an immersion for all $u \in \|U \times B\|$, it is called immersion.

The proof of Proposition 3.1.17 is very similar to the proof of Proposition 3.1.16 and omitted here, see Leites 1980, §2.3.7.

3.2 Families of Supermanifolds

Definition 3.2.1 (Supermanifold) A supermanifold of dimension $m|n$ is a pair $(\|M\|, \mathcal{O}_M)$ consisting of a second countable, Hausdorff topological space $\|M\|$ and a sheaf of superrings $\mathcal{O}_M$ on $\|M\|$ such that every point $p \in \|M\|$ has a neighbourhood that is isomorphic to a superdomain of dimension $m|n$. Maps of supermanifolds are maps of ringed spaces.

Definition 3.2.2 (Coordinates) Let $U \subset M$ be an open superdomain in a supermanifold M and $\varphi: U \to V \subset \mathbb{R}^{m|n}$ an isomorphism. The pullback $\varphi^\# X^A$ of the standard coordinates X^A on V are called local coordinates on M, and φ is called a chart. Let $\varphi_i: U_i \to V_i$ for $i = 1, 2$ be two local charts such that $U_1 \cap U_2 \neq \emptyset$.

Then the functions $\left(\varphi_2^{-1}\right)^{\#} \circ \varphi_1^{\#} X^A$ on $\varphi_2(U_1 \cap U_2) \subseteq V_2$ are called a change of coordinates from V_1 to V_2.

Example 3.2.3 (Split Supermanifolds) Any "classical" smooth manifold $|M|$ of dimension m is a supermanifold of dimension $m|0$, compare Example 3.1.2. Let $E \to |M|$ be a vector bundle over $|M|$ of rank n. Let us denote by $\Gamma\left(\bigwedge E\right)$ the sheaf of sections of the exterior algebra of E. Then $M = (\|M\|, \Gamma\left(\bigwedge E\right))$, where $\|M\|$ denotes the underlying topological space of $|M|$, is a supermanifold of dimension $m|n$. Supermanifolds obtained in this way from sections of an exterior algebra are called split supermanifolds.

To see that M is a supermanifold, let x^a be coordinates on $U \subseteq |M|$, and η^α be a frame of E on the same domain. The sheaf $\Gamma\left(\bigwedge E\right)\big|_U$ is isomorphic to $C^\infty(U, \mathbb{R}) \otimes \bigwedge_n$, where the sections η^α generate the Grassmann algebra. Hence (x^a, η^α) form a set of supercoordinates for M. Let (y^b, θ^β) be a second set of such coordinates. Then the coordinate transformation is of the special form

$$y^b = f^b(x) \qquad\qquad \theta^\beta = \eta^\alpha{}_\alpha f^\beta(x), \tag{3.2.4}$$

that is, y^b does not depend on η and θ^β depends only linearly on η. Consequently, a supermanifold obtained from the sections of an exterior algebra of a vector bundle allows for an atlas, such that all coordinate changes are of the restricted form (3.2.4). As a general coordinate change may have higher terms, it is surprising that by a theorem due to Marjorie Batchelor, every smooth supermanifold is isomorphic to a split one, sec Batchelor (1979). Notice, however, that Batchelor's theorem is not valid for complex supermanifolds (to be studied in Chap. 7 and Part II).

Example 3.2.5 (Projective Superspaces) The real projective superspace of dimension $m|n$, denoted by $\mathbb{P}_{\mathbb{R}}^{m|n}$, is covered by $m+1$ superdomains $U_i = \mathbb{R}^{m|n}$, $i = 1, \ldots, m+1$ with coordinates $X_i^A = (x_i^a, \eta_i^\alpha)$. The coordinate changes φ_{ij} for $i < j$ are given by

$$\left(x_i^1, \ldots, x_i^m\right) = \left(\frac{x_j^1}{x_j^i}, \ldots, \frac{x_j^{i-1}}{x_j^i}, \frac{x_j^{i+1}}{x_j^i}, \ldots, \frac{x_j^{j-1}}{x_j^i}, \frac{1}{x_j^i}, \frac{x_j^j}{x_j^i}, \ldots, \frac{x_j^m}{x_j^i}\right),$$

$$\left(\eta_i^1, \ldots, \eta_i^n\right) = \left(\frac{\eta_j^1}{x_j^i}, \ldots, \frac{\eta_j^n}{x_j^i}\right).$$

Obviously we obtain that $\mathbb{P}_{\mathbb{R}}^{m|0}$ coincides with the classical definition of the m-dimensional projective space. However, an interpretation of $\mathbb{P}_{\mathbb{R}}^{m|n}$ as the set of (even) lines in $\mathbb{R}^{m+1|n}$ will only be possible later once we have introduced families of supermanifolds.

The Definition 3.2.1 of supermanifolds has been used and studied intensively since the 1980s. A lot of geometrical concepts from differential geometry can be generalized to supergeometry, using this definition of supermanifold. For an overview we may refer to Leites (1980); Manin (1988) and Carmeli et al. (2011). As argued in Sect. 3.1, it is however necessary to generalize to families of supermanifolds. Families of supermanifolds are implicitly used for a long time and have appeared explicitly in Deligne and Morgan (1999).

Definition 3.2.6 Let B be a supermanifold and M, N supermanifolds with morphisms $b_M\colon M \to B$ and $b_N\colon N \to B$. The fibered product of M and N over B is a supermanifold $M \times_B N$ with morphisms of supermanifolds $p_M\colon M \times_B N \to M$ and $p_N\colon M \times_B N \to N$, called projection morphisms, such that

i) the following diagram commutes

$$\begin{CD} M \times_B N @>{p_N}>> N \\ @V{p_M}VV @VV{b_N}V \\ M @>{b_M}>> B \end{CD}$$

ii) for all supermanifolds L with morphisms of supermanifolds $f\colon L \to M$ and $g\colon L \to N$ such that $b_N \circ g = b_M \circ f$ there exists a unique morphism of supermanifolds $(f, g)\colon L \to M \times_B N$ such that the following diagram commutes:

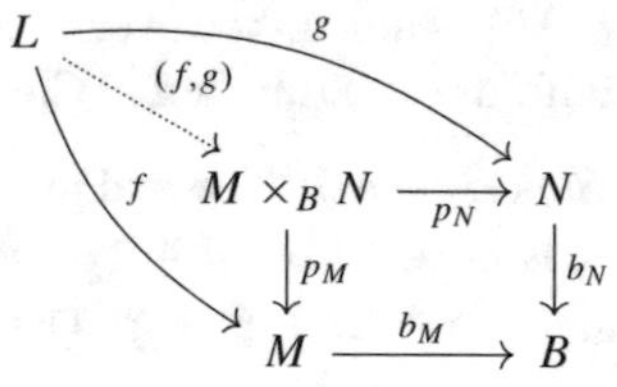

By the universality property of the definition, the fibered product of supermanifolds is unique up to isomorphism. For existence results, see the following Example 3.2.7 and Proposition 3.2.11.

Example 3.2.7 (Cartesian Product) For any supermanifold M there is a unique map $M \to \mathbb{R}^{0|0}$. Consequently, $\mathbb{R}^{0|0}$ is the final object in the category of supermanifolds. The cartesian product $M \times N = M \times_{\mathbb{R}^{0|0}} N$ exists for all supermanifolds M and N. This is a consequence of Lemma 3.1.8.

Definition 3.2.8 (Families of Supermanifolds) A supermanifold M with a submersion $b_M\colon M \to B$ is also called a family of supermanifolds over the base B or shorter a supermanifold over B. By Proposition 3.1.16, for every point $p \in M$ there is a neighbourhood $U \subseteq M$ such that b_M coincides with the projection $U \simeq U_1 \times U_2 \to U_2$ for $U_1 \subseteq \mathbb{R}^{m|n}$ and $U_2 \subseteq B$. In this case we call M a supermanifold of relative dimension $m|n$ over B. If M can be written as a product globally, that is, $M = M' \times B$, we say that M is a trivial family over B.

Let M be a supermanifold over B and M' a supermanifold over B' and $b\colon B \to B'$ a map of supermanifolds. A map of supermanifolds $f\colon M \to M'$ over B is a map of supermanifolds, such that the diagram

$$\begin{array}{ccc} M & \xrightarrow{f} & M' \\ \downarrow{\scriptstyle b_M} & & \downarrow{\scriptstyle b_{M'}} \\ B & \xrightarrow{b} & B' \end{array}$$

commutes. For $B = B'$ and $b = \mathrm{id}_B$, the map f is called a map of supermanifolds over B. The set of all maps of supermanifolds over B from M to M' is denoted $\mathrm{Hom}_B(M, M')$, and the category of supermanifolds over B will be denoted by SMan_B.

The subset of $\mathrm{Hom}_B(M, M')$ that is invertible will be denoted $\mathrm{Diff}_B(M, M')$ and in the case $M = M'$ we will write $\mathrm{End}_B(M) = \mathrm{Hom}_B(M, M)$ and $\mathrm{Diff}_B(M) = \mathrm{Diff}_B(M, M)$. Elements of $\mathrm{Diff}_B(M, M')$ are called diffeomorphisms.

By definition it is possible to cover every supermanifold over B of relative dimension $m|n$ by charts of the form $U \times V$ where $U \subseteq \mathbb{R}^{m|n}$, $V \subseteq B$ and the projection b_M is locally given by the projection to the second factor $U \times V \to V$. Such charts are called relative charts. The pullback of the standard coordinates on U by the chart map are called relative coordinates on M. By restricting the open set $V \subseteq B$ further, one can assume that all coordinate changes between relative coordinates are of the form as in Theorem 3.1.10. If the family is trivial one can find an atlas such that all coordinate changes respect the product structure. In that case the coordinate changes are independent from coordinates of the base B.

Let $f\colon M \to N$ be a map of supermanifolds over B. As M and N can be covered by relative coordinate charts, the map f can be given in coordinates, as in Theorem 3.1.10.

Remark 3.2.9 Every supermanifold M as in Definition 3.2.1 is trivially a family of supermanifolds over the point $\mathbb{R}^{0|0}$. Indeed, the projection $M = M \times \mathbb{R}^{0|0} \to \mathbb{R}^{0|0}$ is a submersion. More generally, for an arbitrary supermanifold B, the product $M \times B$ is a trivial family of supermanifolds over B. This ties in with the general Principle 3.2.12 to be discussed later on.

Remark 3.2.10 An open submanifold $U \subseteq M$ of a family $b_M\colon M \to B$ may not be a supermanifold over B. This is because the map $b_M|_U : U \to B$ is no longer a submersion. For this one needs to restrict the image as well.

The following proposition assures the existence of fibered products over B if one of the maps to B is a submersion. This will have numerous consequences.

Proposition 3.2.11 *Let $b_M \colon M \to B$ be a submersion, and $b \colon B' \to B$ a map of supermanifolds. The fibered product $M \times_B B'$ of M and B' over B exists and is a supermanifold over B'.*

Proof Let us first treat the case, where M is a trivial family $M = M' \times B$. It is obvious that in that case $M \times_B B' = M' \times B'$. Any map $f \colon M = M' \times B \to N = N' \times B$ of trivial families of supermanifolds over B extends to a map $f' \colon M' \times B' \to N' \times B'$ in the following way. By the properties of the product, we know that $f = (\overline{f}, \mathrm{id}_B)$, where $\overline{f} = f \circ p_{N'} \colon M' \times B \to N'$. The map f' is then given by $f' = (\overline{f} \circ b, \mathrm{id}_{B'})$.

Any non-trivial family of supermanifolds M is locally a trivial family of superdomains $U \times V \to V$. The family of superdomains $U \times b^{-1}(V)$ forms a local patch of $M \times_B B'$. The different local patches of $M \times_B B'$ glue together because the glueings from M extend to $M \times_B B'$ as seen in the local case. □

Proposition 3.2.11 shows the existence of the fiber product $M \times_B N$ for families of supermanifolds $b_M \colon M \to B$ and $b_N \colon N \to B$. As $M \times_B N$ fulfills the universal property of the cartesian product in the category SMan_B, we will sometimes drop the index B from the notation and simply write $M \times N$ for the product of the families $M \to B$ and $N \to B$.

Furthermore, for any family of supermanifolds $b_M \colon M \to B$ and a morphism $b \colon B' \to B$ there exists a family of supermanifolds $M' = M \times_B B' \to B'$ by Proposition 3.2.11. Occasionally one might also denote M' as b^*M. This procedure is also called base change, as the base of the family $M \to B$ is being replaced by B'. For any map of supermanifolds $f \colon M \to N$ over B the map $f \times \mathrm{id}_{B'} \colon b^*M \to b^*N$ is a map of supermanifolds over B'. Hence for any map $b \colon B' \to B$ we obtain a covariant functor $b^* \colon \mathsf{SMan}_B \to \mathsf{SMan}_{B'}$.

We have seen in Sect. 3.1, especially in the Examples 3.1.6 and 3.1.14 respectively Examples 3.1.5 and 3.1.12 that a larger base leads to a more complete view on the geometry of supermanifolds. We will consequently assume that any supermanifold is implicitly to be understood as a family of supermanifolds over a base B and any morphism of supermanifolds is a morphism of families of supermanifolds over B. We will furthermore assume that the base of the family is always "large enough", that is, allow arbitrary base changes without notice.

The questions that arises is how geometric properties of a family of supermanifolds M should behave under this base change. Roughly, we should only be interested in properties that are invariant under base change:

Principle 3.2.12 (Base Change) *Any supermanifold M is to be understood implicitly as a family $b_M \colon M \to B$ of supermanifolds. Any map of supermanifolds is to be understood implicitly as a map of supermanifolds over a basis B.*

One should only consider properties of a supermanifolds or maps that are invariant under base change. That is, if a property holds for a supermanifold M or a map $f \colon M \to N$ this property should also hold for b^*M and b^*f, where $b \colon B' \to B$ is any map of supermanifolds.

Properties that are invariant under base change have been called "geometric" in Deligne and Morgan (1999, §2.9). Principle 3.2.12 will hold for all constructions in this work, even though it will not always be formulated explicitly. In practice, most constructions build on the local model of trivial families, glueing and fiber products. Then Principle 3.2.12 is obeyed automatically.

Let us look at examples of properties that are invariant under base change:

Example 3.2.13 (Invariance of the Cartesian Product Under Base Change) Let $b_M\colon M \to B$ and $b_N\colon N \to B$ be families of supermanifolds over B and $b\colon B' \to B$ a map. We want to show that $b^*(M \times N) = b^*M \times b^*N$, where the first product is in the category SMan_B and the second in $\mathsf{SMan}_{B'}$. More explicitly, we have to show that

$$\left(M \times_B B'\right) \times_{B'} \left(N \times_B B'\right) = (M \times_B N) \times_B B'.$$

This claim follows from the associativity of the fiber product and that $B' \times_{B'} P = P$ for any supermanifold P over B'.

Example 3.2.14 (Base Change and Trivial Families) Let $b_M\colon M \to B$ be a trivial family and $b\colon B' \to B$ a map of supermanifolds. Then b^*M is also a trivial family. Hence the property of a family to be trivial is invariant under base change.

In contrast, the property of a family to be non-trivial is not invariant under base change. Indeed, for $M \to B$ a non-trivial family and $b\colon \mathbb{R}^{0|0} \to B$, the family $b^*M \to \mathbb{R}^{0|0}$ is trivial.

Example 3.2.15 (Base Change and Relative Coordinates) Let M be a family of supermanifolds over B and $U \times V$ an relative chart for M with $U \subseteq \mathbb{R}^{m|n}$ and $V \subseteq B$. For any $b\colon B' \to B$ by construction of the fibered product $U \times b^{-1}(V)$ is an relative chart of b^*M and this is as well compatible with composition.

Example 3.2.16 (Base Change and Reduced Space) The reduced space of a supermanifold is not compatible with Principle 3.2.12. For a family of supermanifolds $M \to B$, the reduced space M_{red} is a family over B_{red}. Hence, in general, M_{red} is not a family of supermanifolds over B and the map $i_{red}\colon M_{red} \to M$ is not a map of supermanifolds over B. In Sect. 3.3 we will introduce the concept of underlying even manifold which can be seen as a replacement for the reduced space that is invariant under base change.

Further examples of invariance under base change, to be studied later are given by the tangent bundle ($b^*TM = T(b^*M)$, see Sect. 4.3) and the functoriality of the integral (see Lemma 8.3.6).

3.3 Underlying Even Manifolds

We have argued in Example 3.2.16 that the reduced space is not compatible with the principle of base change, see Principle 3.2.12. In this section, we introduce the concept of the underlying even manifold of a supermanifold that generalizes the concept of reduced space to families of supermanifolds in a way compatible with Principle 3.2.12.

Notice that for a superdomain $\mathbb{R}^{m|n}$ over $\mathbb{R}^{0|0}$ the reduced space is just $\mathbb{R}^m$ with the sheaf of smooth functions on $\mathbb{R}^m$. Consequently for a supermanifold M of dimension $m|n$ over $\mathbb{R}^{0|0}$ the reduced space M_{red} is a manifold of dimension $m|0$. The embedding $i\colon \mathbb{R}^m \hookrightarrow \mathbb{R}^{m|n}$ is given in standard coordinates x^a, η^α on $\mathbb{R}^{m|n}$ by

$$i^\# x^a = x^a \qquad\qquad i^\# \eta^\alpha = 0.$$

Starting in Chap. 11 where we study the relation between super Riemann surfaces and Riemann surfaces, we are interested in families of dimension $m|0$ over B that embed into families of dimension $m|n$. This leads to the following definition:

Definition 3.3.1 Let $M = (\|M\|, \mathcal{O}_M)$ be a family of supermanifolds of dimension $m|n$ over B. A family of supermanifolds $|M| = (\|M\|, \mathcal{O}_{|M|})$ of dimension $m|0$ together with an embedding of families of supermanifolds $i\colon |M| \to M$ which is the identity on the underlying topological space is called an underlying even manifold.

Example 3.3.2 (Underlying Even Manifolds for $\mathbb{R}^{m|n}$) Consider an embedding $i\colon \mathbb{R}^{m|0} \times B \to \mathbb{R}^{m|n} \times B$ such that $\|i\|$ is the identity. Denote the standard coordinates on $\mathbb{R}^{m|0}$ by y^a and the standard coordinates on $\mathbb{R}^{m|n}$ by (x^b, η^β). Then i can be expressed in coordinates:

$$i^\# x^b = g^b(y), \qquad\qquad i^\# \eta^\beta = g^\beta(y). \tag{3.3.3}$$

If one chooses coordinates $L^C = (l^c, \lambda^\gamma)$ on B one can expand the even function $g^b(y)$ further

$$i^\# x^b = g^b(y) = y^b + \sum_{\underline{\nu}\neq 0} \lambda^{\underline{\nu}}{}_{\underline{\nu}}g^b(y, l).$$

Here the zero order term is given by the fact that $\|i\|$ should be the identity. However, the functions ${}_{\underline{\nu}}g^b(y, l)$ and $g^\beta(y) = g^\beta(y, l, \lambda)$ are arbitrary (with the sole exception of prescribed parity). Consequently, underlying even manifolds of the trivial family $\mathbb{R}^{m|n} \times B$ are not at all unique if $B \neq B_{red}$, in contrast to the reduced space.

It is always possible to find relative coordinates $(\tilde{x}^b, \tilde{\eta}^\beta)$ on $\mathbb{R}^{m|n} \times B$ and relative coordinates $\tilde{y}^a$ on $\mathbb{R}^{m|0} \times B$ such that

$$i^\# \tilde{x}^a = \tilde{y}^a, \qquad\qquad i^\# \tilde{\eta}^\beta = 0. \tag{3.3.4}$$

Indeed, using the coordinate transformation

$$\tilde{y}^a = y^b + \sum_{\underline{\nu} \neq 0} \lambda^{\underline{\nu}}{}_{\underline{\nu}} g^b(y, l)$$

on $\mathbb{R}^{m|0} \times B$ and the coordinate change

$$\tilde{x}^b = x^b \qquad\qquad \tilde{\eta}^\beta = -g^\beta(x) + \eta^\beta$$

on $\mathbb{R}^{m|n} \times B$ assures the Eq. (3.3.4). Put differently, there are diffeomorphisms $\xi \in \mathrm{Diff}_B(\mathbb{R}^{m|0} \times B)$ and $\Xi \in \mathrm{Diff}_B(\mathbb{R}^{m|n} \times B)$ such that $\Xi \circ i \circ \xi$ coincides with the standard underlying even manifold of $\mathbb{R}^{m|n} \times B$ given by Eq. (3.3.4).

There are automorphisms Ξ of $\mathbb{R}^{m|n} \times B$ such that $i \circ \Xi = i$. Those can best be expressed in the coordinates $\tilde{x}^b$, $\tilde{\eta}^\beta$ as

$$\Xi^\# \tilde{x}^b = \tilde{x}^b + \tilde{\eta}^\mu f_\mu^b(\tilde{x}, \tilde{\eta}), \qquad\qquad \Xi^\# \tilde{\eta}^\beta = \tilde{\eta}^\mu f_\mu^\beta(\tilde{x}, \tilde{\eta}).$$

The functions f_μ^B are arbitrary functions on $\mathbb{R}^{m|n} \times B$ with appropriate parity.

We will now generalize the results of Example 3.3.2 to families of supermanifolds. To every family of supermanifolds there exists an underlying even manifold (Theorem 3.3.7) that is non-unique if $B \neq B_{red}$. Furthermore, any two underlying even manifolds are diffeomorphic (Corollary 3.3.13). To prove those theorems we need some preparation.

Definition 3.3.5 (Support of a Section) The support of an arbitrary section s of the sheaf $\mathcal{F}$ of rings or modules on M is defined to be the complement of the set where s and the zero section s_0 coincide:

$$\mathrm{supp}\,(s) = \|M\| \setminus \left(\bigcup_{\{U \subseteq \|M\| \text{ open} | s|_U = s_0|_U\}} U \right)$$

The support supp(s) is a closed subset of $\|M\|$ because it is the complement of a union of open sets.

Proposition 3.3.6 (Proposition 4.2.7 in Carmeli et al. 2011) *Let $\{U_i\}_{i \in I}$ be an open cover of the supermanifold M. Then there exists a partition of unity dominated* *by* $\{U_i\}$. *That is, there are even global sections* $\{f_i\}_{i\in I}$ *of* $\mathcal{O}_M$ *such that*

i) $\operatorname{supp} f_i \subseteq U_i$ *for all* $i \in I$ *and the indexed family* $\{\operatorname{supp} f_i\}$ *is locally finite,*
ii) $(f_i)_{red} \geq 0$ *for all* $i \in I$,
iii) $\sum_{i\in I} f_i = 1$.

Proof Let us first assume that all the U_i are coordinate neighbourhoods with compact closure in $\|M\|$. As in the proof of existence of partitions of unity for topological manifolds, see Munkres (2000, Theorem 41.7), we may assume that there is a locally finite cover $\{V_i\}_{i\in I}$ of $\|M\|$ such that the closure $\overline{V_i} \subset U_i$ for all $i \in I$. Furthermore, there exists a locally finite cover $\{W_i\}_{i\in I}$ such that $\overline{W_i} \subset V_i$ for all $i \in I$. For every $i \in I$ there exists a positive bump function $g_i \in C^\infty(\|U_i\|, \mathbb{R})$ such that $\operatorname{supp} g_i \subset V_i$ and $g_i = 1$ on W_i. Since U_i is a coordinate neighbourhood, g_i can be lifted to $\mathcal{O}_M(\|U_i\|)$ and extended to $\mathcal{O}_M(\|M\|)$ by zero. Since $\{V_i\}$ is locally finite, the following sum is well defined

$$g = \sum_{i\in I} g_i .$$

Since all $(g_i)_{red}$ are positive, the function g_{red} is also positive and hence g is invertible. The functions $f_i = g_i/g$ are a partition of unity dominated by $\{U_i\}$.

The general case, of U_i not necessarily coordinate neighbourhoods can be reduced to the previous case by considering a refinement $\{V_j\}$ of $\{U_i\}$ consisting of coordinate neighbourhoods with compact closure. If g_j is a partition of unity dominated by V_j, a partition of unity dominated by $\{U_i\}$ is given by

$$f_i = \sum_{\{j\in J | V_j \subset U_j\}} g_j . \qquad \square$$

Theorem 3.3.7 *Let* $M = (\|M\|, \mathcal{O}_M)$ *be a family of supermanifolds over* B. *Also, let* $\|U_1\| \subseteq \|M\|$ *be a subset (which might also be empty) such that there is an underlying even manifold* $|U_1|$ *with given embedding* $i_U : |U_1| \to U_1$ *and* $U_2 \subset U_1$ *an open subset such that its closure is contained in* U_1. *There exists an underlying manifold* $|M|$ *and an embedding* $i : |M| \to M$ *such that* $|U_1|$ *coincides with* $|M|$ *and* i *with* i_U *over* $\|U_2\|$.

Proof Let V_k be an open cover of the family $b_M : M \to B$ by relative coordinate charts V_k. As M is paracompact, we may assume that V_k is a countable cover, see Munkres (2000, Theorem 30.3), hence $k = 1, \ldots$. Let us write $V_k = F_k \times b_M(V_k)$ with coordinates $X_k^A = (x_k^a, \eta_k^\alpha)$ on F_k. We will denote the coordinate changes as follows:

$$f_{kl}^{\#} X_l^A = f_{kl}^A(X_k) = \sum_{\underline{\nu}} \eta_k^{\underline{\nu}} {}_{\underline{\nu}} f_{kl}^A(x_k) .$$

Here the sum runs over all odd multiindices $\underline{\nu}$ including zero. The manifold $|M|$ that we are going to construct is covered by the same open sets $\|V_k\| = \|F_k\| \times \|b_M(V_k)\|$ and have relative coordinates y_k^a such that $\left(y_k^a\right)_{red} = \left(x_k^a\right)_{red}$. Notice that the coordinate changes $h_{kl}^{\#} y_l^a = h_{kl}^a(y_k)$ need to be constructed in the proof.

We construct a family $b_{|M|}\colon |M| \to B$ of relative dimension $m|0$ and a map $i\colon |M| \to M$ over B inductively. To start the induction we may assume without loss of generality that U_1 is covered by the first j open sets, that is

$$U_1 = \bigcup_{k=1}^{j} V_k.$$

Furthermore, we assume that

$$U_2 \cap \bigcup_{k>j} V_k = \emptyset.$$

If $U_1 = \emptyset$ choose an arbitrary embedding $i|_{V_1}\colon |V_1| \to V_1$ over $b_M(V_1)$ as in Example 3.3.2.

Suppose now that we have the structure of an underlying even manifold together with the embedding i for $\bigcup_{k=0}^{m-1} V_k$. We assume that i is given in the coordinates X_k^A and y_k^a by

$$i^{\#} x_k^a = y_k^a + g_k^a(y_k), \qquad\qquad i^{\#} \eta_k^\alpha = g_k^\alpha(y_k),$$

where g_k^a is an even nilpotent function and g_k^α an odd nilpotent function. We will show that we can extend the underlying even manifold structure and the embedding i to $\bigcup_{k=0}^{m} V_k$. In order to extend the manifold structure we have to give the coordinate changes h_{km}. For g_m^A to describe an extension of the given i we need that the following compatibility conditions hold on $V_k \cap V_m$ for all $k < m$:

$$i^{\#} f_{km}^{\#} X_m^A = h_{km}^{\#} i^{\#} X_m^A \tag{3.3.8}$$

By what has been discussed in Example 3.3.2, we may assume that $g_k^A = 0$ for all $k < m$. Hence the compatibility conditions (3.3.8) read

$$\begin{aligned} {}_0 f_{km}^a(y_k) &= i^{\#} f_{km}^{\#} x_m^a = h_{km}^{\#} i^{\#} x_m^a = h_{km}^{\#} \left(y_m^a + g_m^a(y_m)\right), \\ {}_0 f_{km}^\alpha(y_k) &= i^{\#} f_{km}^{\#} \eta^\alpha = h_{km}^{\#} i^{\#} \eta^\alpha = h_{km}^{\#} g_m^\alpha(y_m). \end{aligned}$$

For $g_m^a = 0$ the first equation can be read as a definition of h_{km}, whereas the second equation specifies g_m^α on $V_m \cap V_k$. However, the function g_m^α may not extend to the whole of V_m because it may be unbounded. Let $\{\sigma, \tau\}$ be a partition of unity dominated by $\{\bigcup_{k=0}^{m-1} V_k, V_m\}$. The function $t^\alpha = \sigma g_m^\alpha$ defined on the set $V_m \cap \bigcup_{k=0}^{m-1} V_k$ can be extended to V_m by zero. We will now construct $\tilde{h}_{kl}$ and an embedding j that coincide with h_{kl} and i respectively on $\bigcup_{k=0}^{m-1} V_k \setminus V_m$ such that

$$j^\# x_m^a = y_m^a, \qquad j^\# \eta_m^\alpha = t^\alpha. \tag{3.3.9}$$

Hence the manifold structure and the embedding j extend to $\bigcup_{k=0}^{m} V_k$.

Let j be in the coordinates X_k^A be given by

$$j^\# x_k^a = y_k^a, \qquad j^\# \eta_k^\alpha = \tilde{g}_k^\alpha(y_k).$$

The coordinate changes $\tilde{h}_{kl}$ are then determined by the compatibility conditions (3.3.8):

$$\sum_{\underline{\nu}} \tilde{g}_k^{\underline{\nu}}(y_k)\, {}_{\underline{\nu}} f_{kl}^a(y_k) = j^\# f_{kl}^\# x^a = \tilde{h}_{kl}^\# j^\# x^a = \tilde{h}_{kl}^\# y^a$$

Notice that $\tilde{h}_{kl}$ differs from h_{kl} only by a nilpotent term dependent on g_k^α. Furthermore, the functions $\tilde{h}_{kl}$ satisfy the cocycle conditions because f_{kl} satisfy them:

$$\tilde{h}_{kl}^\# \tilde{h}_{lp}^\# y_p^a = \tilde{h}_{kl} j^\# f_{lp}^\# x_p^a = j^\# f_{kl}^\# f_{lp}^\# x_p^a = j^\# f_{kp}^\# x_p^a = \tilde{h}_{kp}^\# y_p^a$$

It remains to see that Eq. (3.3.9) determines $\tilde{g}_k^\alpha$ uniquely. We have to expand $\tilde{h}_{km}^\# \left(j^\# \eta_m^\alpha - t^\alpha\right)$ with respect to coordinates $L_k^A = (l_k^a, \lambda_k^\alpha)$ of the base:

$$\begin{aligned}
0 = \tilde{h}_{km}^\# \left(j^\# \eta_m^\alpha - t^\alpha\right) &= j^\# f_{km}^\# \eta_m^\alpha - \tilde{h}_{km}^\# t^\alpha = \sum_{\underline{\nu}} \tilde{g}_k^{\underline{\nu}}\, {}_{\underline{\nu}} f_{km}^\alpha - \tilde{h}_{km}^\# t^\alpha \\
&= \sum_{\underline{\kappa} \neq 0} \lambda^{\underline{\kappa}} \left({}_{\underline{\kappa}0} f_{km}^\alpha + {}_{\underline{\kappa}}\tilde{g}_k{}^\nu\, {}_{0\nu} f_{km}{}^\alpha - {}_{\underline{\kappa}}\left(\sigma\, {}_0 f_{km}{}^\alpha\right) + {}_{\underline{\kappa}} R^\alpha \right) \\
&= \sum_{\underline{\kappa} \neq 0} \lambda^{\underline{\kappa}} \left({}_{\underline{\kappa}}\tilde{g}_k{}^\nu\, {}_{0\nu} f_{km}{}^\alpha + {}_{\underline{\kappa}}\left(\tau\, {}_0 f_{km}{}^\alpha\right) + {}_{\underline{\kappa}} R^\alpha \right)
\end{aligned} \tag{3.3.10}$$

Here the additional leftmost indices of ${}_0 f_{km}^\alpha$, $\tilde{g}_k^\nu$ and σ indicate the λ-dependence. The term ${}_{\underline{\kappa}} R^\alpha$ contains all terms containing ${}_{\underline{\pi}}\tilde{g}_k{}^\alpha$ of order lower than $\underline{\kappa}$. The matrix ${}_{0\nu} f_{km}{}^\alpha$ is invertible because the coordinate change f_{km} is invertible. Hence Eq. (3.3.10) is solvable by recursion. The support of $\tilde{g}_k^\alpha$ is contained in the support of τ. Consequently j and $\tilde{h}_{kl}$ coincide with i and h_{kl} outside of V_m. □

Notice that in this proof $|M|$ is constructed as a family of supermanifolds. It is not clear, a priori, that for two underlying even manifolds $i\colon |M|_1 \to M$ and $j\colon |M|_2 \to M$ the families of supermanifolds $|M|_1$ and $|M|_2$ are isomorphic. Roughly, as their reduced manifolds must coincide, they can only differ in higher nilpotent terms of the coordinate changes. It is a consequence of the following proposition that $|M|_1$ and $|M|_2$ are indeed diffeomorphic.

Proposition 3.3.11 *Let M and N be families of supermanifolds over B with odd dimension zero and $\zeta : M_{red} \to N_{red}$ a diffeomorphism over B_{red}. There exists a diffeomorphism $\xi : M \to N$ of families over B such that $\xi_{red} = \zeta$.*

Proof The proof is by induction over open sets as in Theorem 3.3.7. Let M be covered by countably many coordinate neighbourhoods U_k, for $k = 1, 2, \ldots$ with coordinates x_k^a and coordinate changes h_{kl}. The manifold M_{red} can be covered by the same coordinate neighbourhoods U_k, but the coordinate changes are given by ${}_0h_{kl} = (h_{kl})_{red}$. Similarly, let N be covered by coordinate neighbourhoods V_k with coordinates y_k^a and the corresponding coordinate changes g_{kl}. Then the coordinate changes on N_{red} are given by ${}_0g_{kl} = (g_{kl})_{red}$. We may furthermore assume that $\zeta(U_k) = V_k$.

The maps ζ and the map ξ to be constructed differ only by nilpotent terms:

$$\xi^{\#} y_k^a = \zeta^{\#} y_k^a + r_k^a(x_k)$$

In particular ξ is invertible for any choice of r_k^a because ζ is invertible. It remains to construct the terms r_k^a in a consistent way, that is such that ξ fulfils

$$h_{lk}^{\#} \xi^{\#} y_k^a = \xi^{\#} g_{lk}^{\#} y_k^a \tag{3.3.12}$$

for all k and l wherever h_{lk} and g_{lk} are defined. This is feasible because the reduction of Eq. (3.3.12) yields a corresponding compatibility condition for ζ.

On the first open sets U_1 and V_1 the functions r_1^a can be defined arbitrarily, for example $r_1^a = 0$. Let us now assume that ξ is constructed on $\bigcup_{k=1}^{m-1} U_k$ and show that we can extend ξ to $\bigcup_{k=1}^{m} U_k$. The compatibility condition (3.3.12) defines the functions r_m^a on $U_m \cap \bigcup_{k=1}^{m-1} U_k$. However r_m^a may not extend to the whole of U_m because it may be unbounded. Let $\{\sigma, \tau\}$ be a partition of unity dominated by $\{\bigcup_{k=1}^{m-1} U_k, U_m\}$. Then σr_m^a extends to U_m by zero. Define

$$\tilde{\xi}^{\#} y_m^a = \zeta^{\#} y_m^a + \sigma r_m^a.$$

We will show that there is a diffeomorphism $\tilde{\xi} : \bigcup_{k=1}^{m} U_k \to \bigcup_{k=1}^{m} V_k$ that coincides with ξ on $\bigcup_{k=1}^{m-1} U_k \setminus U_m$. Let $\tilde{\xi}$ be given by

$$\tilde{\xi}^{\#} y_k^a = \zeta^{\#} y_k^a + r_k^a + \tilde{r}_k^a,$$

for nilpotent $\tilde{r}_k^a$. On U_m the terms $\tilde{r}_k^a$ can be obtained using the compatibility condition (3.3.12) for $\tilde{\xi}$:

$$h_{km}^{\#} \tilde{\xi}^{\#} g_{mk}^{\#} y_k^a = \tilde{\xi}^{\#} y_k^a = \zeta^{\#} y_k^a + r_k^a + \tilde{r}_k^a.$$

The support of $\tilde{r}_k^a$ is a subset of the support of τ, hence ξ coincides with $\tilde{\xi}$ on $\bigcup_{k=1}^{m-1} U_k \setminus U_m$. □

Corollary 3.3.13 *Let M be a supermanifold and $i\colon |M|_1 \to M$ and $j\colon |M|_2 \to M$ be two underlying even manifolds. Then $|M|_1$ and $|M|_2$ are diffeomorphic as families of supermanifolds over B.*

Proof The map i and j are the identity on $\|M\|$. By Proposition 3.3.11 the identity can be lifted to a diffeomorphisms $|M|_1 \to |M|_2$. □

Corollary 3.3.14 *Let $i\colon |M| \to M$ be an underlying even manifold and $\Xi\colon M \to M$ be a diffeomorphism. There exists an embedding $j\colon |M| \to M$ and a diffeomorphism $\xi\colon |M| \to |M|$ such that*

$$\Xi \circ i = j \circ \xi.$$

The maps j and ξ are determined up to a $\sigma \in \operatorname{Diff}(|M|)$ such that $\|\sigma\| = \mathrm{id}_{\|M\|}$, that is, for $j' = j \circ \sigma^{-1}$ and $\xi' = \sigma \circ \xi$ it holds also $\Xi \circ i = j' \circ \xi'$.

Proof By Proposition 3.3.11, the diffeomorphism Ξ_{red} can be lifted to diffeomorphism $\xi\colon |M| \to |M|$. Then $j = \Xi \circ i \circ \xi^{-1}$ is an embedding of the underlying even manifold. □

For a given underlying even manifold $i\colon |M| \to M$ one has the following inclusions of sets

$$\{\Xi \in \operatorname{End}(M) \mid i \circ \Xi = i\} \subseteq \{\Xi \in \operatorname{End}(M) \mid \|\Xi\| = \mathrm{id}_{\|M\|}\} \subset \operatorname{End}(M)$$

where the first inclusion is strict if $B \neq B_{red}$. By an inductive method similar to the proof of Theorem 3.3.7 one can show that for any two underlying even manifolds $i\colon |M| \to M$ and $j\colon |M| \to M$ there exists a smooth map $\Xi\colon M \to M$ such that

$$\Xi \circ i = j.$$

Consequently, the quotient

$$\left.\{\Xi \in \operatorname{End}(M) \mid \|\Xi\| = \mathrm{id}_{\|M\|}\}\middle/\{\Xi \in \operatorname{End}(M) \mid i \circ \Xi = i\}\right.$$

classifies the different embeddings of underlying even manifolds. Locally, in a coordinate chart on $|M|$, any element of the quotient set is given by the functions g_k^a and g_k^α on $|M|$ from Eq. (3.3.3). In that case the addition of the functions g_k^a and g_k^α endows the quotient set with a group structure.

Example 3.3.15 Let $|M|$ be a supermanifold of dimension $m|0$ over B and $E \to |M|$ a vector bundle of rank n. The supermanifold $M = (\|M\|, \bigwedge \Gamma(E))$ is a supermanifold of dimension $m|n$ over B. Like in Example 3.2.3, relative coordinates of $|M|$ can be extended to relative coordinates of M. The coordinate changes are of the form

$$y^b = f^b(x), \qquad \theta^\beta = \eta^\alpha{}_\alpha f^\beta(x).$$

Obviously, $|M|$ is the underlying even manifold of M and an embedding is given by $i^\# x^a = x^a$, $i^\# \eta^\alpha = 0$.

We call M a relative split supermanifold. Notice that relative split supermanifolds are more general than the split supermanifolds from Example 3.2.3, since f^b may depend on B.

Any relative split supermanifold allows for a projection $p\colon M \to |M|$, given in the local coordinates above by $p^\# x^a = x^a$. This motivates the following definition:

Definition 3.3.16 Let $i\colon |M| \to M$ be an embedding of the underlying even manifold of M. We say that M is projected relative to i if there exists a projection $p\colon M \to |M|$ such that $p \circ i = \mathrm{id}_{|M|}$.

If M is a supermanifold over $B = \mathbb{R}^{0|0}$ and $i_{red}\colon M_{red} = |M| \to M$, we say that M is projected if it is projected relative to i_{red}.

A related notion is the notion of a function factor. A function factor is the image of $\mathcal{O}_{|M|}$ under $p^\#$ for p a projection. The notion of a projected supermanifold got wide attention in Donagi and Witten (2015), where it was shown that in the holomorphic setting not every supermanifold is projected.

Remark 3.3.17 In the Rogers–DeWitt approach and in the physics-literature one encounters sometimes the related concept of "body manifold" and "body projection", see, for example, Rogers (2007, Chapter 5.4). It associates to a supermanifold M of dimension $m|n$ a supermanifold M_{body} of dimension $m|0$ and a projection $M \to M_{body}$ which is the identity on even points. The body projection seems to be similar to projections. However, projections relative to some embedding i are, in contrast to the body-projection, not at all unique. This is an important difference between the ringed space approach and the Rogers–DeWitt approach to supermanifolds.

3.4 Functor of Points

In this section we compare the approach of families of supermanifolds to the functor of points approach. The functor of points approach for supermanifolds has gained some popularity, in particular due to the work of Christoph Sachse, see, for example, Sachse (2009), based on an earlier preprint of Molotkov (2010). The underlying idea of the functor of points has been used earlier in algebraic geometry and is commonly attributed to Alexander Grothendieck. For an explanation of the functor of points in the setting of algebraic geometry see Eisenbud and Harris (2000, Chapter I.4 and Chapter VI). As the principle to use the functor of points is so well established in algebraic geometry, it has been applied to supergeometry several times, see, for example, Deligne and Morgan (1999, §2.8–§2.9); Carmeli et al. (2011, Chapter 2.4).

In the setting of supergeometry, arguments in favour of the functor of points approach are actually quite similar to the arguments we gave in favour of using families of supermanifolds. In particular the principle of functoriality under base change (Principle 3.2.12) is shared between both approaches. However, the resulting categories of supermanifolds are not isomorphic in all cases.

Definition 3.4.1 Let S and M be supermanifolds over B. An S-point p of M is a map $p: S \to M$ over B.

Consequently, an S-point of the supermanifold M over B is a section of $M \times_B S \to S$. In particular a B-point of M is a section of $M \to B$.

The B-points of a supermanifold often allow for an interpretation of supermanifolds closer to the geometric intuition one might have from classical geometry. For example, as will be seen in Chap. 5, the B-points of the linear super Lie groups correspond to even matrices. Recall that we have already discussed the B-points of $\mathbb{R}^{m|n}$ in Example 3.1.12. Here we will show the geometric interpretation of the B-points of $\mathbb{P}^{m|n}$.

Example 3.4.2 The set of even lines in $\mathcal{O}_B^{m+1|n}$ is in one-to-one correspondence to the B-points of $\mathbb{P}^{m|n}$. Let us denote the trivial family $\mathbb{P}_{\mathbb{R}}^{m|n} \times B \to B$ by $\mathbb{P}_B^{m|n}$. Let v be an even element of the $\mathcal{O}_B$-module $\mathcal{O}_B^{m+1|n}$ and l the even, one-dimensional subspace of $\mathcal{O}_B^{m+1|n}$ spanned by v. For every even $\lambda \in \mathcal{O}_B$ the vector λv spans the same line l and every even, one-dimensional subspace is spanned by some even vector v. Describing the vector v by its coordinates with respect to a basis, we obtain that the collection of even, one-dimensional subspaces of $\mathcal{O}_B^{m+1|n}$ is given by a tuple of $m+1$ even and n odd elements of $\mathcal{O}_B$ up to rescaling. Writing v^A for the coordinates of v, we will write $[v^A] = [v^1 : \ldots : v^{m+1+n}]$ for the line it determines. Hence, $[v^A] = [\lambda v^A]$ for all even $\lambda \in \mathcal{O}_B$. One of the even coordinates, say the i-th of v must be invertible, hence we may assume without loss of generality $v^i = 1$. Then we map the even line determined by v to the B-point of $\mathbb{P}_B^{m|n}$ lying in $U_i = \mathbb{R}^{m|n}$ and given by the coordinates v^A, leaving out v^i.

If two of the even coordinates of v are invertible, the line l can be mapped to two different open coordinate sets of $\mathbb{P}_B^{m|n}$. The glueing of the coordinate patches, as outlined in Example 3.2.5 assures that we obtain a well-defined one-to-one correspondence between the B-points of $\mathbb{P}_B^{m|n}$ and lines l in $\mathcal{O}_B^{m+1|n}$.

As in purely even geometry, one can now verify, that the projective superspaces carry projective coordinates, that is $m+n+1$-tuples $[X^0 : \ldots : ZX^m : \Theta^1 : \ldots : \Theta^n]$ of $m+1$ even and n odd sections of $\mathcal{O}_{\mathbb{P}_B^{m|n}}$ up to identification of $[X^0 : \ldots : X^m : \Theta^1 : \ldots : \Theta^n]$ and $[\lambda X^0 : \ldots : \lambda X^m : \lambda\Theta^1 : \ldots : \lambda\Theta^n]$. Here $\lambda \in \mathcal{O}_{\mathbb{P}_B^{m|n}}$ is invertible. On the open set U_i the projective coordinates are given by $[x_i^1 : \ldots : x_i^i : 1 : x_i^{i+1} : \ldots : x_i^m : \eta_i^1 : \ldots : \eta_i^m]$. Furthermore, one can check that a linear transformation $L \in \mathrm{GL}_B(m+1|n)$ induces an automorphism of $\mathbb{P}_B^{m|n}$. For every invertible $b \in \mathcal{O}_B$, the linear maps L and bL induce the same automorphism of $\mathbb{P}_B^{m|n}$.

The functor of points h_M of a supermanifold M over B is the contravariant functor that sends a supermanifold S to the set of S-points of M:

$$\begin{aligned} h_M \colon \mathsf{SMan}_B &\to \mathsf{Sets} \\ S &\mapsto \mathrm{Hom}_B(S, M) \\ (f \colon S' \to S) &\mapsto \begin{cases} \mathrm{Hom}_B(S, M) \to \mathrm{Hom}_B(S', M) \\ \qquad\qquad p \mapsto p \circ f \end{cases} \end{aligned}$$

The point functors of all supermanifolds form a covariant functor to the category of contravariant functors from SMan_B to Sets:

$$\begin{aligned} h_* \colon \mathsf{SMan}_B &\to \mathsf{Sets}^{\mathsf{SMan}_B^{op}} \\ M &\mapsto h_M \end{aligned}$$

The functor h_* is a faithfull embedding by the Yoneda-Lemma, see, for example, Mac Lane (1998, Chapter III.2). Several questions arise now:

i) *Redundancy*: A point functor determines a supermanifold. But is it possible to identify the supermanifold uniquely by less data?
ii) *Representability*: Given a functor $F \in \mathsf{Sets}^{\mathsf{SMan}_B^{op}}$, is it the point functor of a supermanifold M, that is, $F = h_M$?
iii) *Extension of the category*: Is there a subcategory of $\mathsf{Sets}^{\mathsf{SMan}_B^{op}}$ that extends the category of supermanifolds in a geometrically interesting way?

While the first and second question are very well studied for $\mathsf{SMan} = \mathsf{SMan}_{\mathbb{R}^{0|0}}$, see, for example, Sachse (2009); Carmeli et al. (2011) and Fioresi and Zanchetta (2017), the third question allows for different answers. In Molotkov (2010) and Sachse (2009) the category SMan is extended to include also infinite dimensional manifolds, while in Alldridge et al. (2014) a large category of superspaces for which the chart theorem holds is constructed.

As we have seen before, the S-points of a supermanifold M over $\mathbb{R}^{0|0}$ correspond to sections of the trivial family $M \times S \to S$. Consequently, in any functor of points approach to SMan, trivial families of supermanifolds and the Principle 3.2.12 appear. Non-trivial families of supermanifolds usually do not appear in the reformulation of supergeometry in terms of the functor of points. However, for example, the group $\mathrm{Diff}(M)$ has been realized as an infinite dimensional super Lie group in Sachse (2009). The S-points of the super Lie group $\mathrm{Diff}(M)$ are given by diffeomorphisms of the trivial family $M \times S \to S$ over S, that is they take the additional parameters from S into account.

To summarize, the functor of points approach to supermanifolds is a reformulation of the theory of supermanifolds that allows for certain geometrically interesting extensions but usually considers only trivial families. In this work we will need however non-trivial families, for Example in (9.1.5), but not necessarily the extensions that the functor of points approach offers. A unified approach, that is a functorial reformulation of non-trivial families of supermanifolds, seems to be possible and desirable but is beyond the scope of this work.

Chapter 4
Vector Bundles

The goal of this chapter is to explain the generalisations of vector bundles and, in particular, tangent bundles to families of supermanifolds. Vector bundles are fiber bundles where the typical fiber is a vector space. In supergeometry, the relation between a super vector space and the corresponding linear supermanifold is slightly more complicated than in ordinary differential geometry because there are no points corresponding to odd elements of the vector space. Consequently, the theory of sections of a vector bundle is more complicated than expected, and the more algebraic approach via locally free modules is to be preferred sometimes.

In Sect. 4.1 we set up the notation for fiber bundles over families of supermanifolds.

Section 4.2 introduces vector bundles as fiber bundles where the typical fiber is a vector space. Furthermore, the category equivalence between vector bundles over M and locally free sheaves over $\mathcal{O}_M$ is demonstrated.

In the third section, the tangent bundle of a family of supermanifolds is introduced and tangent maps are defined via the locally free sheaf of derivations of the structure sheaf.

In the fourth section we treat the theory of connections in vector bundles and in particular the equivalence with covariant derivatives. This is mainly a preparation for Chap. 6, where connections with additional geometric properties will be treated from the point of view of frame bundles.

Besides setting the notation, this chapter shows how vector bundles and in particular the tangent bundle are constructed in a way that is compatible with base extensions. Furthermore, the theory of connections on super vector bundles could not be found in the literature.

E. Keßler, *Supergeometry, Super Riemann Surfaces and the Superconformal Action Functional*, Lecture Notes in Mathematics 2230,
https://doi.org/10.1007/978-3-030-13758-8_4

4.1 Fiber Bundles

Recall that by Principle 3.2.12 all supermanifolds are implicitly to be understood as families of supermanifolds over a base B. All maps are maps of families of supermanifolds over B. In particular, the fiber product of families of supermanifolds M and N over B is denoted by $M \times N = M \times_B N$. As in ordinary differential geometry, a fiber bundle is a supermanifold that is locally a product of supermanifolds:

Definition 4.1.1 Let M and F be supermanifolds. The supermanifold $E = M \times F$ together with the projection $\pi : E \to M$ on the first factor is called a trivial fiber bundle with typical fiber F. Let $f : M \to N$ be a map, $\pi_E : E = M \times F \to M$ and $\pi_{E'} : E' = N \times F' \to N$ trivial fiber bundles over M and N respectively. A map $g : E \to E'$ is called a homomorphism of trivial fiber bundles over f, if g can be written as $g = (f \circ \pi_E, h)$ where $h : M \times F \to F'$:

$$\begin{array}{ccc} E = M \times F & \xrightarrow{g=(f\circ\pi_E,h)} & E' = N \times F' \\ \downarrow{\scriptstyle \pi_E} & & \downarrow \\ M & \xrightarrow{f} & N \end{array}$$

The fiber bundle homomorphism g is a fiber bundle isomorphism if g and f are isomorphisms of supermanifolds.

A fiber bundle E over M with typical fiber F is a map of supermanifolds $\pi : E \to M$ such that there is an open cover U_i of M and local trivialisation isomorphisms $\phi_i : \pi^{-1}(U_i) \to U_i \times F$ such that $\pi : \pi^{-1}(U_i) \to U_i$ is a trivial fiber bundle with typical fiber F.

Let $\pi_E : E \to M$ and $\pi_{E'} : E' \to N$ be fiber bundles over M and N with typical fibers F and F' respectively. A map of supermanifolds $g : E \to E'$ is called a homomorphism of fiber bundles over $f : M \to N$ if it is a homomorphism of trivial fiber bundles when restricted to trivializing open sets on both sides.

The following is a reformulation of Definition 4.1.1:

Proposition 4.1.2 *Let $\pi : E \to M$ be a fiber bundle with typical fiber F and U_i a covering of M such that there are isomorphisms of supermanifolds $\phi_i : U_i \times F \to \pi^{-1}(U_i) \subset E$. The maps*

$$\phi_{ij} = \phi_j^{-1} \circ \phi_i\Big|_{U_i \cap U_j} : \left(U_i \cap U_j\right) \times F \to \left(U_j \cap U_i\right) \times F$$

are isomorphisms of trivial fiber bundles and are called cocycles or glueing functions of E. They fulfill the Čech cocycle conditions

$$\phi_{ii} = \mathrm{id}, \qquad \phi_{ij} = \phi_{kj} \circ \phi_{ik}, \tag{4.1.3}$$

wherever all three of them are defined.

The fiber bundle E is completely determined by the typical fiber F, a covering U_i of M, and glueing functions ϕ_{ij} as above fulfilling the cocycle condition (4.1.3).

Definition 4.1.4 Let $\pi : E \to M$ be a fiber bundle over M and $U \subset M$. A section of E over U is a map of supermanifolds $s: U \to \pi^{-1}(U)$ such that $\pi \circ s = \mathrm{id}_U$. The sections of E form a sheaf on $\|M\|$, denoted by $\mathrm{Sec}\,(E)$.

Definition 4.1.5 Let $f: M \to N$ be a morphism of supermanifolds over B and E a fiber bundle with typical fiber F over N. The pullback $f^*E = M \times_N E$ is a fiber bundle over M with the same typical fiber F, called the pullback bundle. The projection of the fiberproduct $f^*E = M \times_N E$ to E is a morphism of fiber bundles over f, denoted by $\overline{f}: f^*E \to E$.

If $p: B \to N$ is a B-point of N, the pullback p^*E is also called fiber of E over p.

Let s be a section of E. Then we call the section $f^*s = (\mathrm{id}_M, f \circ s)$ of f^*E the pullback of s along f.

Let $g: E \to E'$ be a fiber bundle homomorphism over f. By definition of the pullback bundle, the map g factorizes into $g = \overline{f} \circ \overline{g}$, where $\overline{g} = (\pi_E, g): E \to f^*E'$ is a fiber bundle homomorphism over id_M and $\overline{f}$ does not depend on g.

4.2 Vector Bundles and Sections

Definition 4.2.1 A trivial fiber bundle $\pi : M \times \mathbb{R}^{p|q} \to M$ is called a trivial vector bundle. A homomorphism of trivial vector bundles over $f: M \to N$ is a map $(f \circ \pi, l)$ where $l: M \times \mathbb{R}^{p|q} \to \mathbb{R}^{r|s}$ is a family of linear maps between linear superspaces. A general vector bundle of rank $p|q$ is a fiber bundle $\pi : E \to M$ whose typical fiber is the linear supermanifold $\mathbb{R}^{p|q}$ and such that all cocycles are isomorphisms of trivial vector bundles. Vector bundle homomorphisms are locally homomorphisms of trivial vector bundles.

Thus vector bundles are particular fiber bundles; hence Proposition 4.1.2 and Definition 4.1.5 apply. In particular, the pullback of a vector bundle is a vector bundle of the same rank and every vector bundle homomorphism factorizes over the pullback bundle. For a non-trivial example of a vector bundle, the tangent bundle, see Sect. 4.3.

Example 4.2.2 Given vector bundles E and F over M, one can construct vector bundles ΠE, $E^\vee$, $\mathrm{Ber}\, E$, $E \oplus F$, $E \otimes F$, $\mathrm{Hom}(E, F)$ over M with the expected properties. As an example we give the explicit construction of ΠE: Suppose the vector bundle E is of rank $m|n$, trivial over the open sets U_i and glued by the linear functions ϕ_{ij}. Then the vector bundle ΠE is the vector bundle of rank $n|m$ that is trivial over U_i and glued by the linear (even) maps $\Pi\left(\phi_{ij}\right)$.

Example 4.2.3 (The Sheaf $\mathrm{Sec}\,(E)$*)* Let first $E = M \times \mathbb{R}^{p|q}$ be a trivial vector bundle over the supermanifold M. Any section s of E is given by a map

$\overline{s}\colon M \to \mathbb{R}^{p|q}$. Denote the standard coordinates on the fiber $\mathbb{R}^{p|q}$ by E^B and define sections e_A of E by setting

$$\overline{e}_A^{\#} E^B = \begin{cases} 1 & A = B \\ 0 & \text{else} \end{cases} . \tag{4.2.4}$$

Then any section s can by described by $\overline{s} = \overline{s}^{\#} E^A e_A$. This yields an additive structure on $\operatorname{Sec}(E)$ by setting $s + t = \left(\overline{s}^{\#} E^A + \overline{t}^{\#} E_A\right) e_A$. The additive structure generalizes to non-trivial vector bundles, because the glueing maps are linear maps.

However, multiplication of $\overline{s}$ by a function $f \in \mathcal{O}_M$ will yield a section $f \cdot s$ only if f is even, because the map $\overline{s}\colon M \to \mathbb{R}^{p|q}$ has to be even. This can be seen as a more general instance of the difference between super vector spaces and linear supermanifolds as discussed before in Example 3.1.12.

Hence, in order to have a tensor calculus similar to classical differential geometry it will be necessary to enlarge the sheaf of sections:

Definition 4.2.5 Let $\pi\colon E \to M$ be a vector bundle. We denote by $\Gamma(E)$ the sheaf $\operatorname{Sec}(E) \oplus \operatorname{Sec}(\Pi E)$. From Example 4.2.3 it is clear that $\Gamma(E)$ is a sheaf of $\mathcal{O}_M$-modules by setting $\Gamma(E)_0 = \operatorname{Sec}(E)$ and $\Gamma(E)_1 = \operatorname{Sec}(\Pi E)$.

Suppose the vector bundle E is trivial over M. For any trivialization, the sections e_A, defined in Eq. (4.2.4), define a basis for $\Gamma(E)$. In the non-trivial case, one can choose a local basis. The elements of such a (local) basis are also referred to as frames. Any section $s \in \Gamma(E)$ can then be written (locally) as an $\mathcal{O}_M$-linear combination of the basis vectors, $s = s^A e_A$. We are going to use the same conventions as for the linear algebra of modules.

Recall that a sheaf of $\mathcal{O}_M$-modules $\mathcal{E}$ is called free of rank $m|n$ if $\mathcal{E} \simeq \mathcal{O}_M^{m|n} = \mathcal{O}_M^m \oplus \Pi \mathcal{O}_M^n$. A sheaf of $\mathcal{O}_M$-modules $\mathcal{E}$ is called locally free of rank $m|n$ if M has an open cover $\{U_i\}_{i\in I}$ such that $\mathcal{E}|_{U_i}$ is a free $\mathcal{O}_M|_{U_i}$-module. We denote by $\mathsf{FreeMod}_M$ and $\mathsf{LFreeMod}_M$ the categories of free and locally free $\mathcal{O}_M$-modules respectively, where the morphisms are given by linear maps. We denote by $\mathsf{FreeMod}_M^{ev}$ and $\mathsf{LFreeMod}_M^{ev}$ the categories of free and locally free sheaves over $\mathcal{O}_M$ respectively, where the morphisms are given by even linear maps.

Lemma 4.2.6 *The section functor*

$$\begin{aligned} \Gamma\colon \mathsf{TrivVB}_M &\to \mathsf{FreeMod}_M^{ev} \\ E &\mapsto \Gamma(E) \\ \left(f\colon E \to E'\right) &\mapsto \left(s\colon U \to E \oplus \Pi E \mapsto (f \oplus \Pi f) \circ s\colon U \to E' \oplus \Pi E'\right) \end{aligned}$$

is an isomorphism of categories between the category TrivVB_M *of trivial vector bundles over M and the category* $\mathsf{FreeMod}_M^{ev}$ *of free* $\mathcal{O}_M$*-modules with even linear maps.*

Proof From the discussion before it is clear that the choice of a set of frames e_A gives a basis for $\Gamma(E)$. Hence $\Gamma(E)$ is free and its dimension coincides with the rank of E. As the dimension classifies trivial vector bundles and sheaves of free modules completely, it is clear that the section functor is bijective on the objects of TrivVB_M and $\mathsf{FreeMod}_M^{ev}$ respectively.

It remains to show that homomorphisms between vector bundles correspond bijectively to linear maps between free $\mathcal{O}_M$-modules. Let $l\colon M\times\mathbb{R}^{p|q}\to M\times\mathbb{R}^{r|s}$ be a homomorphisms between trivial vector bundles. By definition l is fiberwise linear, that is, given by a map $\bar{l}\colon M\times\mathbb{R}^{p|q}\to\mathbb{R}^{r|s}$. With respect to coordinates E^A and F^B of the fibers $\mathbb{R}^{p|q}$ and $\mathbb{R}^{r|s}$ respectively, the map $\bar{l}$ can be written as

$$\bar{l}^{\#}F^B = E^A\,{}_AL^B,$$

where ${}_AL^B$ is an even matrix of functions from $\mathcal{O}_M$. Let e_A and f_B be the frames corresponding to E^A and F^B, then l acts on the frame e_A by $\langle e_A, l\rangle = {}_AL^B f_B$, because

$$\overline{\langle e_A, l\rangle}^{\#}F^B = \bar{e}_A^{\#}\left(\bar{l}^{\#}F^B\right) = {}_AL^B.$$

Hence, any fiberwise linear map l corresponds to an even $\mathcal{O}_M$-linear map between free $\mathcal{O}_M$-modules. Conversely, any even linear map between free $\mathcal{O}_M$-modules defines a fiberwise linear map between vector bundles. □

Applying Lemma 4.2.6 to trivializing open subsets yields immediately the following proposition (compare Balduzzi et al. 2011).

Proposition 4.2.7 *The section functor is an isomorphism of categories between the category* VB_M *of vector bundles over* M *and the category* $\mathsf{LFreeMod}_M^{ev}$ *of locally free sheaves of* $\mathcal{O}_M$*-modules with even linear maps.*

Hence vector bundles on supermanifolds can be described either as particular fiber bundles or as locally free sheaves of $\mathcal{O}_M$-modules. In the following we will use both descriptions interchangeably.

Remark 4.2.8 The section functor respects linear algebra. For example:

$$\begin{aligned}\Gamma\left(E^\vee\right) &= (\Gamma(E))^\vee\\ \Gamma(\Pi E) &= \Pi(\Gamma(E))\\ \Gamma(E\oplus F) &= \Gamma(E)\oplus\Gamma(F)\\ \Gamma(E\otimes F) &= \Gamma(E)\otimes_{\mathcal{O}_M}\Gamma(F)\end{aligned}$$

Remark 4.2.9 Let $f\colon M\to N$ be a map of supermanifolds and $\mathcal{E}$ a locally free sheaf of $\mathcal{O}_N$-modules. Recall that the inverse image sheaf $f^*\mathcal{E}$ is defined as

$$f^*\mathcal{E} = \mathcal{O}_M\otimes_{f^{-1}\mathcal{O}_N} f^{-1}\mathcal{E},$$

see for example Hartshorne (1977, Chapter II.5). For any vector bundle E over N it holds

$$f^*\Gamma(E) = \Gamma\left(f^*E\right).$$

To show this isomorphism it suffices again to show the local case when E is trivial over M. Then $\Gamma(E)$ is generated by the frame e_A of sections of E. The sections f^*e_A form an $\mathcal{O}_N$-linear basis of $\Gamma(f^*E)$.

Remark 4.2.10 In the approach presented here the category VB_M is isomorphic to $\mathsf{LFreeMod}_M^{ev}$. An odd morphism between the locally free sheaves $\mathcal{E}$ and $\mathcal{F}$ does not correspond to some bundle map between supermanifolds. If a treatment as a map between bundles of supermanifolds is desired it can be reformulated as an even morphism between $\mathcal{E}$ and $\Pi\mathcal{F}$. Alternatively, if a treatment of even and odd morphisms on the same geometric footing would be necessary, one could do so by defining vector bundles of rank $r|s$ to have fibre $\mathbb{R}^{r|s} \oplus \mathbb{R}^{s|r}$ and proceed similarly to the definition of $\Gamma(E)$.

4.3 Tangent Bundle

In this section we define the tangent bundle TM of a supermanifold M via the locally free sheaf of derivations $\operatorname{Der}_{\mathcal{O}_B}(\mathcal{O}_M)$ and Proposition 4.2.7.

Example 4.3.1 Let $X^A = (x^a, \eta^\alpha)$ be relative coordinates on $\mathbb{R}^{m|n} \times B$. The partial derivatives in coordinate directions ∂_{X^A} are defined by

$$\partial_{x^a}\left(b\eta^{\underline{\nu}} f(x)\right) = b\eta^{\underline{\nu}}\partial_{x^a} f(x) \qquad\qquad \partial_{\eta^\alpha}\left(bf(x)\eta^\beta\right) = bf(x)\delta_\alpha^\beta$$

for $b \in \mathcal{O}_B$, $f(x) \in \left(\mathcal{O}_{\mathbb{R}^{m|n}\times B}\right)_{red}$ and any $\mathbb{Z}_2$-multiindex $\underline{\nu}$. Imposing additivity and the Leibniz rule, the partial derivatives ∂_{X^A} extend to $\mathcal{O}_{\mathbb{R}^{m|n}\times B}$. Hence ∂_{X^A} are $\mathcal{O}_B$-linear derivations of $\mathcal{O}_{\mathbb{R}^{m|n}\times B}$. In Proposition 4.3.2 below we will show that any derivation $Y \in \operatorname{Der}_{\mathcal{O}_B}(\mathcal{O}_{\mathbb{R}^{m|n}\times B})$ is a linear combination of ∂_{X^A}.

Proposition 4.3.2 *The sheaf* $\operatorname{Der}_{\mathcal{O}_B}(\mathcal{O}_M)$ *is a locally free sheaf of* $\mathcal{O}_M$*-modules.*

Proof Let $U \subseteq M$ be a coordinate patch with relative coordinates $X^A = (x^a, \eta^\alpha)$. Any derivation Y is an $\mathcal{O}_M|_U$-linear combination of ∂_{X^A}. To confirm this let

$$Y^A = Y(X^A)$$

and remark that the derivations Y and $Y^A\partial_{X^A}$ coincide on all polynomial functions in the coordinates. The equality of Y and $Y^A\partial_{X^A}$ on arbitrary supersmooth functions can be achieved with the help of Hadamard's Lemma, as in ordinary differential geometry. □

Definition 4.3.3 The vector bundle whose sheaf of sections is $\mathrm{Der}_{\mathcal{O}_B}(\mathcal{O}_M)$ is called tangent bundle of M and denoted by TM. Let M be of dimension $m|n$. The tangent bundle is a vector bundle of rank $m|n$ over M. The sections of TM are also called vector fields and form a Lie algebra, see Example 2.11.4.

The dual bundle $T^\vee M = (TM)^\vee$ is called cotangent bundle and its sheaf of sections is denoted $\Omega(M) = \Gamma\left(T^\vee M\right)$. It is the degree one part of its exterior algebra $\Omega^\bullet(M) = \bigwedge^\bullet(\Omega(M))$. Elements of $\Omega^\bullet(M)$ are called differential forms.

Remark 4.3.4 Let us call to mind that the exterior algebra is not bounded if $n > 0$. There are elements of arbitrary high degree. Furthermore the Berezinian $\mathrm{Ber}\, T^\vee M$ is not part of this sequence (likewise if $n > 0$).

The results on modules of derivations from Sect. 2.9 carry over directly to the corresponding sheaves on supermanifolds. In particular, the exterior derivative $d\colon \mathcal{O}_M \to \Omega(M)$ is given by its action on vector fields M:

$$\langle X, df\rangle = Xf$$

It follows that the dual of the coordinate vector fields ∂_{X^A} is given by dX^A. The exterior derivative can be continued to a derivation $d\colon \Omega^k(M) \to \Omega^{k+1}(M)$ as in Example 2.9.5.

Any map of supermanifolds $f\colon M \to N$ the differential of $f^\#$ gives a linear map between locally free sheaves

$$df^\#\colon f^*\mathrm{Der}_{\mathcal{O}_B}(\mathcal{O}_N)^\vee = \mathcal{O}_M \otimes_{f^{-1}\mathcal{O}_N} f^{-1}\mathrm{Der}_{\mathcal{O}_B}(\mathcal{O}_N)^\vee \to \mathrm{Der}_{\mathcal{O}_B}(\mathcal{O}_M)^\vee,$$

see Example 2.9.4. We will call the dual of $df^\#$ differential of f. The differential is given by

$$\begin{aligned} df\colon \Gamma(TM) &\to \Gamma\left(f^*TN\right) \\ X &\mapsto Xf^\#. \end{aligned}$$

In local coordinates X^A of M and Y^B of N the differential is given by

$$df\left(\partial_{X^A}\right) = \left(\partial_{X^A} f^\# Y^B\right) f^*\partial_{Y^B}.$$

Notice that we have already used the coordinate expressions for the differential in the inverse function Theorem 3.1.15. The tangent map of f is the map $Tf\colon TM \to TN$ induced by df.

Let now $f: M \to N$ and $g: N \to L$ be maps of supermanifolds over B and X^A, Y^B and Z^C be local coordinates for M, N and L respectively. By the chain rule we obtain

$$\begin{aligned} d(g \circ f)\left(\partial_{X^A}\right) &= \left(\partial_{X^A} f^{\#} g^{\#} Z^C\right) f^* g^* \partial_{Z^C} \\ &= \left(\partial_{X^A} f^{\#} Y^B\right) f^{\#}\left(\partial_{Y^B} g^{\#} Z^C\right) f^* g^* \partial_{Z^C} = \left(\partial_{X^A} f^{\#} Y^B\right) f^*\left(dg \partial_{Y^B}\right) \\ &= \left(df \circ f^* dg\right)\left(\partial_{X^A}\right), \end{aligned}$$

where $f^* dg\colon f^*TN \to f^*g^*TL$ is the pullback of dg along f. For the tangent map, we obtain the simpler $T(g \circ f) = Tg \circ Tf$.

Note that in most cases, the image $dfX \in \Gamma(f^*TN)$ of a section $X \in \Gamma(TM)$ does not arise from a section of TN. Even if there is a section $X^N \in \Gamma(TN)$ such that $dfX = f^*X^N$, the vector field X^N does not have to be unique. However, if the image of df can be realized as sections of TN, the map df preserves the Lie algebra structure. More precisely, let $X, Y \in \Gamma(TM)$ and $X^N, Y^N \in \Gamma(TN)$ such that $dfX = f^*X^N$ and $dfY = f^*Y^N$, then

$$df[X, Y] = f^*[X^N, Y^N]. \tag{4.3.5}$$

Indeed, for any $F \in \mathcal{O}_N$,

$$\begin{aligned} \left(f^*[X^N, Y^N]\right) F &= f^{\#} X^N Y^N F - (-1)^{p(X^N)p(Y^N)} f^{\#} Y^N X^N F \\ &= X f^{\#} Y^N F - (-1)^{p(X^N)p(Y^N)} Y f^{\#} X^N F \\ &= \left(XY - (-1)^{p(X)p(Y)} YX\right) f^{\#} F = df[X, Y]F \end{aligned}$$

One particular case such that any dfX arises from a vector field on the target is the case when f is a diffeomorphism, as was shown in Theorem 3.1.15.

Let now $\alpha \in \Omega(N)$ and $f: M \to N$. Define the form $\alpha_f \in \Omega(M)$ via its application to vector fields $X \in \Gamma(TM)$ by

$$\left\langle X, \alpha_f \right\rangle = \left\langle dfX, f^*\alpha \right\rangle.$$

The map $\alpha \mapsto \alpha_f$ is a map of sheaves of modules $\Omega(N) \to \Omega(M)$ over $f^{\#}$, but not a vector bundle morphism over f. By construction it is a contravariant functor. That is, for a map $g: M' \to M$ we have $\alpha_{g \circ f} = \left(\alpha_f\right)_g$.

Remark 4.3.6 Note that in other texts on differential geometry what we call α_f is often denoted $f^*\alpha$, see, for example, Jost (2011) and Kobayashi and Nomizu (1996). As we need it frequently, we prefer to reserve the notation $f^*\alpha$ for the element of $f^*\Omega(N)$ and warn about the possible confusion.

In the following, we also need frequently:

Definition 4.3.7 Let E be a vector bundle over M. The sheaf of differential forms on M with values in E, that is, sections of $\bigwedge^\bullet T^\vee M \otimes E$ will be denoted by $\Omega^\bullet(M, E)$. In the case of a trivial vector bundle $E = M \times V$ we may also write $\Omega^\bullet(M, V)$.

Let $X \in \Gamma\left(\bigwedge TM\right)$ and $\alpha \in \Omega^\bullet(M, E)$. We define $\langle X, \alpha\rangle \in \Gamma(E)$ by the linear extension of $\langle X, \overline{\alpha} \otimes e\rangle = \langle X, \overline{\alpha}\rangle\, e$ for $\overline{\alpha} \in \Omega^\bullet(M)$ and $e \in \Gamma(E)$.

4.4 Connections on Vector Bundles

In this section we generalize the theory of connections on vector bundles to supermanifolds. Definitions and statements carry over from ordinary differential geometry to super differential geometry. The challenge is to formulate the proofs accordingly. Covariant derivatives, their curvature and torsion are standard in differential geometry, see, for example, Kobayashi and Nomizu (1996). The description of connections on vector bundles as a splitting of the short exact sequence 4.4.2 can be found, for instance, in Tolksdorf (in preparation).

Lemma 4.4.1 *Let $\pi : E \to M$ be a vector bundle of rank $r|s$ over the supermanifold M. The kernel of the differential of $d\pi : TE \to \pi^*TM$ is given by π^*E. Thus there is a short exact sequence of vector bundles over E:*

$$0 \longrightarrow \pi^*E \xrightarrow{\iota_{\pi^*E}} TE \xrightarrow{d\pi} \pi^*TM \longrightarrow 0 \tag{4.4.2}$$

Proof Let $\{U_i\}$ be a cover by coordinate charts of M that trivializes E. That is, $E|_{U_i} = U_i \times \mathbb{R}^{r|s}$ and the standard basis of $\mathbb{R}^{r|s}$ induces local sections F_{iA} of $E|_{U_i}$. Let us denote the coordinates on U_i by X_i^A and the standard coordinates on $\mathbb{R}^{r|s}$ by E_i^A. The product $E|_{U_i} = U_i \times \mathbb{R}^{r|s}$ induces a local direct sum decomposition $TE|_{U_i} = \pi^*E|_{U_i} \oplus \pi^*TU_i$. Consequently the sections $\partial_{E_i^A} = \pi^*F_{iA}$ together with $Y_{iB} = \pi^*\partial_{X_i^A}$ form a local base for the free module of sections $\Gamma\left(TE|_{U_i}\right)$. As π is locally the projection on the second factor it is clear that $d\pi : \Gamma\left(TE|_{U_i}\right) \to \pi^*TU_i$ is given as a $\mathcal{O}_{E|_{U_i}}$-linear map between free modules by

$$\begin{aligned} d\pi : TE|_{U_i} &\to \pi^*TU_i \\ Y_{iB} &\mapsto \pi^*\partial_{X_i^B}\,. \\ \pi^*F_{iA} &\mapsto 0 \end{aligned}$$

The kernel of $d\pi$ is obviously generated by $\pi^* F_{iA}$ and thus the map $\iota_{\pi^* E}$ is locally given by

$$\begin{aligned} \iota_{\pi^* E} \colon \pi^* E|_{U_i} &\to TE|_{U_i} \\ \pi^* F_A &\mapsto \pi^* F_A. \end{aligned}$$

This proves the statement locally.

Let us now study how the local descriptions glue to a global one. On the intersection of the trivializing sets, we have the coordinate change $\phi_{ij} \colon U_i \cap U_j \to U_j \cap U_i$. On the level of E the change of coordinates is given by $\Phi_{ij} = (\phi_{ij}, g_{ij})$, where g_{ij} is a family of invertible linear maps

$$g_{ij} \colon U_i \cap U_j \times \mathbb{R}^{r|s} \to \mathbb{R}^{r|s}.$$

The map g_{ij} transforms the frames of E by $F_{iA} = \left(g_{ij}^{\#} E_j^B\right) F_{jB}$. Consequently, the frames on TE transform by

$$\begin{aligned} d\Phi_{ij} Y_{iA} &= \left(\partial_{X_i^A} \phi_{ij}^{\#} X_j^B\right) \partial_{X_j^B} + \left(\partial_{X_i^A} g_{ij}^{\#} E_j^B\right) \partial_{E_j^B}, \\ d\Phi_{ij} \partial_{E_i^A} &= \left(\partial_{E_i^A} g_{ij}^{\#} E_j^B\right) \partial_{E_j^B}. \end{aligned}$$

In particular the subspace generated by $\pi^* F_{iA}$ and $\pi^* F_{jB}$ is the same, and the sequence (4.4.2) is proven globally. □

Remark 4.4.3 The proof of 4.4.1 has also shown that TE is a vector bundle over TM with typical fiber $\mathbb{R}^{r|s} \times \mathbb{R}^{r|s}$ and glueing function (g_{ij}, g_{ij}). Furthermore $\pi^* TM = E \times_M TM$ is also a vector bundle over TM and the tangent map $d\pi$ is a linear map over TM.

Definition 4.4.4 A splitting of the short exact sequence (4.4.2) is called a connection on E. Such a splitting gives a direct sum decomposition $TE = \pi^* E \oplus \pi^* TM$ or equivalently a projection $p_{\pi^* E} \colon TE \to \pi^* E$ that is left inverse to $\iota_{\pi^* E}$, that is $p_{\pi^* E} \circ \iota_{\pi^* E} = \mathrm{id}_{\pi^* E}$. Still equivalently, there is an inclusion $\iota_{\pi^* TM} \colon \pi^* TM \to TE$ such that $d\pi \circ \iota_{\pi^* TM} = \mathrm{id}_{\pi^* TM}$.

The connection is called linear if the inclusion $\iota_{\pi^* TM}$ is also linear over TM.

Definition 4.4.5 (see Deligne and Morgan 1999, §3.6) A covariant derivative ∇ on the vector bundle E is an additive $\mathcal{O}_B$-linear morphism $\nabla \colon \Gamma(E) \to \Omega^1(M, E)$ of $\mathcal{O}_M$-modules obeying the Leibniz identity:

$$\nabla f v = df \otimes v + f \nabla v$$

For a tangent vector field X we write $\nabla_X v = \langle X, \nabla v \rangle$, compare Definition 4.3.7.

Remark 4.4.6 Covariant derivatives do not have a parity. While one might call a covariant derivative ∇ even if $\nabla_X v$ has parity $p(X) + p(v)$, it is impossible to define odd covariant derivatives. Indeed, the condition $p(\nabla_X v) = p(X) + p(v) + 1$ contradicts the Leibniz-identity. The reason is that covariant derivatives do not form a linear space, but rather an affine space, as we will see in Sect. 6.4. A similar remark applies to connections on vector bundles.

Proposition 4.4.7 *For a vector bundle $\pi : E \to M$ of rank $r|s$ the following structures are equivalent:*

i) A covariant derivative ∇ on E.
ii) A set of local gauge potentials: Let $\{U_i\}_{i\in I}$ be an open cover of M that trivializes E and $g_{ij} : U_i \cap U_j \times \mathbb{R}^{r|s} \to \mathbb{R}^{r|s}$ the corresponding glueing functions, see Proposition 4.1.2. Local gauge potentials are $A_i \in \Omega^1(U_i, \mathrm{GL}(r|s))$ such that on $U_i \cap U_j$ the following holds:

$$A_i = {g_{ij}}^{-1} A_j g_{ij} + {g_{ij}}^{-1} dg_{ij} \tag{4.4.8}$$

iii) A linear connection $TE = \pi^* E \oplus \pi^* TM$ on E.

Proof We prove the equivalence i)⇔ii) first and afterwards the equivalence iii) ⇔ii).

The description of a covariant derivative in local frames leads to local gauge potentials, as in point ii). Let F_{iA} be a frame in E over the open set $U_i \subset M$. The action of the covariant derivative ∇ on F_{iA} can be written as

$$\nabla F_{iA} = {A_i}_A{}^B F_B$$

where $A_i \in \Omega^1(U_i, \mathrm{GL}(r|s))$. If now $F_{iA} = \left(g_{ij}\right)_A{}^B F_{jB}$, then,

$$\begin{aligned}
(A_i)_A{}^C F_{iC} &= \nabla F_{iA} = \nabla\left(\left(g_{ij}\right)_A{}^B F_{jB}\right) \\
&= \left(g_{ij}\right)_A{}^B \nabla F_{jB} + \left(dg_{ij}\right)_A{}^B F_{jB} \\
&= \left(\left(g_{ij}\right)_A{}^B (A_j)_B{}^C + \left(dg_{ij}\right)_A{}^C\right)\left(g_{ij}^{-1}\right)_C{}^D F_{iD}
\end{aligned}$$

which is the written out version of the following expression:

$$A_i = {g_{ij}}^{-1} A_j g_{ij} + {g_{ij}}^{-1} dg_{ij}$$

Consequently, any covariant derivative leads to a set of gauge potentials and conversely. This shows the equivalence i)⇔ii).

We will now show that the local gauge potentials give rise to a linear connection by constructing $\iota_{\pi^*TM} : \pi^*TM \to TE$ explicitly. In the notation from

Lemma 4.4.1, the A_i give rise to a linear connection locally via

$$\begin{aligned} \iota_{\pi^* TM} \colon \pi^* TU_i &\to TE|_{U_i} \\ \pi^* \partial_{X_i^A} &\mapsto Y_{iA} + E_i^B{}_B \left(\left\langle \partial_{X_i^A}, A_i \right\rangle \right)^C \partial_{E_i^C} \end{aligned} \tag{4.4.9}$$

This local definition of $\iota_{\pi^* TM}$ glues to a global, linear connection on TE, because the terms in the transformation of (4.4.9) that are proportional to the derivatives of g_{ij} do cancel. Conversely, any linear connection is of the form (4.4.9). □

Remark 4.4.10 As in "classical" differential geometry, connections on vector bundles E and F over M induce connections on $E^\vee$, $E \oplus F$, $E \otimes F$. This is best seen using covariant derivatives. Let ∇^E be the associated covariant derivative on E, ∇^F on F and so on. Then the following formulas define covariant derivatives on $E^\vee$, $E \oplus F$ and $E \otimes F$ respectively

$$\begin{aligned} d \langle X, \alpha \rangle &= \left\langle \nabla^E X, \alpha \right\rangle + \left\langle X, \nabla^{E^\vee} \alpha \right\rangle, \\ \nabla^{E \oplus F} X \oplus Y &= \nabla^E X \oplus \nabla^F Y, \\ \nabla^{E \otimes F} X \otimes Y &= \nabla^E X \otimes Y + X \otimes \nabla^F Y, \end{aligned}$$

where X is a section of E, Y a section of F and α a section of $E^\vee$.

Definition 4.4.11 The curvature tensor $R^\nabla \in \Omega^2(M, \operatorname{End} E)$ of a covariant derivative ∇ on the vector bundle $\pi : E \to M$ is given by

$$R^\nabla(X, Y)s = \nabla_X \nabla_Y s - (-1)^{p(X)p(Y)} \nabla_Y \nabla_X s - \nabla_{[X,Y]} s$$

where X and Y are pure sections of TM, whereas s is a section of E.

It is an easy calculation that the curvature tensor R^∇ is indeed a tensor, that is, $\mathcal{O}_M$-linear in its arguments:

$$\begin{aligned} R^\nabla(X, Y) f s &= (-1)^{p(Y)p(f)} R^\nabla(X, fY)s = (-1)^{(p(Y)+p(X))p(f)} R^\nabla(fX, Y)s \\ &= (-1)^{(p(Y)+p(X))p(f)} f R^\nabla(X, Y)s \end{aligned}$$

Furthermore R^∇ is anti-symmetric in the first two arguments

$$R^\nabla(X, Y)s = -(-1)^{p(X)p(Y)} R^\nabla(Y, X)s.$$

Definition 4.4.12 The torsion tensor T^∇ of a covariant derivative ∇ on the tangent bundle $TM \to M$ is defined for pure sections X and Y of TM by

$$T^\nabla(X, Y) = \nabla_X Y - (-1)^{p(X)p(Y)} \nabla_Y X - [X, Y]$$

One can check that T^∇ is indeed a tensor and is anti-symmetric:

$$T^\nabla(X,Y) = -(-1)^{p(X)p(Y)}T^\nabla(Y,X)$$

Remark 4.4.13 The additional sign prefactors in the definition of R^∇ and T^∇ compared to the classical formulas are the natural generalization of symmetric tensors to supersymmetric ones. In addition for any other choice of sign prefactors the resulting R^∇ and T^∇ would not be tensors.

Proposition 4.4.14 (Existence of Pullback Connection) *Let $f\colon M \to N$ be a map of supermanifolds, $E \to N$ a vector bundle and ∇ a covariant derivative on E. There is a unique connection ∇^{f^*E} on f^*E such that for all sections $s \in \Gamma(E)$ and $X \in \Gamma(TM)$*

$$\nabla_X^{f^*E} f^*s = \left\langle dfX, f^*(\nabla s)\right\rangle.$$

Here $f^*(\nabla s)$ is a section of $f^*\left(\Omega^1(N,E)\right)$.

For the curvature of ∇^{f^*E} it holds for all $X, Y \in \Gamma(TM)$

$$R^{\nabla^{f^*E}}(X,Y) = \left(f^*R^\nabla\right)(dfX, dfY).$$

In the case $E = TN$, we have furthermore

$$\left(f^*T^\nabla\right)(dfX, dfY) = \nabla_X^{f^*E} dfY - (-1)^{p(X)p(Y)}\nabla_Y dfX - df[X,Y].$$

Most of the time we will study connections or covariant derivatives on vector bundles that enjoy compatibility conditions with additional structures, such as metrics or almost complex structures. The most convenient way to formulate those compatibility conditions is in the language of principle bundles to be studied below in Chap. 6. In Sect. 6.3 we will show how connections on principal bundles induce connections on associated vector bundles and study their curvature and torsion in Sects. 6.6 and 6.7 respectively. This will lead to a geometric interpretation and further results on the curvature and torsion tensors introduced here. For examples of connections we refer to Sects. 6.9, 6.10 and Chap. 10.

Chapter 5
Super Lie Groups

This chapter gives an introduction to the theory of super Lie groups. However, the choice of topics is restricted to what will be needed for the theory of principal bundles in Chap. 6. Super Lie groups are of interest to physics as symmetry groups. An early mathematically rigorous treatment of super Lie groups is presented in Kostant (1977).

The first section of this chapter gives the definition of a super Lie group and shows that the general linear group $\mathrm{GL}(m|n)$ and several subgroups are super Lie groups.

The second section treats representations and actions of super Lie groups. Several examples needed later on are listed and the notions of equivariant and invariant map are given.

The third section introduces the Lie superalgebras of the linear groups that we are interested in.

The results in this chapter parallel the classical theory of Lie groups and are quite well known. They can be found, for example, in Carmeli et al. (2011) and Deligne and Morgan (1999) that are quite close to the language we use here, as well as in Bartocci et al. (1991) and Tuynman (2004). For a more specialized text, we refer to Fioresi and Gavarini (2012).

The theory of super Lie groups profits from the language of families of supermanifolds that we advocate here in several ways. Even though the linear groups that we will study here, are trivial families of supermanifolds, the presence of additional odd parameters from the base B allows for a quite geometric and intuitive interpretation. For example, the Lie superalgebras of the super Lie groups will be modules over $\mathcal{O}_B$ and thus in the case of the linear groups be made up of matrices with possibly odd entries. Similarly, B-points of the linear super Lie groups can be understood as matrices with entries from $\mathcal{O}_B$.

E. Keßler, *Supergeometry, Super Riemann Surfaces and the Superconformal Action Functional*, Lecture Notes in Mathematics 2230,
https://doi.org/10.1007/978-3-030-13758-8_5

5.1 Definition and Examples

Definition 5.1.1 A super Lie group over the supermanifold B is a supermanifold G over B together with maps $m\colon G \times_B G \to G$, $i\colon G \to G$ over B and a section $e\colon B \to G$ satisfying

i) associativity of the multiplication m

$$\begin{array}{ccc} G \times_B G \times_B G & \xrightarrow{\mathrm{id}_G \times m} & G \times_B G \\ \downarrow{\scriptstyle m \times \mathrm{id}_G} & & \downarrow{\scriptstyle m} \\ G \times_B G & \xrightarrow{\quad m \quad} & G \end{array}$$

ii) multiplication with the Identity e

$$\begin{array}{ccc} G & \xrightarrow{(\mathrm{id}_G, \hat{e})} & G \times_B G \\ \downarrow{\scriptstyle (\hat{e}, \mathrm{id}_G)} & \searrow{\scriptstyle \mathrm{id}_G} & \downarrow{\scriptstyle m} \\ G \times_B G & \xrightarrow{\quad m \quad} & G \end{array}$$

Here $\hat{e}\colon G \to G$ is the composition of $p_G\colon G \to B$ and e.

iii) inverse property of i

$$\begin{array}{ccc} G & \xrightarrow{(\mathrm{id}_G, i)} & G \times_B G \\ \downarrow{\scriptstyle (i, \mathrm{id}_G)} & \searrow{\scriptstyle \hat{e}} & \downarrow{\scriptstyle m} \\ G \times_B G & \xrightarrow{\quad m \quad} & G \end{array}$$

In short, G is a group object in the category SMan_B of supermanifolds over B. One verifies that for any morphism $p : B' \to B$ we get a super Lie group p^*G over B'.

A smooth map $f\colon G \to H$ between super Lie groups G and H is called a homomorphism of super Lie groups if the map f commutes with multiplication, that is, if the following diagram is commutative:

$$\begin{array}{ccc} G \times_B G & \xrightarrow{m_G} & G \\ \downarrow{\scriptstyle f \times_B f} & & \downarrow{\scriptstyle f} \\ H \times_B H & \xrightarrow{m_H} & H \end{array}$$

Example 5.1.2 (Trivial Group) The trivial group is given by the supermanifold B over B together with $m = i = e = \mathrm{id}_B$.

Example 5.1.3 ($\mathrm{GL}(m|n)$ *as a Super Lie Group)* The general linear group $\mathrm{GL}_R(m|n) \subset \mathrm{Mat}_R(m|n)$ is the subset of the even invertible square matrices over the superring R, see Definition 2.3.3. By Corollary 2.3.5, an even square matrix is invertible if its upper right and lower left block are.

In the case $R = \mathcal{O}_B$ we can construct a super Lie group, denoted $\mathrm{GL}_B(m|n)$, such that its B-points correspond bijectively to the invertible square matrices in $\mathrm{Mat}_{\mathcal{O}_B}(m|n)$. Indeed, let us consider coordinates on the linear supermanifold $\mathrm{Mat}_B(m|n) = \mathbb{R}^{m^2+n^2|2mn} \times B$, arranged in matrix form

$$\begin{pmatrix} {}_a x^b & {}_a \eta^\beta \\ {}_\alpha \eta^b & {}_\alpha x^\beta \end{pmatrix},$$

where x represent the even coordinates and η the odd ones. As in Example 3.1.12, B-points of the real linear supermanifold $\mathrm{Mat}_B(m|n)$ correspond to even square matrices in $\mathrm{Mat}_{\mathcal{O}_B}(m|n)$. $\mathrm{GL}_B(m|n)$ is the open submanifold of $\mathrm{Mat}_B(m|n)$, given by restriction to the open subset of $\| \mathrm{Mat}_B(m|n)\|$ where the determinant of the upper left and lower right block do not vanish. It is of no surprise that arranged in matrix form the effect of matrix multiplication and forming of an inverse can just be read off and seen to be smooth. The identity morphism is the B-point corresponding to the identity matrix. Hence, $\mathrm{GL}_B(m|n)$ is a super Lie group over B whose B-points correspond to elements of $\mathrm{GL}_{\mathcal{O}_B}(m|n)$.

Since $\mathrm{GL}_B(m|n) = \mathrm{GL}_{\mathbb{R}^{0|0}}(m|n) \times B$ is a trivial family of supermanifolds, we will drop the index B from the notation. Thus, we will write $\mathrm{GL}(m|n)$ for the super Lie group and assume, as always, implicitly that we have extended the base as necessary.

Example 5.1.4 The super Lie group $\mathrm{GL}(m|n)$ from Example 5.1.3 has, of course, several interesting sub Lie groups:

- the group of orthogonal matrices $\mathrm{O}(m|2n)$, see Definition 2.10.9,
- the group of matrices that commute with a given almost complex structure, see Sect. 2.12,
- the group of unitary matrices $\mathrm{U}(m|n)$, see Definition 2.12.7.

Here by sub Lie group of G we understand a super Lie group H with an immersion $j\colon H \to G$ which is a homomorphism of super Lie groups such that $\|j\|$ is injective. To prove that these subgroups of $\mathrm{GL}(m|n)$ are actually sub Lie groups, one has to show that the subgroup structure on the level of B-points induces an immersed submanifold. Intuitively, this uses that the equations defining the particular subgroup in $\mathrm{GL}(m|n)$ are actually smooth equations.

As an example, let us construct the manifold structure and the immersion j for $\mathrm{O}(m|2n)$. The subgroup $\mathrm{O}_{\mathcal{O}_B}(m|2n)$ of $\mathrm{GL}_{\mathcal{O}_B}(m|2n)$ consists of the matrices A such that $b(Av, Aw) = b(v, w)$ for all $v, w \in \mathcal{O}_B^{m|2n}$, where b is the standard bilinear form from Example 2.10.2. With respect to the standard basis of $\mathcal{O}_B(m|2n)$ the bilinear form b can be seen as an even symmetric matrix. The module of even symmetric matrices has rank $\frac{1}{2}\left(m(m+1) + 2n(2n-1)\right)|2mn$. We denote the corresponding linear superdomain by $S^2_B(m|2n)$, its coordinates by ${}_A S_B$ and the map induced from $b(A\cdot, A\cdot)$ by $f\colon \mathrm{GL}_B(m|2n) \to S^2_B(m|2n)$. In the local coordinates ${}_A X^B$ the map f is given by $f^\# {}_A S_B = {}_A X^C {}_C b_D {}^D X_B$. Its reduced differential is easily seen to be surjective at all points. Consequently, by the characterization of submersions in Proposition 3.1.16, we know that locally around any $p \in \|\operatorname{GL}(m|2n)\|$, the map f is can be understood as the projection on the first factor in $V \times_B U \to V$, where V is an open neighbourhood around $f_{red}(p)$ and U a superdomain of dimension

$$\begin{aligned} m^2 + 4n^2 - \frac{1}{2}\left(m(m+1) + 2n(2n-1)\right) &|4mn - 2mn \\ = \frac{1}{2}(m(m-1) + 2n(2n+1))&|2mn. \end{aligned}$$

Hence for any $p \in \|\operatorname{GL}(m|2n)\|$ such that $f_{red}(p) = b$, the superdomain U gives a coordinate neighbourhood for $\operatorname{O}(m|2n)$. Since the splitting provided by the characterization of submersions depends smoothly on the point p, the local coordinate patches glue together to give $\operatorname{O}(m|2n)$ a manifold structure. The immersion j is locally given by the map $j|_U : U \to U \times V$ whose image is in the fiber above b. Using those local coordinates, it is obvious that the immersion j is also a group homomorphism.

5.2 Lie Group Actions

Definition 5.2.1 A homomorphism of super Lie groups $\rho\colon G \to \operatorname{GL}(V)$ is called a linear representation of G on V.

Example 5.2.2 (Trivial Representation) Any group G can be represented on any vector space V via

$$G \longrightarrow B \xrightarrow{e} \operatorname{GL}(V).$$

Example 5.2.3 (Defining Representation) Let G be a subgroup of $\operatorname{GL}(V)$. The inclusion $i\colon G \to \operatorname{GL}(V)$ is a Lie group homomorphism called the defining representation.

Example 5.2.4 (Linear Algebra of Representations) Let $\rho\colon G \to \operatorname{GL}(V)$ be a representation of G on V. There are induced representations $\rho^\vee$ on $V^\vee$, $\operatorname{Ber}\rho$ on $\operatorname{Ber} V$, and $\bigwedge \rho$ on $\bigwedge V$ by applying dualization, Berezinian and products on $\operatorname{GL}(V)$.

Given representations ρ of G on V and σ of G on W there are induced representations of G on $V \oplus W$ and $V \otimes W$ denoted $\rho \oplus \sigma$ and $\rho \otimes \sigma$.

Definition 5.2.5 (See Balduzzi et al. 2009, Def. 3.1) A morphism of supermanifolds $a : G \times M \to M$ over B is called a (left-) action of G on M if it satisfies

$$a \circ (m \times \mathrm{id}_M) = a \circ (\mathrm{id}_G \times a), \qquad a \circ \left(\hat{e}, \mathrm{id}_M\right) = \mathrm{id}_M .$$

Given a G-action on M we call M a G-space.

We say that G acts freely on M if for any B-points $m\colon B \to M$ and $g\colon B \to G$ we have that $a \circ (g \times m) = m$ implies $g = e$. Right actions can be defined analogously.

Example 5.2.6 (Action of the Trivial Group) The trivial group B acts on any supermanifold M over B with a given by the isomorphism $B \times_B M = M$.

Example 5.2.7 (Left and Right Action of G on Itself) Any super Lie group acts on itself from the left by multiplication: $L\colon G \times G \to G$. The axioms of a group action are obvious. Let $g\colon B \to G$ be a B-point of G and $\hat{g}\colon G \to B \to G$ the composition of the map g with the projection to the base B. Then we call L_g the composition

$$G \xrightarrow{(\hat{g},\mathrm{id})} G \times G \xrightarrow{m} G.$$

It acts on another B-point h of G as $L_g h = gh$.

Analogously does G act on itself from the right by multiplication: $R\colon G \times G \to G$. We define R_g analogously and get $R_g h = hg$.

Example 5.2.8 (Adjoint Action of G on Itself) The action of G on itself given by the composition of

$$G \times G \xrightarrow{(\mathrm{id},i)\times \mathrm{id}} G \times G \times G \xrightarrow{\mathrm{id}\times\sigma_{23}} G \times G \times G \xrightarrow{\mathrm{id}\times m} G \times G \xrightarrow{m} G$$

is called the adjoint action of G, denoted by Ad. Here σ_{23} is the permutation of the second and third factor in the product. Given two B-points g and h of G, define Ad_g as the composition

$$G \xrightarrow{(\hat{g},\mathrm{id})} G \times G \xrightarrow{\mathrm{Ad}} G$$

Then it holds that $\mathrm{Ad}_g = L_g \circ R_{g^{-1}}$ and $\mathrm{Ad}_g h = ghg^{-1}$. The neutral element e is a fixed point of this action.

Example 5.2.9 (Action on the Tangent Bundle) Let a be an action of the super Lie group G on the supermanifold M. Then G also acts on TM and this action is a bundle map over a by the following construction:

$$\begin{array}{ccc}
TG \times TM & \xrightarrow{Ta} & TM \\
{\scriptstyle s_0\times\mathrm{id}}\uparrow\ \downarrow{\scriptstyle \pi_{TG}\times\mathrm{id}} & & \downarrow{\scriptstyle \mathrm{id}} \\
G \times TM & \xrightarrow{a_{TM}} & TM \\
\downarrow{\scriptstyle \mathrm{id}\times\pi_{TM}} & & \downarrow{\scriptstyle \pi_{TM}} \\
G \times M & \xrightarrow{a} & M
\end{array}$$

Here s_0 is the zero section and thus the action on TM is defined as $a_{TM} = Ta \circ (s_0 \times \mathrm{id})$. To see that this is indeed an action consider the diagram

$$\begin{array}{ccccc}
TM & \xrightarrow{(0,\mathrm{id})} & TG \times TM & \xrightarrow{Ta} & TM \\
\downarrow{\scriptstyle \mathrm{id}} & & {\scriptstyle s_0 \times \mathrm{id}} \uparrow \ \downarrow {\scriptstyle \pi_{TG} \times \mathrm{id}} & & \downarrow{\scriptstyle \mathrm{id}} \\
TM & \xrightarrow{(e,\mathrm{id})} & G \times TM & \xrightarrow{a_{TM}} & TM \\
\downarrow{\scriptstyle \pi_{TM}} & & \downarrow{\scriptstyle \mathrm{id} \times \pi_{TM}} & & \downarrow{\scriptstyle \pi_{TM}} \\
M & \xrightarrow{(e,\mathrm{id})} & G \times M & \xrightarrow{a} & M
\end{array}$$

As the lower line is the identity, also the upper line is the identity. It follows that the middle line is the identity as well. Associativity can be read off in just the same way from the appropriate diagram.

With the help of the following Example 5.2.10 one can extend a_{TM} to actions on $T^{\vee}M$, $\mathrm{Ber}\, TM$, $\bigwedge^{\bullet} TM$.

Example 5.2.10 (Actions on Vector Bundles) As a generalization of the preceding Example 5.2.9 we now look at actions on vector bundles or families of representations. Let a be an action of G on M and $\pi : E \to M$ a vector bundle over M. An action a_E of G on E is called a vector bundle action over a or a family of representations, if it fits into the following commutative diagram

$$\begin{array}{ccc}
G \times E & \xrightarrow{a_E} & E \\
\downarrow{\scriptstyle \mathrm{id} \times \pi} & & \downarrow{\scriptstyle \pi} \\
G \times M & \xrightarrow{a} & M
\end{array}$$

and is locally of the form $a_E = (a, r) \colon G \times M \times V \to M \times V$. Here r is a family of representations of G indexed by M on V, the typical fiber of E.

By applying linear algebra to families of representations over a, one can construct families of representations over a on $E^{\vee}$, $\mathrm{Ber}\, E$, $\bigwedge E$, $E \oplus F$, $E \otimes F$ out of families of representations over a on vector bundles E and F (see Example 5.2.4).

Example 5.2.11 (Adjoint Representation of G on Its Lie Algebra) Restrict the action Ad_{TG} on TG to the fiber above e. As e is a fixed point of Ad it gives a linear representation of G on its Lie algebra $\mathfrak{g} = T_eG$:

$$\mathrm{ad} \colon G \times T_eG \to T_eG$$

Example 5.2.12 (Inverse Action) Given any action a on M define the action by the inverse a^i as the composition

$$G \times M \xrightarrow{i \times \mathrm{id}_M} G \times M \xrightarrow{a} M$$

This is of course an action of G on M as well.

Definition 5.2.13 (γ-Equivariant Maps) Let a_M be an action of G on M and a_N and action of H on N. Let furthermore $\gamma: G \to H$ be a Lie group homomorphism and $f: M \to N$ a homomorphism of supermanifolds. The map f is called γ-equivariant if the following diagram is commutative:

$$
\begin{array}{ccc}
G \times M & \xrightarrow{\gamma \times f} & H \times N \\
\downarrow{\scriptstyle a_M} & & \downarrow{\scriptstyle a_N} \\
M & \xrightarrow{\ \ f\ \ } & N
\end{array}
$$

An id_G-equivariant map is also called G-equivariant.

Lemma 5.2.14 *Let* $s: M \to G \times M$ *be a section of the projection* $p_M: G \times M \to M$. *Any other section* t *of* p_M *can be written as* $t = (m \times \mathrm{id}_M)(g, s)$ *for some function* $g: M \to G$. *Let* G *act on* $G \times M$ *by multiplication from the left,* a_N *be an action of* H *on* N *and* $f: G \times M \to N$ *a* γ-equivariant map. Denote $f^s: M \to N$ the map $f \circ s$ and analogously $f^t = f \circ t$. Then $f^t = a_N(\gamma \circ g, f^s)$ and f is completely determined by s and f^s.

Proof The transformation property of f^t is a consequence of the γ-equivariance of f:

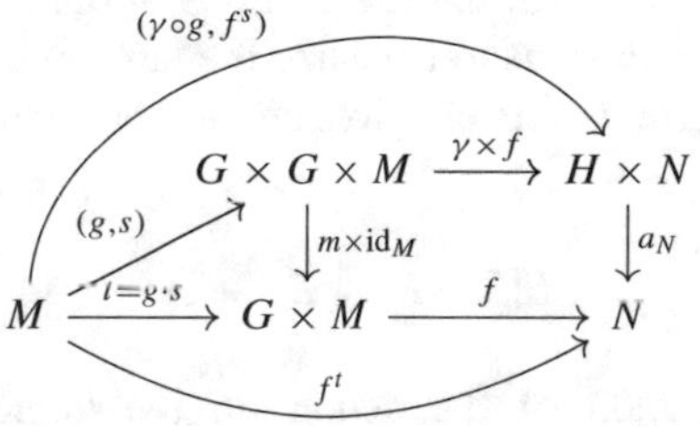

The section f^s is of the form (id_M, g) for some map $g: M \to G$. Let $s_{\mathrm{id}}: M \to M \times G$ be the identity section. Then $f^{s_{\mathrm{id}}} = a_N(\gamma \circ g^{-1}, f^s)$ and $f^{s_{\mathrm{id}}}$ determines f, again by the γ-equivariance of f:

$$
\begin{array}{ccc}
G \times G \times M & \xrightarrow{\gamma \times f} & H \times N \\
{\scriptstyle \mathrm{id}_G \times s_{\mathrm{id}}}\Big\uparrow \Big\downarrow{\scriptstyle m \times \mathrm{id}_M} & \nearrow{\scriptstyle \gamma \times f^{s_{\mathrm{id}}}} & \Big\downarrow{\scriptstyle a_N} \\
G \times M & \xrightarrow{\ \ f\ \ } & N
\end{array}
$$

That is, $f = a_N \circ (\gamma \times f^{s_{\mathrm{id}}})$. $\qquad\square$

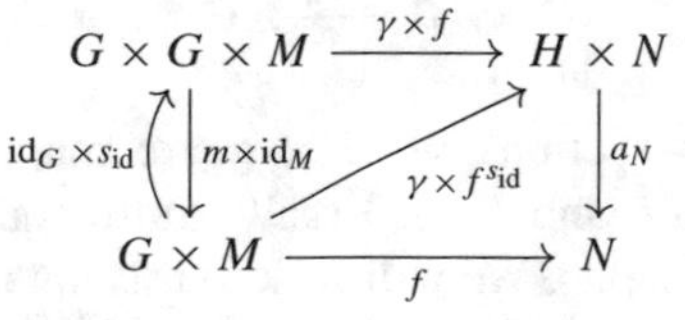

Definition 5.2.15 (G-Invariant Maps) Let a be an action of G on M and $f\colon M \to N$ a homomorphism of supermanifolds. The map f is called G-invariant if the following diagram is commutative:

$$\begin{array}{ccc} G \times M & \xrightarrow{a} & M \\ \downarrow{\scriptstyle p_M} & & \downarrow{\scriptstyle f} \\ M & \xrightarrow{f} & N \end{array}$$

Lemma 5.2.16 *Let G act on $G \times M$ by multiplication from the left and $f\colon G \times M \to N$ a G-invariant map. There exists a map $\overline{f}\colon M \to N$ such that $f = \overline{f} \circ p_M$ where p_M is the projection on the second factor $p_M\colon G \times M \to M$.*

Proof This is a special case of Lemma 5.2.14 where $H = B$ is the trivial group. Let $s\colon M \to G \times M$ be any section of p_M. The map $\overline{f} = f^s = f \circ s\colon M \to N$ does not depend on s by the transformation formula in Lemma 5.2.14. □

5.3 Lie Algebra of a Lie Group

Let us denote the tangent space of the Lie group G at the B-point e by $\mathfrak{g} = e^*TG$. The multiplication on G allows us to trivialize the tangent bundle of G. Indeed, the left action L of the Lie group on itself gives an isomorphism when restricted to the identity:

$$L|_e : G \times_B B \to G$$

Consequently, the restriction of the action on the tangent bundles L_{TG} (see Example 5.2.9) gives an isomorphism λ of vector bundles over $L|_e$:

$$\begin{array}{ccc} G \times \mathfrak{g} & \xrightarrow{\lambda} & TG \\ \downarrow & & \downarrow \\ G \times_B B & \xrightarrow{L|_e} & G \end{array}$$

By construction, constant sections of $G \times \mathfrak{g}$ are mapped to sections that are equivariant with respect to L and L_{TG}. Usually those equivariant vector fields are called left-invariant vector fields. We will stick to this notation.The inverse of λ is a $\mathfrak{g}$-valued form on G, called the Maurer–Cartan form ζ^{MC}.

Of course, using R instead of L one would obtain another trivialization of TG. The constant sections of $G \times \mathfrak{g}$ are then mapped to right-invariant vector fields.

Remark 5.3.1 Let us give two reformulations of the definition of left-invariant vector fields.

Recall from Example 5.2.7 that for any B-point $g\colon B \to G$, the multiplication with g from the left is an isomorphism L_g of supermanifolds. Its tangent map TL_g is the restriction of L_{TG} to g. A vector field X on G is then left-invariant, if and only if $X \circ L_g = TL_g \circ X$. This is the definition of left-invariant vector fields to be found in most books on differential geometry, for example, Baum (2009).

If one considers the vector field X as a derivation, the equation $L_{TG} \circ (\mathrm{id}_G \times X) = X \circ L$ specialises to

$$(\mathbb{1} \otimes X)\, m^\# = m^\# X.$$

On the right hand side, we use that as a map $L = m$. On the left-hand side, the differential operator $\mathbb{1} \otimes X$ corresponds to the section (s_0, X) of $TG \times TG$ that appears in the definition of L_{TG}. This is the definition to be found in texts on supergeometry, for example in Carmeli et al. (2011).

The left- resp. right invariant vector fields form an $\mathcal{O}_B$-module by scalar multiplication. Furthermore, they are closed under the Lie-bracket, as the following little calculation shows for left invariant vector fields.

$$\begin{aligned}
m^\#[X, Y]f &= m^\# \left(XY - (-1)^{p(X)p(Y)} YX\right) f \\
&= (\mathbb{1} \otimes X)\, m^\# Y f - (-1)^{p(X)p(Y)} (\mathbb{1} \otimes Y)\, m^\# X f \\
&= \Big((\mathbb{1} \otimes X)(\mathbb{1} \otimes Y) - (-1)^{p(X)p(Y)} (\mathbb{1} \otimes Y)(\mathbb{1} \otimes X)\Big)\, m^\# f \\
&= (\mathbb{1} \otimes [X, Y])\, m^\# f
\end{aligned}$$

Definition 5.3.2 We call the Lie superalgebra $\mathfrak{g}$ of left invariant vector fields the Lie superalgebra of G. As an $\mathcal{O}_B$ module the Lie superalgebra $\mathfrak{g}$ is isomorphic to e^*TG.

Example 5.3.3 The Lie algebra $\mathfrak{gl}(m|n)$ of $\mathrm{GL}(m|n)$ is isomorphic to $\mathrm{Mat}(m|n)$ with the Lie algebra structure given by the commutator (see Example 2.11.2). Indeed, by construction $\mathrm{GL}(m|n)$ is a full-dimensional, open submanifold of the linear space $\mathrm{Mat}(m|n)$, so the tangent space at the identity is $\mathrm{Mat}(m|n)$. It remains to demonstrate that the Lie algebra structure of $\mathfrak{gl}(m|n)$ is given by the commutator of matrices. Let us consider the standard coordinates ${}_A X^B$ for $\mathrm{Mat}(m|n)$ and two matrices ${}_A M^B$, and ${}_A N^B$ in $e^*T\,\mathrm{GL}(m|n) = \mathfrak{gl}(m|n)$. By construction, the corresponding left invariant vector fields are given by

$$M = \sum_{A,B} {}_A X^C {}_C M^B \partial_{{}_A X^B} \qquad\qquad N = \sum_{D,E} {}_D X^F {}_F N^E \partial_{{}_D X^E}$$

and their commutator is calculated to be

$$\begin{aligned}
[M,N] &= MN - (-1)^{p(M)p(N)} NM \\
&= \sum_{A,B,D,E} {}_A X^C \, {}_C M^B \partial_{{}_A X^B} \, {}_D X^F \, {}_F N^E \partial_{{}_D X^E} \\
&\quad - (-1)^{p(M)p(N)} \sum_{A,B,D,E} {}_A X^C \, {}_C N^B \partial_{{}_A X^B} \, {}_D X^F \, {}_F M^E \partial_{{}_D X^E} \\
&= \sum_{A,E} {}_A X^C \left({}_C M^F \, {}_F N^E - (-1)^{p(M)p(N)} \, {}_C N^F \, {}_F M^E \right) \partial_{{}_A X^E}
\end{aligned}$$

which is the left invariant vector field to the commutator of ${}_A M^B$, and ${}_A N^B$. This proves the claim.

Proposition 5.3.4 *Let $f \colon G \to H$ be a homomorphism of super Lie groups. The tangent map Tf induces a homomorphism $f_* \colon \mathfrak{g} \to \mathfrak{h}$ between the Lie superalgebras of G and H.*

Proof Any homomorphism of Lie groups maps the identity to the identity. Consequently Tf induces a map $f_* \colon e_G^* TG \to e_H^* TH$. It remains to be shown that f_* is a Lie algebra homomorphism, that is, $f_*[X, Y] = [f_* X, f_* Y]$. As the differential preserves Lie-brackets, see Eq. (4.3.5), it remains to show that Tf maps left-invariant vector fields to left-invariant vector fields. As left-invariant vector fields correspond to constant sections of the trivial bundle $G \times \mathfrak{g}$, the claim follows from the commutativity of the following diagram.

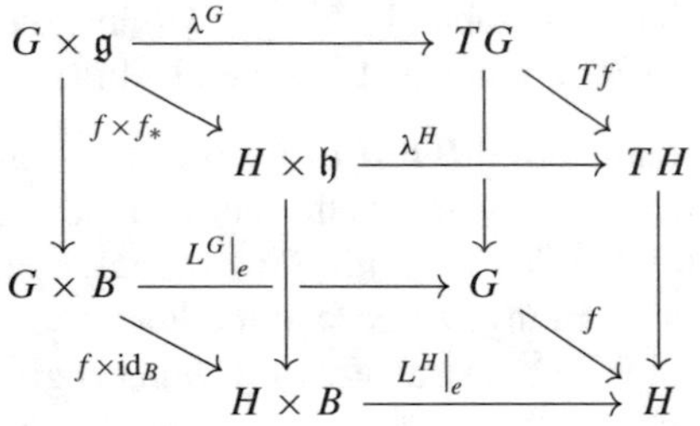

Commutativity of the diagram follows from the properties of the Lie group homomorphism f. □

Example 5.3.5 The Lie algebra of $O(m|2n)$ denoted $\mathfrak{o}(m|2n)$ is given by all matrices l such that

$$b(l(v), w) + (-1)^{p(l)p(v)} b(v, l(w)) = 0$$

for all vectors v, w and the standard bilinear form b. As $\mathrm{O}(m|2n)$ is a sub Lie group of $\mathrm{GL}(m|2n)$, by Proposition 5.3.4 the Lie algebra $\mathfrak{o}(m|2n)$ is a sub Lie algebra of $\mathfrak{gl}(m|2n)$ given by the differential of the inclusion $\mathrm{O}(m|2n) \to \mathrm{GL}(m|2n)$. One can show that the even part of the Lie algebra $\mathfrak{o}(m|2n)$ is given by block matrices of the following form

$$\left\{ \begin{pmatrix} A & B \\ C & D \end{pmatrix} \middle| \; A^T + A = 0,\, C^T J + B = 0,\, D^T J + JD = 0 \right\} \subset \mathfrak{gl}(m|2n),$$

whereas the odd part is given by

$$\left\{ \begin{pmatrix} A & B \\ C & D \end{pmatrix} \middle| \; A^T - A = 0,\, C^T J + B = 0,\, D^T J - JD = 0 \right\} \subset \mathfrak{gl}(m|2n).$$

A similar argument works for the Lie algebra $\mathfrak{u}(m|n)$ of $\mathrm{U}(m|n)$.

Chapter 6
Principal Fiber Bundles

In this chapter we treat the theory of principal bundles and connections on them for families of supermanifolds. The most important examples of principal bundles are frame bundles of vector bundles. Many extra structures on vector bundles, such as metrics or almost complex structures can actually be formulated in terms of a reduction of the structure group of the frame bundle of the vector bundle. The theory of connections on principal bundles sheds light on properties of covariant derivatives that are compatible with such extra structures.

As principal bundles and covariant derivatives are quite important in physics to model gauge theories, several generalizations thereof incorporating anti-commuting variables have appeared. Mathematical treatments of super principal bundles and connections on them are rare. The books Tuynman (2004) and Bartocci et al. (1991) contain a chapter each on principal bundles in their respective theory of supermanifolds. For the theory of families of supermanifolds we use here, some claims can be found in Deligne and Morgan (1999). As the theory of reduction of the structure group of the frame bundle will play such a prominent role in the second part and has not be treated in the other texts, we give a rather complete treatment here. We put emphasis on how connections on reductions of the frame bundle induce extra structure on the associated covariant derivatives in associated vector bundles.

The results and presentation in this chapter follow quite closely what is standard in classical differential geometry. Indeed, the presentation here owes a lot to the books Baum (2009), Kobayashi and Nomizu (1996), Lawson and Michelsohn (1989), Tolksdorf (in preparation). However, the classical proofs often do not extend to the super case directly, as they are most of the time formulated in terms of points of the manifolds. Instead, the proofs had to be reformulated to use the language of sheaves used to describe supermanifolds.

E. Keßler, *Supergeometry, Super Riemann Surfaces and the Superconformal Action Functional*, Lecture Notes in Mathematics 2230,
https://doi.org/10.1007/978-3-030-13758-8_6

6.1 Definition and Basic Properties

Definition 6.1.1 Let G be a super Lie group over B with multiplication m. A trivial principal G-bundle over a supermanifold M is the manifold $P = M \times_B G$ together with the projection $\pi\colon M \times_B G \to M$ and the right-action $a\colon P \times_B G \to P$ given by $a = \mathrm{id}_M \times m$.

Let now $\gamma\colon G \to H$ a homomorphism of Lie groups, $P = M \times_B G$ a trivial principal G-bundle over M and $Q = N \times_B H$ a trivial principal H-bundle over the supermanifold N, and $f\colon M \to N$ a map of supermanifolds. A γ-morphism of trivial principal bundles is a γ-equivariant map $g\colon P \to Q$ such that the following diagram is commutative:

$$\begin{array}{ccc} P & \xrightarrow{g} & Q \\ \downarrow & & \downarrow \\ M & \xrightarrow{f} & N \end{array}$$

A general principal bundle is a fiber bundle $P \to M$ with typical fiber G that is constructed from cocycles which are isomorphisms of trivial principal bundles, compare Definition 4.1.1 and Proposition 4.1.2. The right action of G on the trivial fiber bundles extends to a global right action $a_P\colon P \times G \to P$.

Let $P \to M$ be a principal G-bundle over M and $Q \to N$ a principal H-bundle over N and $f\colon M \to N$ a map of supermanifolds and $\gamma\colon G \to H$ a Lie group homomorphism. A map $g\colon P \to Q$ is called a γ-morphism of principal bundles over f if its restriction to trivializing open sets on both sides it is a γ-morphism of trivial principal bundles.

Remark 6.1.2 The group action a_P acts fiberwise, that is, the projection $\pi\colon P \to M$ is G-invariant. This is obvious, as the action a_P is over trivializing neighbourhoods of the form $\mathrm{id}_M \times m$.

The group action a_P acts transitively on the fibers. Let s and t be local sections of P on an open set U. Then there exists a unique map $g\colon U \to G$ such that $s = a_P \circ (t, g)$. This follows in a trivialization from the existence of an inverse in G.

Furthermore the group action a_P is a free G-action on P. That is for any B-points p of P and g of G, the equality $a_P \circ (p \times g) = p$ implies $g = e$. This can also best be seen for the local case where it reduces to the uniqueness of the unity element of G.

Lemma 6.1.3 *Let $P = M \times G$ and $Q = N \times H$ be trivial principal bundles and $g\colon P \to Q$ a γ-morphism of fiber bundles over $f\colon M \to N$. Then there exists a map $h\colon M \to H$ such that g is given as the composition*

$$P = M \times G \xrightarrow{(f,h)\times\gamma} N \times H \times H \xrightarrow{\mathrm{id}_N \times m_H} Q = N \times H.$$

Proof As g is a map between trivial principal bundles over f, it can be written as $g = (f, \overline{h})$, where $\overline{h}\colon M \times G \to H$. Let $s = (\mathrm{id}_M, e_G)$ be the identity section of P. Define $h = \overline{h} \circ s$. The claimed identity follows from the γ-equivariance of g, see Lemma 5.2.14. □

Proposition 6.1.4 *Let $s\colon M \to P$ be a section of the principal bundle P. Then P is a trivial principal bundle and there is an isomorphism of trivial principal bundles $\phi\colon M \times G \to P$ that maps s to the identity section of $M \times G$:*

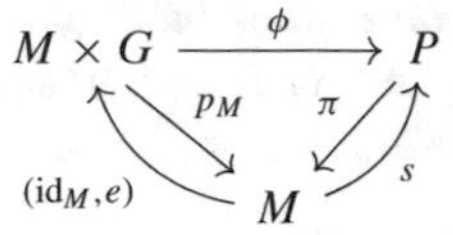

Let furthermore t be another section of P and $\tilde{\phi}\colon M \times G \to P$ the corresponding isomorphism that maps t to the identity section. By Lemma 6.1.3 the isomorphism of trivial principal bundles $\tilde{\phi}^{-1} \circ \phi$ can be written as left multiplication with a function $g\colon M \to G$. Then it holds that $t = s \cdot g^{-1}$.

Proof We construct the isomorphism ϕ of principal bundles by $\phi = a_P \circ (s \times \mathrm{id}_G)$. By the properties of the action a_P, the map ϕ is G-equivariant. It can be seen that ϕ is an isomorphism by restricting to open sets $U \subset M$ such that P is trivial over U. □

Corollary 6.1.5 *Let a trivialization of P be given by open sets U_i of M and glueing functions ϕ_{ij}. Over each U_i there is an isomorphism $\phi_i\colon U_i \times G \to P|_{U_i}$ such that $\phi_{ij} = \phi_j^{-1} \circ \phi_i$. Thus there are local sections $s_i\colon U_i \to P$ given by $s_i = \phi_i \circ (\mathrm{id}_{U_i}, e)$. Vice-versa, by Proposition 6.1.4, local trivializations can be given in terms of local sections s_i.*

By Lemma 6.1.3 the cocycles ϕ_{ij} can be written as left multiplication with a function $g_{ij}\colon U_i \cap U_j \to G$. Then it holds that $s_j = s_i \cdot g_{ij}^{-1}$. The maps g_{ij} fulfill a multiplicative cocycle property, similar to Baum (2009, Definition 2.5).

Example 6.1.6 (Frame Bundle) Let $E \to M$ be a vector bundle of rank $r|s$ and U_i a cover of M that trivializes E, that is, $E|_{U_i}$ is isomorphic to $U_i \times \mathbb{R}^{r|s}$. Let F_{iA} be the corresponding standard local frame over U_i. We know that all other even frames are related to F_{iA} by an even linear transformation from $\mathrm{GL}(r|s)$. That is, one can give the set of all local frames over U_i the following supermanifold structure: $U_i \times \mathrm{GL}(r|s)$. In particular over $U_i \cap U_j$, it holds that $F_{iA} = \left(g_{ij}\right)_A{}^B F_{jB}$, where $g_{ij}\colon U_i \cap U_j \to \mathrm{GL}(r|s)$ is a family of linear maps. By Corollary 6.1.5, the data of U_i and g_{ij} is sufficient to construct a principle fiber bundle with structure group $\mathrm{GL}(r|s)$, called the frame bundle of E. In the case $E = TM$ we also speak of the frame bundle of M.

Any local section of P corresponds to an even local frame of E and vice versa.

The following is a generalization of Lemma 5.2.16 in the context of principal bundles.

Proposition 6.1.7 *For any $\overline{f}: M \to N$, the map $f = \overline{f} \circ \pi: P \to N$ is G-invariant. Conversely, let $f: P \to N$ be a G-invariant map. Then there exists a map $\overline{f}: M \to N$ such that $f = \overline{f} \circ \pi$.*

Proof Let U_i be a cover of M that trivializes P. The associated local sections s_i give rise to the maps $f^{s_i} = f \circ s_i: U_i \to N$. By Lemma 5.2.16 the maps f^{s_i} are invariant under change of trivialization, and thus glue together to a well-defined map $\overline{f}: M \to N$. □

Corollary 6.1.8 *Let $E \to M$ be a fiber bundle over M and $\pi: P \to M$ a G-principal bundle over M. Then any G-invariant section s of $\pi^* E$ is of the form $s = \pi^* \overline{s}$ for some section $\overline{s}: M \to E$.*

6.2 Associated Fiber Bundles

Definition 6.2.1 (Associated Fiber Bundle) Let P be a G-principal bundle over M and $a_F: G \times F \to F$ an action of G on the supermanifold F. Let U_i be a trivializing cover over M of P and ϕ_{ij} the corresponding cocycles. Recall from Lemma 6.1.3 that ϕ_{ij} can be written as left-multiplication with some function g_{ij}. We define a fiber bundle E over M with typical fiber F by the cocycles $\sigma_{ij} = (\mathrm{id}_{U_i \cap U_j}, \overline{\sigma}_{ij})$. Here the map $\overline{\sigma}_{ij}: U_i \cap U_j \times F \to F$ is given by $\overline{\sigma}_{ij} = a_F \circ \left(g_{ij} \times \mathrm{id}_F\right)$. The fiber bundle E is called the a_F-associated bundle to P, also denoted as $E = P \times_{a_F} F$.

If the action a_F is given by a linear representation ρ of G on a vector space $F = V$, the resulting fiber bundle is actually a vector bundle denoted by $E = P \times_\rho V$.

Example 6.2.2 Let E be a vector bundle of rank $(r|s)$ over M. Let P be the frame bundle of E and $\rho: \mathrm{GL}(r|s) \to \mathrm{GL}(r|s)$ the representation given by the identity. Then $P \times_\rho \mathbb{R}^{r|s}$ is isomorphic to E.

Example 6.2.3 (Frames of Associated Vector Bundles) Let P be a principal G-bundle over M and $\rho: G \to \mathrm{GL}(m|n)$ a representation on the linear group of the vector space $V = \mathbb{R}^{m|n}$. A section s of P gives rise to a frame F_A of $E = P \times_\rho V$ and vice versa. Without loss of generality, we can assume that $P = M \times G$ and the section s is given by the identity section. In that case the vector bundle $E = M \times \mathbb{R}^{m|n}$ is trivial. Let us denote the standard frame of $\mathbb{R}^{m|n}$ by F_A. As in Sect. 4.2, the corresponding constant sections of E are given by constant maps $F_A: M \to V \oplus \Pi V$. By Lemma 5.2.14, they give rise to equivariant maps $\overline{F}_A: P \to V \oplus \Pi V$, such that $\overline{F}_A \circ s = F_A$.

More generally we have:

Proposition 6.2.4 *Let $a_F: G \times F \to F$ be an action of G on the supermanifold F and $i: G \to G$ the inverse map on G. There is a one-to-one correspondence between i-equivariant maps $f: P \to F$ and sections s of $P \times_{a_F} F$.*

Proof Let U_j be a trivializing cover for P, and t_j the corresponding trivializing sections with $t_k \cdot g_{kj}^{-1} = t_j$. The map $f^{t_j} = f \circ t_j$ gives rise to local sections $s|_{U_j} = (\mathrm{id}_{U_j}, f^{t_j})$ of $P \times_{a_F} F|_{U_j} = U_j \times F$. Conversely the map f can be reconstructed from f^{t_j} and the i-equivariance by Lemma 5.2.14. It remains to show that the sections $s|_{U_j}$ do not depend on t_j and that they glue together to a well-defined section s of P. By the i-invariance of f we obtain that $f^{t_j} = a_N(g_{kj}, f^{t_k})$ which is precisely the required transformation behaviour for sections of $P \times_{a_F} F$. □

6.3 Connections

Let $\pi : P \to M$ be a principal G-bundle over the B-supermanifold M. We call the fiberwise kernel of $d\pi$ vertical subbundle $VP \subset TP$.

$$0 \longrightarrow VP \longrightarrow TP \xrightarrow{d\pi} \pi^*TM \longrightarrow 0 \tag{6.3.1}$$

The subbundle VP is invariant under the action of a_{TP} because a acts fiberwisely. Consequently the bundle $HP = \pi^*TM = {}^{TP}\!/_{VP}$, called the horizontal bundle, inherits a G-action which we will call a_{HP}.

Lemma 6.3.2 *There is a vector bundle isomorphism*

$$\lambda \colon P \times \mathfrak{g} \to VP \subset TP,$$

that is, VP is a trivial vector bundle with typical fiber $\mathfrak{g}$. Furthermore the inverse adjoint action $\mathrm{ad}^i : G \times \mathfrak{g} \to \mathfrak{g}$ extends to a right action of G on $P \times \mathfrak{g}$ and the map λ is equivariant that is the following diagram is a commutative diagram of bundle maps over $a \colon P \times G \to P$.

$$\begin{array}{ccc}
P \times \mathfrak{g} \times G & \xrightarrow{\lambda \times \mathrm{id}_G} & VP \times G \\
\big\downarrow{\scriptstyle a \times \mathrm{ad}^i} & & \big\downarrow{\scriptstyle a_{TP}} \\
P \times \mathfrak{g} & \xrightarrow{\lambda} & VP
\end{array}$$

Proof Define λ as the composition of the upper row of the diagram

$$\begin{array}{ccccccc}
P \times \mathfrak{g} & \xrightarrow{s_0 \times \mathrm{id}_{\mathfrak{g}}} & TP \times \mathfrak{g} & \longrightarrow & TP \times TG & \xrightarrow{Ta} & TP \\
& \searrow & \big\downarrow & & \big\downarrow & & \big\downarrow \\
& & P & \xrightarrow{(\mathrm{id}_P, e)} & P \times G & \xrightarrow{a} & P
\end{array}$$

By definition of a principal bundle, the action a acts fiberwise, so the image of λ is in VP. To see that λ is a vector bundle isomorphism, we may work locally, such that $P = M \times G$. The vertical bundle is then given by $TG \to M \times G$ and the claim reduces to the trivialization of TG by multiplication from the right, as in Sect. 5.3.

To see that λ is indeed an equivariant bundle map over a consider the diagram

$$\begin{array}{ccccccc}
P\times\mathfrak{g}\times G & \xrightarrow{s_0\times \mathrm{id}_{\mathfrak{g}}\times \mathrm{id}_G} & TP\times\mathfrak{g}\times G & \longrightarrow & TP\times TG\times G & \xrightarrow{Ta\times\mathrm{id}_G} & TP\times G\\
\downarrow{\scriptstyle a\times \mathrm{ad}^i} && \downarrow{\scriptstyle a_{TP}\times\mathrm{ad}^i} && \downarrow{\scriptstyle a_{TP}\times \mathrm{Ad}^i_{TG}} && \downarrow{\scriptstyle a_{TP}}\\
P\times\mathfrak{g} & \xrightarrow{s_0\times\mathrm{id}_{\mathfrak{g}}} & TP\times \mathfrak{g} & \longrightarrow & TP\times TG & \xrightarrow{Ta} & TP
\end{array}$$

This is a diagram where every square commutes. The first square commutes, as the zero section is G-invariant, the second square commutes by the definition of ad^i and to see that the third square commutes one goes back to the definitions of a_{TP} and Ad^i_{TG}:

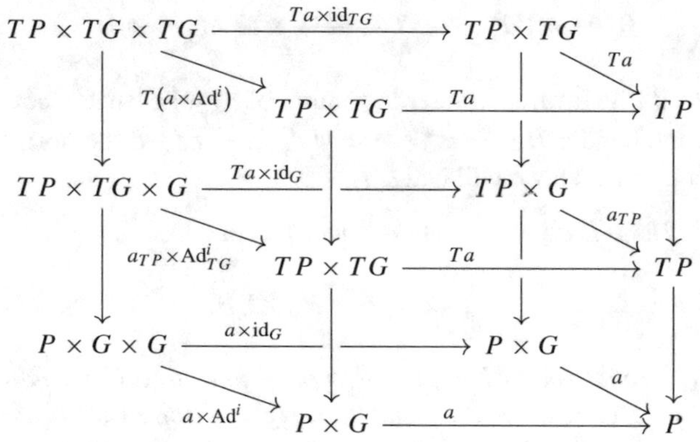

As the lower level of the diagram is commutative, also the upper level is. This in turn makes also the middle level commutative. □

Remark 6.3.3 To see how this fits into the description of connections in G-principal bundles consider a B-point p of P and define the map

$$\mathcal{R}_p = a \circ (p \times \mathrm{id}_G) \colon G \to P,$$

and for each choice of B-point g of G a map

$$\mathcal{R}_g = a \circ (\mathrm{id}_P \times g) \colon P \to P.$$

The invariance of VP is then expressed as

$$V_{pg}P = d\mathcal{R}_g V_p P.$$

The fiber of λ over the point p is then the differential of $\mathcal{R}_p$ at the point e, and the image of a constant section h of $P \times \mathfrak{g}$ under λ is called fundamental vector field $\widetilde{h} = \lambda \circ h$. The equivariance of the map λ is then in this notation that

$$d\mathcal{R}_g \widetilde{h} = \widetilde{\mathrm{ad}(g^{-1})h}.$$

Definition 6.3.4 A G-equivariant splitting of the short exact sequence

$$0 \longrightarrow VP \stackrel{\iota}{\longrightarrow} TP \stackrel{d\pi}{\longrightarrow} HP \longrightarrow 0 \tag{6.3.5}$$

is called a connection. Such an equivariant splitting is given either by

(i) an inclusion of $i_H \colon HP \to TP$ which is right inverse to $d\pi$, that is, $d\pi \circ i_H = \mathrm{id}_{HP}$ and invariant under the G-action on TP, or $i_H \circ a_{HP} = a_{TP} \circ (i_H \times \mathrm{id}_G)$ or
(ii) a projection $p_V \colon TP \to VP$ which is invariant under the G-action, that is, $p_V \circ a_{TP} = a_{VP} \circ (p_V \times \mathrm{id}_G)$ and left inverse to ι, or $p_V \circ \iota = \mathrm{id}_{VP}$.

Let us identify VP with $P \times \mathfrak{g}$ using Lemma 6.3.2 and call the vertical projector p_V of a connection, viewed as a differential form with values in $\mathfrak{g}$, connection form denoted $\omega \in \Omega(P, \mathfrak{g})$. Let us work out the additional properties of ω.

Since ω gives a splitting of the short exact sequence (6.3.5), we have that ω is left-inverse to $\iota \colon \Gamma(P \times \mathfrak{g}) \to \Gamma(TP)$, that is

$$\omega \circ \iota = \mathrm{id}_{\Gamma(P \times \mathfrak{g})}. \tag{6.3.6}$$

For the G-equivariance of ω recall that $T(P \times G) = p_P^* TP \oplus p_G^* TG$. The action of a_{TP} on ω is given by the form $\omega_a = a^*\omega \circ da|_{p_P^* TP \oplus 0}$ on $P \times G$. Here ω_a is an section of $p_P^* T^\vee P \otimes \mathfrak{g}$. Then the G-equivariance of ω is given by

$$a^*\omega \circ da|_{p_P^* TP} = \mathrm{id}_{p_P^* TP} \otimes \mathrm{ad}^i \circ p_P^* \omega, \tag{6.3.7}$$

where the composition on the right hand side is to apply the representation ad^i to the $\mathfrak{g}$-factor of the form.

Example 6.3.8 (Connections on a Trivial Principal Bundle) Let $P = M \times G$ be a trivial principal G-bundle. We denote the projections on the factors by $\pi_M \colon M \times G \to M$ and $\pi_G \colon M \times G \to G$. As P is trivial, the tangent bundle TP possesses a canonical direct sum decomposition $TP = \pi_M^* TM \oplus \pi_G^* TG$. The vertical bundle of P is given by $\pi_G^* TG \hookrightarrow \pi_M^* TM \oplus \pi_G^* TG = TP$, and the canonical splitting $TP = \pi_M^* TM \oplus \pi_G^* TG$ determines a canonical connection. Clearly, not every connection is the canonical connection.

Let us now study the connection form of a given connection. Every section of TP is an $\mathcal{O}_P = \mathcal{O}_{M\times G}$ linear combination of vector fields $\pi_M^* X \oplus 0$ and $0 \oplus \pi_G^* g$, where $X \in \Gamma(TX)$ and $g \in \Gamma(TG)$ is right-invariant. We can identify the right-invariant vector field g with an element $\tilde{g} \in \mathfrak{g}$. Equation (6.3.6) implies $\langle 0 \oplus \pi_G^* g, \omega\rangle = \pi_G^* g$.

The vector field $\pi_M^* X \oplus 0$ is G-invariant, that is, $da\left(p_P^* \pi_M^* X \oplus 0\right) = a^*\left(\pi_M^* X \oplus 0\right)$. Hence by Eq. (6.3.7),

$$a^*\langle \pi_M^* X \oplus 0, \omega\rangle = \left\langle p_P^*\left(\pi_M^* X \oplus 0\right), a^*\omega \circ da\right\rangle = \mathrm{ad}^i \circ p_P^* \langle \pi_M^* X \oplus 0, \omega\rangle .$$

Put differently, the section $\langle \pi_M^* X \oplus 0, \omega\rangle$ is given by an ad^i-equivariant map from $P = M \times G$ to $\mathfrak{g}$. According to Lemma 5.2.14, such equivariant map is determined by its values along a section $s\colon M \to P = M \times G$, that is, by $s^*\langle \pi_M^* X \oplus 0, \omega\rangle = \langle X \oplus 0, s^*\omega\rangle$.

By Proposition 6.1.4, we can assume without loss of generality that s is given by the identity section $s = (\mathrm{id}_M, e)$. In that case, $ds(X) = X \oplus 0$ and $\langle X \oplus 0, s^*\omega\rangle = \langle X, \omega_s\rangle$ where $\omega_s = s^*\omega \circ ds$ denotes the pullback of ω along s.

The connection form ω can be reconstructed from ω_s by setting

$$\langle \pi_M^* X \oplus 0, \omega\rangle = \mathrm{ad}^i \circ \langle X, \omega_s\rangle , \qquad \langle 0 \oplus \pi_G^* g\rangle = \pi_G^* g.$$

Consequently, the space of connections on $P = M \times G$ is given by $\Omega(M, \mathfrak{g})$, where the canonical connection corresponds to the choice $\omega_s = 0$.

In Example 6.3.8 we have seen that the connection form ω on a trivial principal bundle is completely determined by a differential form on the base manifold M with values in $\mathfrak{g}$. Since P is locally trivial, a connection on P can equivalently described by a set of local connection forms. Let (U_i, s_i) be a trivializing cover of P with open sets $U_i \subset M$ and local sections $s_i : U_i \to P$. According to Corollary 6.1.5 the isomorphisms between $P|_{U_i}$ and $P|_{U_j}$ is given by left multiplication with $g_{ij} : U_i \cap U_j \to G$. Then on the overlap $U_i \cap U_j$ it holds that

$$s_i = s_j \cdot g_{ij} = a(s_j, g_{ij}). \tag{6.3.9}$$

Let ω be a connection form and call

$$\omega_{s_i} = s_i^* \omega \circ ds_i$$

the local connection forms. Hence, ω_{s_i} is a $\mathfrak{g}$-valued form on U_i. To see how ω_{s_i} transforms into ω_{s_j} we calculate the derivative of (6.3.9) first

$$ds_i = d\big(a\left(s_j, g_{ij}\right)\big) = \left(\left(s_j, g_{ij}\right)^* da\right) \circ \left(ds_j \oplus dg_{ij}\right)$$

and thus

$$\begin{aligned}\omega_{s_i} &= s_i^*\omega \circ ds_i = (s_j, g_{ij})^* a^*\omega \circ \left((s_j, g_{ij})^* da\right) \circ (ds_j \oplus dg_{ij}) \\ &= (s_j, g_{ij})^* \left(a^*\omega \circ da\right) \circ (ds_j \oplus 0) \\ &\quad + (s_j, g_{ij})^* \left(a^* (\omega \circ \iota) \circ \left(0 \oplus \zeta^{MC} \circ dg_{ij}\right)\right) \\ &= (s_j, g_{ij})^* \left(\mathrm{id}_{p_P^* TP} \otimes \mathrm{ad}^i \circ p_P^*\omega\right) \circ (ds_j \oplus 0) + g_{ij}^*\zeta^{MC} \circ dg_{ij} \\ &= \mathrm{id}_{TM} \otimes \mathrm{ad}^i_{g_{ij}} \circ \omega_{s_j} + \zeta^{MC}_{g_{ij}}\end{aligned}$$

Here $\zeta^{MC}_{g_{ij}} = g_{ij}^*\zeta^{MC} \circ dg_{ij}$ denotes the pullback of the Maurer–Cartan form along g_{ij}. It is a $\mathfrak{g}$-valued forms on $U_i \cap U_j$. With this notation we have shown:

Proposition 6.3.10 *There is a 1-to-1 correspondence between connections on P and local connection forms ω^{s_i} over the trivialization (U_i, s_i) of P that transform according to*

$$\omega_{s_i} = \mathrm{id}_{TM} \otimes \mathrm{ad}^i_{g_{ij}} \circ \omega_{s_j} + \zeta^{MC}_{g_{ij}} \tag{6.3.11}$$

Corollary 6.3.12 *Let $\rho\colon G \to \mathrm{GL}(r|s)$ be a representation of G and $E = P \times_\rho \mathbb{R}^{r|s}$ the associated vector bundle. Then a connection on P induces a connection on E.*

Proof Let the G-principal bundle P be given by an open cover U_i of M, local sections $s_i\colon U_i \to P|_{U_i}$ and glueing functions $g_{ij}\colon U_i \cap U_j \to G$. The corresponding trivialization of E is given by $U_i \times \mathbb{R}^{r|s}$ and the glucing functions $\rho \circ g_{ij}$. Let now ω be a connection on P. The local connection forms ω_{s_i} induce local gauge potentials

$$A_i = \rho_* \circ \omega_{s_i}$$

for a connection on E. The transformation formula (6.3.11) induces the correct transformation behaviour of the A_i, see Eq. (4.4.8). □

6.4 The Affine Space of Connections

Proposition 6.4.1 *The space of connections is an affine space over the vector space of sections of $T^\vee M \otimes P \times_{\mathrm{ad}} \mathfrak{g}$.*

Proof Given two connections in terms of connection forms ω and $\tilde{\omega}$ it is clear that their difference $\alpha = \tilde{\omega} - \omega$ is a section of $T^\vee P \otimes \mathfrak{g}$ such that

$$a^*\alpha \circ da|_{p_P^* TP} = \mathrm{id}_{p_P^* TP} \otimes \mathrm{ad}^i \circ p_P^*\alpha, \qquad \alpha \circ \iota = 0.$$

Contrary, given a connection form ω and a form α fulfilling the above conditions, the form $\omega + \alpha$ is also a connection form. The space of forms α is by the following Lemma 6.4.4 isomorphic to $T^\vee M \otimes P \times_{\mathrm{ad}} \mathfrak{g}$. □

Definition 6.4.2 Recall from Definition 4.3.7 that differential forms on P with values in a trivial vector bundle $P \times V \to P$ are denoted by $\Omega^\bullet(P, V)$. For a given connection on P, a form $\alpha \in \Omega^k(P, V)$ is called horizontal if $\alpha(X_1, \ldots, X_k) = 0$ if any of the vector fields X_i is vertical.

Consider the dual of the short exact sequence (6.3.1):

$$0 \longrightarrow \pi^* T^\vee M \xrightarrow{d\pi^\vee} T^\vee P \longrightarrow VP^\vee \longrightarrow 0$$

The image of $d\pi^\vee$ is given by 1-forms on P that vanish on vertical vector fields. A horizontal differential form is thus a section of $\pi^* \bigwedge^\bullet T^\vee M$.

Definition 6.4.3 Suppose that ρ is a representation of G on V. According to Definition 5.2.13, a section $\alpha \in \Omega^k(P, V)$ is called G-equivariant if

$$a^*\alpha \circ \bigwedge^k da\Big|_{p_P^* TP} = \mathrm{id}_{p_P^* TP} \otimes \rho \circ p_P^* \alpha.$$

The sheaf of horizontal G-equivariant k-forms is denoted by $\Omega^k(P, V)_{hor}^{(G,\rho)}$.

Lemma 6.4.4 *Let ρ be a representation of G on V and $E = P \times_\rho V$. The sheaf of horizontal G-equivariant k-forms $\Omega^k(P, V)_{hor}^{(G,\rho)}$ is an $\mathcal{O}_M$-module.*

The $\mathcal{O}_M$-module $\Omega^k(P, V)_{hor}^{(G,\rho^i)}$ of global horizontal G-equivariant k-forms with values in V is linearly isomorphic to the $\mathcal{O}_M$-module $\Omega^k(M, E)$ of global k-forms on M with values in E.

Proof By Proposition 6.1.7, any function $\overline{f} \in \mathcal{O}_M$ extends to a G-invariant function $f \in \mathcal{O}_P$. The product $f\alpha$ of a G-invariant function f with a form $\alpha \in \Omega^k(P, V)_{hor}^{(G,\rho^i)}$ is again a G-equivariant k-form with values in V. This explains the $\mathcal{O}_M$-module structure.

Since α is horizontal it is an element of $\pi^*\Omega^k(M) \subset \Omega^k(P)$. Sections of $\pi^* \bigwedge^\bullet TM$ are linear combinations of pullback sections of the form $\pi^* X$ for $X \in \Gamma\left(\bigwedge^k TM\right)$. Since $\pi^* X$ is G-invariant, we have from the ρ^i-equivariance of α that

$$a^* \langle \pi^* X, \alpha \rangle = \left\langle \pi^* X, a^*\alpha \circ \bigwedge^k da \right\rangle = \rho^i \circ p_P^* \langle \pi^* X, \alpha \rangle .$$

Hence, if $\langle \pi^* X, \alpha \rangle$ is even, it is an i-equivariant map from P to V. By Proposition 6.2.4, such i-equivariant maps are in bijection with even sections of E. If

$\langle \pi^* X, \alpha \rangle$ is odd, the same argument applies to ΠV and odd sections of E. Thus we can associate to α the form $\overline{\alpha} \in \Omega^k(M, E)$ such that $\langle X, \overline{\alpha} \rangle = \langle \pi^* X, \alpha \rangle$.

Let us additionally give an explicit local expression for $\overline{\alpha}$. Assume that s is a local section of P over the open set $U \subset M$. Then, by Proposition 6.1.4, $P|_U \simeq M \times G$ where s is the identity section and $E|_U \simeq M \times V$. By Lemma 5.2.14, the equivariant map $\langle \pi^* X, \alpha \rangle$ is completely determined by its values along s, that is by $s^* \langle \pi^* X, \alpha \rangle = \langle X, \alpha_s \rangle$. Here $\alpha_s = s^* \alpha \circ \bigwedge^k ds$ is the pullback of α along s. The differential ds gives the inclusion $\Omega^k(M) \hookrightarrow \Omega^k(P)$.

Vice versa, given $\overline{\alpha} \in \Omega^k(M, E)$, there is a unique horizontal, ρ^i-equivariant form α on P with values in V given by

$$\left\langle \pi^* X, \alpha \right\rangle = \rho^i \circ \pi^* \left\langle X, \overline{\alpha} \right\rangle . \qquad \square$$

The tangential of any representation $\rho\colon G \to \mathrm{GL}(V)$ yields a map of Lie-algebras $\rho_*\colon \mathfrak{g} \to \mathfrak{gl}(V) = \mathrm{End}(V)$. Consequently, $\rho_* \circ \mathrm{ad}$ is a representation of G on $\mathfrak{gl}(V)$ and we obtain an induced morphism of vector bundles

$$\rho_*\colon P \times_{\mathrm{ad}} \mathfrak{g} \to \mathrm{End}(P \times_\rho V),$$

which we will denote also by ρ_*. Let now ω and $\omega + \alpha$ be two connections on P. The induced connections on $P \times_\rho V$, see Corollary 6.3.12 differ by $\rho_* \alpha \in \Omega^1(M, \mathrm{End}\, E)$. Depending on the representation ρ the form $\rho_* \alpha$ may have additional geometric properties, see, for example, Sects. 6.9 and 6.10.

6.5 Exterior and Covariant Derivative

Let $\pi : P \to M$ be a G-principal bundle with a given connection ω. The corresponding horizontal projector is called $h = i_H \circ d\pi$. Let furthermore $\rho\colon G \to \mathrm{GL}(V)$ be a representation of G on V and $E = P \times_\rho V$ the associated vector bundle.

Definition 6.5.1 (Covariant Exterior Derivative) The $\mathcal{O}_B$-linear map

$$\begin{aligned} D_\omega\colon \Omega^k(P, V) &\to \Omega^{k+1}(P, V) \\ \alpha &\mapsto d\alpha \circ \bigwedge\nolimits^{k+1} h \end{aligned}$$

is called covariant exterior derivative associated to the connection ω.

Proposition 6.5.2 *The covariant exterior derivative associated to the connection ω maps equivariant differential forms to horizontal equivariant differential forms:*

$$D_\omega\colon \Omega^k(P, V)^{(G,\rho^i)} \to \Omega^{k+1}(P, V)^{(G,\rho^i)}_{hor}$$

Furthermore it holds for all $\alpha \in \Omega^k(P, V)^{(G,\rho^i)}_{hor}$ *that*

$$D_\omega \alpha = d\alpha + \rho_*(\omega) \wedge \alpha. \tag{6.5.3}$$

Here the second summand is defined via

$$\rho_*(\overline{\omega} \otimes g) \wedge \overline{\alpha} \otimes v = (-1)^{p(g)p(\alpha)} (\overline{\omega} \wedge \overline{\alpha}) \otimes \rho_*(g) v$$

for any differential forms $\overline{\omega}$ *and* $\overline{\alpha}$ *and* $g \in \mathfrak{g}$ *and* $v \in V$.

Proof By construction D_ω is horizontal. Note that for a ρ^i-equivariant form α also $d\alpha$ is ρ^i-equivariant. The ρ^i-equivariance of D_ω follows from the equivariance of h:

$$\begin{aligned} a^* (D_\omega \alpha) \circ \bigwedge^k da &= a^* \left(d\alpha \circ \bigwedge^{k+1} h \right) \circ \bigwedge^k da \\ &= a^* d\alpha \circ \bigwedge^k da \circ p_P^* \bigwedge^{k+1} h \\ &= \left(\mathrm{id}_{p_P^* TP} \otimes \rho^i \right) \circ p_P^* \, d\alpha \circ p_P^* \bigwedge^{k+1} h \\ &= \left(\mathrm{id}_{p_P^* TP} \otimes \rho^i \right) \circ p_P^* D_\omega \alpha. \end{aligned}$$

We will show Eq. (6.5.3) here only in the case $k = 1$, for higher degree forms, the proof is analogous. The proof works by contraction with vector fields X and Y on P which by linearity can be assumed to be either vertical or horizontal. If both vector fields X and Y are horizontal, by definition of $D_\omega \alpha$ it holds that

$$\iota_X \iota_Y D_\omega \alpha = \iota_X \iota_Y \, d\alpha$$

and $\iota_X \omega = \iota_Y \omega = 0$ by definition of the connection form ω.

If any of the vector fields X or Y is vertical, by definition, we have that $\iota_X \iota_Y D_\omega \alpha = 0$. We have thus to show that in those cases the right hand side of Eq. (6.5.3) is also zero.

In the case that both X and Y are vertical, we know that $\iota_X \iota_Y (\rho_*(\omega) \wedge \alpha) = 0$. Consequently, we have to show that $\iota_X \iota_Y \, d\omega = 0$. To this end, we use Eq. (2.9.7):

$$\iota_X \iota_Y \, d\alpha = X (\iota_Y \alpha) - (-1)^{p(X)p(Y)} Y (\iota_X \alpha) - \iota_{[X,Y]} \alpha = 0$$

We have used that the commutator of two vertical vector fields is again vertical, as can be seen from the local description in Sect. 6.3.

Let us now turn to the case where only one of the vectors is vertical. Without loss of generality we may assume that X is vertical. By linearity, we may furthermore assume that X is a constant section of $P \times \mathfrak{g}$. In that case, the commutator of X and Y vanishes, as one can check, once again, in the local description of P. Consequently, using Eq. (2.9.7), it holds that

$$\iota_X \iota_Y \, d\alpha = X\,(\iota_Y \alpha) - (-1)^{p(X)p(Y)} Y\,(\iota_X \alpha) - \iota_{[X,Y]}\alpha = X\,(\iota_Y \alpha)\,.$$

The map $f = \iota_Y \alpha \colon P \to V \oplus \Pi V$ is equivariant, $f \circ a = \rho^{-1} \oplus \Pi \rho^{-1} \circ (f \times \mathrm{id}_G)$. Taking the derivative with respect to G yields

$$\iota_X \iota_Y \, d\alpha = X\,(\iota_Y \alpha) = -\iota_X \rho_*(\omega) \iota_Y \alpha,$$

which cancels

$$\iota_X \iota_Y \,(\rho_*(\omega) \wedge \alpha) = \iota_X \rho_*(\omega) \iota_Y \alpha.$$

Hence the right hand side of Eq. (6.5.3) vanishes and the claim is proven. □

Proposition 6.5.4 (Product Rule for d_ω) *The map D_ω induces an $\mathcal{O}_B$-linear map*

$$d_\omega \colon \Omega^k(M, E) \to \Omega^{k+1}(M, E).$$

such that the product rule

$$d_\omega\,(\alpha \wedge \beta) = d\alpha \wedge \beta + (-1)^k \alpha \wedge d_\omega \beta$$

holds for any $\alpha \in \Omega^k(M)$ and $\beta \in \Omega^l(M, E)$. In particular the for the case $k = 0$ and $l = 1$ the covariant exterior derivative d_ω is a covariant derivative:

$$\nabla^\omega = d_\omega \colon \Gamma(E) \to \Omega^1(M, E)$$

which coincides with the covariant derivative on E associated to the connection on E from Corollary 6.3.12.

Proof The product rule follows from the corresponding product rule for D_ω. It is sufficient to check locally whether ∇^ω coincides with d_ω on $\Gamma(E)$. Let $s \colon U \subseteq M \to P$ be a local section and F_A the corresponding local frame of E. Denote furthermore by $\overline{F}_A \colon P \to V \oplus \Pi V$ the corresponding G-equivariant map. By Proposition 6.5.2 it holds that

$$D_\omega \overline{F}_A = \rho_* \omega \wedge \overline{F}_A$$

which pulls back to $\rho_* \circ \omega^s$ on U. □

6.6 Curvature

In this section we will introduce the curvature F^ω of a connection ω on a G-principal bundle $P \to M$. The curvature measures to what extent the operator D_ω fails to be exact. The curvature can also be expressed in terms of covariant derivatives in associated fiber bundles.

Definition 6.6.1 The exterior covariant derivative of the connection form $\omega \in \Omega(P, \mathfrak{g})$ is called curvature form of ω:

$$F^\omega = D_\omega \omega \in \Omega^2(P, \mathfrak{g})_{hor}^{(G, \mathrm{ad}^i)}$$

Proposition 6.6.2 *For the curvature F^ω we have the following identities:*

(i) the structure equation: $F^\omega = d\omega + \frac{1}{2}[\omega, \omega]$
(ii) the Bianchi identity: $D_\omega F^\omega = 0$
(iii) For any horizontal form $\alpha \in \Omega^k(P, V)_{hor}^{(G, \rho^i)}$ *it holds that*

$$D_\omega D_\omega \alpha = \rho_*(F^\omega) \wedge \alpha.$$

The commutator in the structure equation is the commutator of differential forms with values in the Lie algebra $\mathfrak{g}$. For two such forms $\alpha = \overline{\alpha} \otimes g$ and $\beta = \beta \otimes h$ the commutator is defined as

$$[\alpha, \beta] = (-1)^{p(g)p(\beta)} \overline{\alpha} \wedge \overline{\beta} \otimes [g, h].$$

Proof of Proposition 6.6.2 As in the proof of Eq. (6.5.3), we will prove the structure equation by assuming that X and Y are either vertical or horizontal. Let us first assume that both X and Y are horizontal. As ω vanishes on horizontal vector fields, we have that

$$\iota_X \iota_Y F^\omega = \iota_X \iota_Y D_\omega \omega = \iota_X \iota_Y \, d\omega = \iota_X \iota_Y \left(d\omega + \frac{1}{2}[\omega, \omega] \right).$$

If any of the vector fields X and Y is vertical it holds that $\iota_X \iota_Y F^\omega = 0$, as F^ω is horizontal. By linearity, we may assume that the vertical vector fields are actually constant sections of $P \times \mathfrak{g}$. Suppose that X is horizontal and Y is vertical. In that case the commutator $[\omega, \omega]$ vanishes and the exterior derivative $d\omega$ vanishes by Eq. (2.9.7). Suppose now that both vector fields are vertical and constant, that is, there are B-points g and h of $\mathfrak{g}$ such that $X = \lambda(g)$ and $Y = \lambda(h)$. By Eq. (2.9.7) it holds that

$$\begin{aligned} \iota_X \iota_Y \, d\omega &= X(\iota_Y \omega) - (-1)^{p(X)p(Y)} Y(\iota_X \omega) - \iota_{[X,Y]}\omega \\ &= -\iota_{[X,Y]}\omega = -[g, h] = -[\iota_X \omega, \iota_Y \omega] = -\frac{1}{2}\iota_X \iota_Y [\omega, \omega]. \end{aligned}$$

Consequently, the right-hand side of the structure equation vanishes also in this case and the first claim is proven.

The Bianchi identity is an easy consequence of the structure equation. By definition, D_ω is the composition of the differential d with the horizontal projector h. Consequently

$$D_\omega F^\omega = \left(dF^\omega\right) \circ h = (d\,d\omega + [d\omega, \omega]) \circ h = [d\omega \circ h, \omega \circ h] = 0$$

Let now $\alpha \in \Omega^k(P, V)^{(G,\rho^i)}$. Applying Eq. (6.5.3) twice yields

$$\begin{aligned} D_\omega D_\omega \alpha &= d\left(d\alpha + \rho_*(\omega) \wedge \alpha\right) + \rho_*(\omega) \wedge \left(d\alpha + \rho_*(\omega) \wedge \alpha\right) \\ &= \rho_*(d\omega) \wedge \alpha + \rho_*(\omega) \wedge \rho_*(\omega) \wedge \alpha \\ &= \rho_*\left(d\omega + \frac{1}{2}[\omega, \omega]\right) \wedge \alpha = \rho_*\left(F^\omega\right) \wedge \alpha \end{aligned}$$

□

Corollary 6.6.3 *By Lemma 6.4.4 the curvature* $F^\omega \in \Omega^2(P, \mathfrak{g})_{hor}^{(G,\mathrm{ad}^i)}$ *induces a curvature form* $F^\omega \in \Omega^2(M, P \times_{\mathrm{ad}} \mathfrak{g})$. *Denote* $R = \rho_* F^\omega \in \Omega^2(M, \mathrm{End}\, E)$, *where* $\rho\colon G \to \mathrm{GL}(V)$ *is a representation and* $E = P \times_\rho V$. *The following identities hold:*

(i) R coincides with R^{∇^ω}, *see* Definition 4.4.11
(ii) the Bianchi identity $d_\omega R = 0$
(iii) for any form $\alpha \in \Omega^k(M, E)$

$$d_\omega d_\omega \alpha = R \wedge \alpha$$

Proof We only need to prove point (i), the other statements then follow from Propositions 6.5.4 and 6.6.2. It suffices to check the identity locally, consequently, we may assume without loss of generality that $P = M \times G$. The identity section of P is denoted by s. We may conclude from Proposition 6.6.2 and Lemma 6.4.4 that for all $X, Y \in \Gamma(TM)$ and $Z \in \Gamma(E)$:

$$\begin{aligned} \langle X \wedge Y, R\rangle Z &= \left\langle X \wedge Y, \rho_* F_s^{\omega}\right\rangle Z \\ &= \rho_* \left\langle X \wedge Y, s^* \left(d\omega + \frac{1}{2}[\omega, \omega]\right) \circ \bigwedge^2 ds \right\rangle Z \\ &= \rho_* \Big(X \langle Y, \omega_s\rangle - (-1)^{p(X)p(Y)} Y \langle X, \omega_s\rangle \\ &\qquad - \langle [X, Y], \omega_s\rangle + [\langle X, \omega_s\rangle, \langle Y, \omega_s\rangle] \Big) Z \\ &= \left([\nabla_X^\omega, \nabla_Y^\omega] - \nabla_{[X,Y]}^\omega\right) Z \end{aligned}$$

□

6.7 Torsion

Let P be a principal G-bundle over M and $\rho\colon G \to \mathrm{GL}(m|n)$ a representation such that $P \times_\rho \mathbb{R}^{m|n} = TM$.

Definition 6.7.1 (Soldering Form) The form $\theta \in \Omega^1(P, \mathbb{R}^{m|n})_{hor}^{(G,\mathrm{ad}^i)}$ associated to $\mathrm{id}_{TM} \in \Omega^1(M, TM)$ by Lemma 6.4.4 is called the soldering form.

Definition 6.7.2 Let ω be a connection on P. The form

$$\Theta^\omega = D_\omega \theta \in \Omega^2(P, \mathbb{R}^{m|n})_{hor}^{(G,\mathrm{ad}^i)}$$

is called the torsion form of ω.

Proposition 6.7.3 *The tensor $T^\omega \in \Omega^2(M, TM)$ associated to the torsion form Θ^ω with the help of Lemma 6.4.4 coincides with the torsion tensor T^{∇^ω} from Definition 4.4.12.*

Proof It suffices to check the claim locally, that is, without loss of generality, we may assume that $P = M \times G$. We denote by s the identity section, and by F_A the corresponding frame of TM. By Proposition 6.5.2, it holds for any vector fields X and Y on P that

$$\begin{aligned}\langle X \wedge Y, D_\omega \theta\rangle &= \langle X \wedge Y, d\theta + \rho_* \omega \wedge \theta\rangle \\ &= X \langle Y, \theta\rangle - (-1)^{p(X)p(Y)} Y \langle X, \theta\rangle - \langle [X, Y], \theta\rangle \\ &\quad + \langle X, \rho_*\omega\rangle \langle Y, \theta\rangle + (-1)^{p(X)p(Y)} \langle Y, \rho_*\omega\rangle \langle X, \theta\rangle .\end{aligned}$$

Consequently, it holds for Θ^ω_s:

$$\begin{aligned}\left\langle F_A \wedge F_B, \Theta^\omega_s \right\rangle &= -[F_A, F_B] + \nabla_{F_A} F_B + (-1)^{p(A)p(B)} \nabla_{F_B} F_A \\ &= \left\langle F_A \wedge F_B, T^{\nabla^\omega} \right\rangle\end{aligned}$$

□

Remark 6.7.4 By the results of Sect. 6.4 any other connection $\tilde{\omega}$ on can be written as

$$\tilde{\omega} = \omega + \alpha$$

for an $\alpha \in \Omega^1(P, \mathbb{R}^{m|n})_{hor}^{(G,\mathrm{ad}^i)}$. For the resulting covariant derivatives it holds that

$$\nabla^{\tilde{\omega}} = \nabla^\omega + A.$$

where $A = \rho_*\alpha \in \Omega^1(M, P \times_{\mathrm{ad}} \mathfrak{g}) \subseteq \Omega^1(M, \operatorname{End} TM)$. It follows that

$$\left\langle X \wedge Y, T^{\nabla^{\tilde{\omega}}} \right\rangle = \left\langle X \wedge Y, T^{\nabla^{\omega}} \right\rangle + \langle X, A \rangle\, Y - (-1)^{p(X)p(Y)} \langle Y, A \rangle\, X.$$

We notice that only the anti-symmetric $A \in \Omega^2(M, TM) \cap \Omega^1(M, P \times_{\mathrm{ad}} \mathfrak{g})$ do change the torsion. In the case, where P is the frame bundle of a vector bundle E, any torsion tensor can be realized by choosing an appropriate anti-symmetric A. However the resulting affine linear map from the space of connections to $\Omega^2(M, TM)$ containing the torsion tensor is in general neither surjective, nor injective. This will be made explicit in the following Sects. 6.9, 6.10 and used in Chap. 10.

Curvature and Torsion are not completely unrelated. A straightforward calculation using the definition of curvature and torsion as well as the Jacobi identity yields:

Lemma 6.7.5 *For any connection on the tangent bundle $TM \to M$ and any vector fields X, Y, Z on M it holds:*

$$\begin{aligned} R(X,Y)Z &+ (-1)^{p(Z)(p(X)+p(Y))} R(Z,X)Y + (-1)^{p(X)(p(Y)+p(Z))} R(Y,Z)X \\ &= \nabla_X T(Y,Z) + (-1)^{p(Z)(p(X)+p(Y))} \nabla_Z T(X,Y) \\ &\quad + (-1)^{p(X)(p(Y)+p(Z))} \nabla_Y T(Z,X) + T(X,[Y,Z]) \\ &\quad + (-1)^{p(Z)(p(X)+p(Y))} T(Y,[Z,X]) + (-1)^{p(X)(p(Y)+p(Z))} T(Z,[X,Y]) \end{aligned}$$

6.8 Pullback of Connections and Reductions of the Structure Group

Let $f\colon M \to N$ be a map between supermanifolds, $\kappa\colon H \to G$ be a super Lie group homomorphism and P a G-principal bundle over M and Q an H-principal bundle over N. Recall that every κ-morphism $\lambda\colon Q \to P$ over f decomposes into a κ-morphism $\overline{\lambda}$ of principal bundles over M and a morphism $\overline{f}$ of G-principal bundles over f:

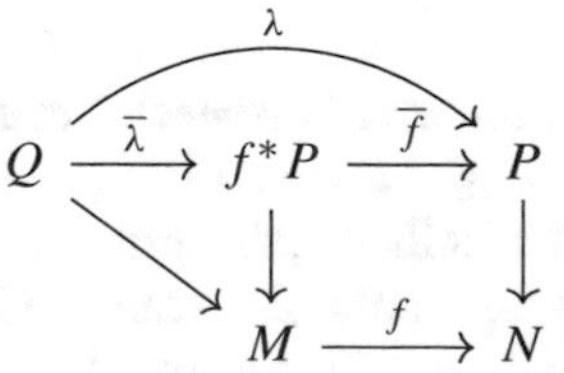

$$\begin{array}{ccccc} & & \overset{\lambda}{\frown} & & \\ Q & \xrightarrow{\overline{\lambda}} & f^*P & \xrightarrow{\overline{f}} & P \\ & \searrow & \downarrow & & \downarrow \\ & & M & \xrightarrow{f} & N \end{array} \tag{6.8.1}$$

In this section we are explaining under which conditions a connection on P induces a connection on f^*P and Q.

Proposition 6.8.2 (Existence of Pullback Connection) *Let P be a G-principal bundle over N and f^*P its pullback along $f\colon M \to N$. Any connection on P induces a connection on f^*P, called the pullback connection.*

Proof Let us denote the projections by $\pi\colon P \to N$ and $\sigma\colon f^*P \to M$ and the bundle map by $\overline{f}\colon f^*P \to P$. Then $\pi \circ \overline{f} = f \circ \sigma$. Let the connection on P be given by the connection form ω_P. We have the following diagram of vector bundles over f^*P:

$$\begin{array}{ccccccccc}
 & & & \overset{\omega_{f^*P}}{\curvearrowleft} & & & & & \\
0 & \longrightarrow & f^*P \times \mathfrak{g} & \longrightarrow & T f^*P & \xrightarrow{d\sigma} & \sigma^*TM & \longrightarrow & 0 \\
 & & \Vert & & \big\downarrow d\overline{f} & & \big\downarrow \sigma^* df & & \\
0 & \longrightarrow & \overline{f}^*(P \times \mathfrak{g}) & \longrightarrow & \overline{f}^*TP & \xrightarrow{\overline{f}^* d\pi} & \overline{f}^*\pi^*TN & \longrightarrow & 0 \\
 & & & \underset{\overline{f}^*\omega_P}{\curvearrowleft} & & & & &
\end{array}$$

Hence $\omega_{f^*P} = \overline{f}^*\omega_P \circ d\overline{f}$ yields a splitting of the upper short exact sequence. From the equivariance of $\overline{f}$ it follows that ω_{f^*P} is indeed a connection form. $\square$

Proposition 6.8.3 (Compatibility of Pullbacks) *Let $f\colon M \to N$ be a map of supermanifolds, P a principal G-bundle over N. Let $\rho\colon G \to \mathrm{GL}(V)$ be a representation. Then it holds that*

$$f^*\left(P \times_\rho V\right) = \left(f^*P\right) \times_\rho V$$

and the connections on $f^*\left(P \times_\rho V\right)$ given by the pullback of the linear connection on $P \times_\rho V$ (see Proposition 4.4.14) and the linear connection associated to the pullback connection on f^*P coincide.

This settles the case of pullback connections. We now turn to the case $f = \mathrm{id}_M$ in (6.8.1).

Definition 6.8.4 Let $\kappa\colon H \to G$ be the inclusion of a subgroup H in G, P a G-principal bundle and Q an H-principal bundle over M. A κ-morphism $Q \to P$ of principal bundles is also called reduction of the structure group.

This construction is most often used in the context of vector bundles with additional structure.

Example 6.8.5 (Bundle of Orthonormal Frames) Let $E \to M$ be a vector bundle of rank $r|s$ with metric g, and $U \subseteq M$ a trivializing open set. Then any orthonormal frame can be reached from a specified orthonormal frame e_A by a transformation in $\mathrm{O}(r|s)$. Thus the set of orthonormal frames can be given the structure of a trivial $\mathrm{O}(r|s)$-principal bundle over U with inclusion $U \times \mathrm{O}(r|s) \to U \times \mathrm{GL}(m|n)$.

This construction globalises to give the principal bundle of orthonormal frames as a reduction of the structure group of the frame bundle of E.

Example 6.8.6 (Complex Frame Bundle) Let now M be a complex manifold of dimension $m|n$. The frame bundle of the underlying real manifold allows for a reduction of the structure group to $\mathrm{GL}_{\mathbb{C}}(m|n) \subseteq \mathrm{GL}_{\mathbb{R}}(2m|2n)$. Indeed, M can be covered by complex coordinate systems, inducing complex coordinate frames on the tangent bundles. The conversion between those is given by a complex matrix, so that one can construct a principal $\mathrm{GL}_{\mathbb{C}}(m|n)$ bundle over M together with an inclusion into the real frame bundle and whose associated vector bundle is isomorphic to the tangent bundle.

In non-super differential geometry H-reductions of a G-principal bundle $P \to M$ are in one-to-one correspondence to sections of $P \times_G {}^{G}\!/_{H}$, see Kobayashi and Nomizu (1996, Proposition I.5.6). An analogous result is to be expected for supermanifolds. For the construction of quotients of super Lie groups, see Carmeli et al. (2011, Chapter 9), Alldridge et al. (2016). However, we will not pursue this issue, since the construction of the quotient ${}^{G}\!/_{H}$ as a supermanifold with G-action is a subtle issue and will not be needed here. Instead we turn to the question when a given connection on P reduces to a given reduction of the structure group.

Proposition 6.8.7 *Let $\kappa\colon H \to G$ be a Lie group homomorphism, and $\lambda\colon Q \to P$ a κ-morphism of principal bundles. A connection on P induces a connection on Q, if the map $\kappa_*\colon \mathfrak{h} \to \mathfrak{g}$ is injective and the connection form $\omega\colon TP \to P \times \mathfrak{g}$ takes actually values in $P \times \kappa_*\mathfrak{h}$.*

Proof The statement follows from inspection of the following diagram:

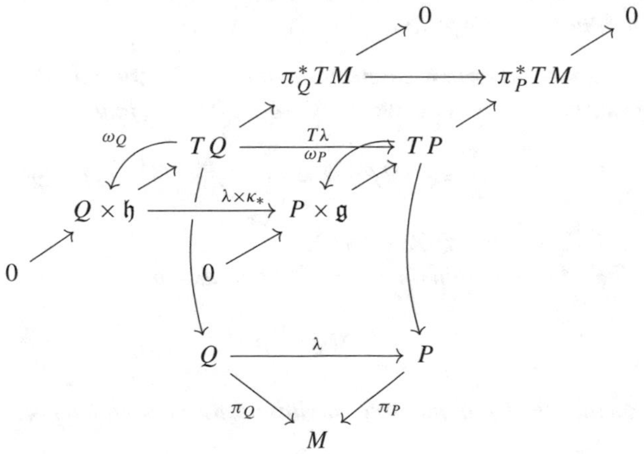

One would like to construct the map $\omega_Q = (\lambda \times \kappa_*)^{-1} \circ \omega_P \circ T\lambda$. With the conditions that κ_* is injective, and that the image of ω_P lies in $P \times \kappa_*\mathfrak{h} \subset P \times \mathfrak{g}$, the inverse $(\lambda \times \kappa_*)^{-1}$ can be taken. The equivariance properties of ω_Q are automatic. $\square$

6.9 Metric Connections

Definition 6.9.1 Let E be a vector bundle over M. An even, symmetric, non-degenerate bilinear, positive bilinear form on $\Gamma(E)$ with values in $\mathcal{O}_M$ is called metric on E. Notice that we call b positive, if it is positive in the sense of Lemma 2.10.8 in any trivializing neighbourhood of E. In the case of $E = TM$, we also speak of a metric on M.

Remark 6.9.2 In Lemma 2.10.8, the existence of square roots is used. Notice that for every invertible $f \in \mathcal{O}_M$ there are two distinct square roots for either f or $-f$. However, -1 does not possess a square root.

Consequently, one might introduce metrics with different signatures on the even part of E. In this work, we will only study positive metrics.

Remark 6.9.3 An analogous comment to Remark 4.3.6 applies. Let b be a metric on the vector bundle E over N and $f: M \to N$ a map of supermanifolds. Then f^*b is a symmetric non-degenerate bilinear form on f^*E.

Thus, the notation used in this work collides with the established notations in differential geometry in the case $E = TN$. For example in Jost (2011), Kobayashi and Nomizu (1996), f^*b denotes $f^*b \circ (df \otimes df)$. We denote the latter by b_f.

Lemma 6.9.4 *Let E be a vector bundle of rank $p|2q$ over the supermanifold M. Recall that the frame bundle of E is a $\mathrm{GL}(p|2q)$-principal bundle over M. A metric g on the vector bundle E is equivalent to a reduction of the frame bundle of E to a bundle with structure group $\mathrm{O}(p|2q)$. The reduction of the frame bundle to $\mathrm{O}(p|2q)$ is called the bundle of orthonormal frames.*

Proposition 6.9.5 *Let E over M be a vector bundle with metric g. For a connection on the frame bundle of E, the following are equivalent:*

(i) The connection reduces to a connection on the orthonormal frame bundle.
(ii) For the induced covariant derivative ∇ on E, it holds that

$$X\left(g\left(Y, Z\right)\right) = g\left(\nabla_X Y, Z\right) + (-1)^{p(X)p(Y)} g\left(Y, \nabla_X Z\right)$$

for all $X \in \Gamma(TM)$ and $Y, Z \in \Gamma(E)$.
(iii) For the induced connection on $E^\vee \otimes E^\vee$ it holds that

$$\nabla g = 0.$$

If the connection fulfills any of the three conditions, the connection is called a metric connection.

Proof Suppose that the connection reduces to a connection on the orthonormal frame bundle. Then for any local orthonormal frame F_A of E, we have $\nabla_X F_A = \langle X, \omega_s \rangle_A{}^B F_B$, where $\langle X, \omega_s \rangle_A{}^B$ is a function with values in $\mathfrak{o}(p|2q)$. Hence, by Example 5.3.5, we have

$$\begin{aligned} 0 &= X\left(g(F_A, F_B)\right) \\ &= g\left(\langle X, \omega_s \rangle_A{}^C F_C, F_B\right) + (-1)^{p(X)p(A)} g\left(F_A, \langle X, \omega_s \rangle_B{}^C F_C\right), \end{aligned}$$

which is equivalent to (ii). Conversely, (ii) implies that the connection form takes values in $\mathfrak{o}(p|2q)$. Since the inclusion $\mathfrak{o}(p|2q) \hookrightarrow \mathfrak{gl}(p|2q)$ is injective, the connection reduces to a connection on the orthonormal frame bundle by Proposition 6.8.7. This shows the equivalence of (i) and (ii).

To show the equivariance of (ii) and (iii), we use the product rule of connections:

$$X\left(g\left(Y, Z\right)\right) = \left(\nabla_X g\right)(Y, Z) + g\left(\nabla_X Y, Z\right) + (-1)^{p(X)p(Y)} g\left(Y, \nabla_X Z\right). \qquad \square$$

By Corollary 6.6.3, the curvature tensor R of a metric connection is a two form with values in the anti-symmetric endomorphisms of E. Hence,

$$\begin{aligned} R\left(X, Y\right) Z &= -(-1)^{p(X)p(Y)} R\left(Y, X\right) Z, \\ g\left(R\left(X, Y\right) Z, W\right) &= -(-1)^{p(Z)(p(X)+p(Y))} g\left(Z, R\left(X, Y\right) W\right). \end{aligned}$$

We are now turning to metric connections on the tangent bundle $TM \to M$ and a question raised in Remark 6.7.4. In Remark 6.7.4 we noticed that the affine linear map that associates to a connection ω on TM its torsion $T^\omega \in \Omega^2(M, TM)$ is neither surjective nor injective in general. However, for metric connections this map is actually a bijection.

Proposition 6.9.6 *Let $\langle \cdot, \cdot \rangle$ be a metric on M. For any $T \in \Omega^2(M, TM)$ there is a unique metric connection ω such that the torsion tensor T^ω coincides with T. The connection is implicitly given by the Koszul formula*

$$\begin{aligned} 2\left\langle \nabla_X Y, Z \right\rangle = {} & X\left\langle Y, Z \right\rangle - (-1)^{p(Z)(p(X)+p(Y))} Z\left\langle X, Y \right\rangle \\ & + (-1)^{p(X)(p(Y)+p(Z))} Y\left\langle Z, X \right\rangle + \left\langle T(X, Y) + [X, Y], Z \right\rangle \\ & + (-1)^{p(Z)(p(X)+p(Y))} \left\langle T(Z, X) + [Z, X], Y \right\rangle \\ & - (-1)^{p(X)(p(Y)+p(Z))} \left\langle T(Y, Z) + [Y, Z], X \right\rangle. \end{aligned}$$

Proof For the pure vector fields consider the compatibility condition of the covariant derivative with the metric and its cyclic permutations:

$$\begin{aligned} X\langle Y, Z\rangle &= \langle \nabla_X Y, Z\rangle + (-1)^{p(X)p(Y)}\langle Y, \nabla_X Z\rangle \\ (-1)^{p(Z)(p(X)+p(Y))} Z\langle X, Y\rangle &= (-1)^{p(Z)(p(X)+p(Y))}\langle \nabla_Z X, Y\rangle \\ &\quad + (-1)^{p(Y)p(Z)}\langle X, \nabla_Z Y\rangle \\ (-1)^{p(X)(p(Y)+p(Z))} Y\langle Z, X\rangle &= (-1)^{p(X)(p(Y)+p(Z))}\langle \nabla_Y Z, X\rangle \\ &\quad + (-1)^{p(X)(p(Y)+p(Z))+p(Y)p(Z)}\langle Z, \nabla_Y X\rangle \end{aligned}$$

Taking their alternating sum

$$\begin{aligned} &X\langle Y, Z\rangle - (-1)^{p(Z)(p(X)+p(Y))} Z\langle X, Y\rangle + (-1)^{p(X)(p(Y)+p(Z))} Y\langle Z, X\rangle \\ &= \left\langle \nabla_X Y + (-1)^{p(X)p(Y)}\nabla_Y X, Z\right\rangle \\ &\quad - (-1)^{p(Z)(p(X)+p(Y))}\left\langle \nabla_Z X - (-1)^{p(X)p(Z)}\nabla_X Z, Y\right\rangle \\ &\quad + (-1)^{p(X)(p(Y)+p(Z))}\left\langle \nabla_Y Z - (-1)^{p(Y)p(Z)}\nabla_Z Y, X\right\rangle \\ &= \langle 2\nabla_X Y - T(X, Y) - [X, Y], Z\rangle \\ &\quad - (-1)^{p(Z)(p(X)+p(Y))}\langle T(Z, X) + [Z, X], Y\rangle \\ &\quad + (-1)^{p(X)(p(Y)+p(Z))}\langle T(Y, Z) + [Y, Z], X\rangle \end{aligned}$$

one gets the claim by reordering. □

As in ordinary differential geometry, the space of metric connections has a canonical connection, the one with vanishing torsion. Lemma 6.7.5 implies additional symmetries of its curvature tensor:

Corollary 6.9.7 *For the vector bundle $TM \to M$ with metric $\langle\cdot,\cdot\rangle$ there is a unique metric and torsion-free connection, called Levi-Civita connection. Its covariant derivative ∇ is implicitly given by*

$$\begin{aligned} 2\langle \nabla_X Y, Z\rangle &= X\langle Y, Z\rangle - (-1)^{p(Z)(p(X)+p(Y))} Z\langle X, Y\rangle \\ &\quad + (-1)^{p(X)(p(Y)+p(Z))} Y\langle Z, X\rangle + \langle [X, Y], Z\rangle \\ &\quad + (-1)^{p(Z)(p(X)+p(Y))}\langle [Z, X], Y\rangle - (-1)^{p(X)(p(Y)+p(Z))}\langle [Y, Z], X\rangle . \end{aligned}$$

The curvature tensor of the Levi-Civita connection has the following additional symmetries

$$R(X,Y)Z + (-1)^{p(Z)(p(X)+p(Y))} R(Z,X)Y + (-1)^{p(X)(p(Y)+p(Z))} R(Y,Z)X = 0,$$

$$\langle R(X,Y)Z, W\rangle = (-1)^{(p(X)+p(Y))(p(Z)+p(W))} \langle R(Z,W)X, Y\rangle .$$

The covariant derivative of the Levi-Civita connection on supermanifolds and its curvature tensor have been studied already in Goertsches (2008).

6.10 Almost Complex Connections

Definition 6.10.1 Let $E \to M$ be a vector bundle. An automorphism $\mathrm{I} \in \mathrm{Aut}(E)$ such that $\mathrm{I}^2 = -\mathrm{id}_E$ is called almost complex structure on E. In the case $E = TM$ we call I an almost complex structure on M.

It is an immediate consequence of this definition that any vector bundle with an almost complex structure must be of even rank, that is, of rank $2p|2q$ for some $p, q \in \mathbb{N}_0$. Furthermore, an almost complex structure gives the vector bundle the structure of a complex vector bundle. The standard almost complex structure from Example 2.12.2 gives an almost complex structure on trivial vector bundles of even rank. Recall that the group $\mathrm{GL}_{\mathbb{C}}(p|q)$ is the subgroup of $\mathrm{GL}_{\mathbb{R}}(2p|2q)$ that commutes with the standard almost complex structure.

Lemma 6.10.2 *Let E be a vector bundle of rank $2p|2q$ over M. An almost complex structure* I *on E is equivalent to a reduction of the structure group of the frame bundle of E to* $\mathrm{GL}_{\mathbb{C}}(m|n)$. *The reduced frame bundle is called the bundle of (almost) complex frames.*

Proposition 6.10.3 *Let E over M be a vector bundle with almost complex structure* I. *For a connection on the frame bundle of E the following are equivalent:*

(i) The connection reduces to a connection on the bundle of complex frames.
(ii) The induced covariant derivative ∇ on E fulfils

$$\nabla \mathrm{I} e = \mathrm{I} \nabla e$$

for all sections e of E.
(iii) For the induced covariant derivative on Aut(E) *it holds that*

$$\nabla \mathrm{I} = 0.$$

The proof of Proposition 6.10.3 is analogous to the proof of Proposition 6.9.5. It is a corollary of this characterization that for connection ω compatible with the almost complex structure I it holds that

$$R(X, Y)\,\mathrm{I} = \mathrm{I}\,R(X, Y).$$

Let us now characterize the torsion of connections compatible with an almost complex structure on M.

Definition 6.10.4 The Nijenhuis tensor of an almost complex structure I on M is defined by

$$N_{\mathrm{I}}(X, Y) = [\mathrm{I}X, \mathrm{I}Y] - \mathrm{I}[\mathrm{I}X, Y] - \mathrm{I}[X, \mathrm{I}Y] - [X, Y]$$

for all vector fields X and Y.

Remark 6.10.5 The Nijenhuis tensor of an almost complex structure has the following properties:

$$\begin{aligned} N_{\mathrm{I}}(X, Y) &= -(-1)^{p(X)p(Y)} N_{\mathrm{I}}(Y, X) \\ N_{\mathrm{I}}(\mathrm{I}X, \mathrm{I}Y) &= -N_{\mathrm{I}}(X, Y) \\ -\mathrm{I}\,N_{\mathrm{I}}(X, Y) &= N_{\mathrm{I}}(\mathrm{I}X, Y) = N_{\mathrm{I}}(X, \mathrm{I}Y) \end{aligned}$$

Proposition 6.10.6 *For the torsion of a connection ω that is compatible with a given almost complex structure the following relation between the torsion tensor T and the Nijenhuis-tensor holds:*

$$T(\mathrm{I}X, \mathrm{I}Y) - \mathrm{I}\,T(\mathrm{I}X, Y) - \mathrm{I}\,T(X, \mathrm{I}Y) - T(X, Y) = -N_{\mathrm{I}}(X, Y). \qquad (6.10.7)$$

Proof For a given connection ω that is compatible with the almost complex structure, Eq. (6.10.7) is a consequence of $\nabla^{\omega}\mathrm{I} = 0$ and the definition of the torsion tensor of a covariant derivative:

$$T(X, Y) = \nabla_X Y - (-1)^{p(X)p(Y)} \nabla_Y X - [X, Y]. \qquad \square$$

It is an interesting corollary that the existence of a torsion-free almost complex connection implies that the Nijenhuis-tensor vanishes and hence the integrability of I in the sense of the following Chap. 7.

Chapter 7
Complex Supermanifolds

In this chapter the theory of smooth families of complex supermanifolds is introduced. Families of complex supermanifolds are locally given by $\mathbb{C}^{m|n}$ and patched by smooth families of holomorphic coordinate changes. Consequently, every smooth family of complex supermanifolds has an underlying (real) family of smooth supermanifolds with an almost complex structure. However, not every smooth family of supermanifolds with almost complex structure lead to a smooth family of complex supermanifolds. A "super" version of the Newlander–Nirenberg-Theorem, originally due to McHugh (1989), Vaintrob (1988) applies to families of supermanifolds.

In the first section we treat the local theory of superholomorphic functions and vector fields. The second section then establishes the notion of smooth family of complex supermanifolds and the appropriate version of the Newlander–Nirenberg-Theorem. The third section gives formulas to compare the commutators of real and complex vector fields that will be needed later.

7.1 Local Theory

The development in this chapter adapts the treatment of Huybrechts (2005, Chapter I) to our needs. Recall that $\mathbb{C}^m$ arises from $\mathbb{R}^{2m}$ with the standard almost complex structure. The coordinate functions $z^1, \dots, z^m$ can be decomposed in real and imaginary part $z^a = x^a + \mathrm{i}y^a$, where $x^1, \dots, x^m, y^1, \dots, y^m$ are real coordinates of $\mathbb{R}^{2m}$. The sheaf of holomorphic functions $\mathcal{H}_{\mathbb{C}^m}$ is the subsheaf of $\mathcal{O}_{\mathbb{R}^{2m}} \otimes \mathbb{C}$ containing all functions $f = u + \mathrm{i}v$ satisfying the Cauchy–Riemann equations

$$\partial_{x^a} u = \partial_{y^a} v, \qquad \partial_{y^a} u = -\partial_{x^a} v.$$

E. Keßler, *Supergeometry, Super Riemann Surfaces and the Superconformal Action Functional*, Lecture Notes in Mathematics 2230,
https://doi.org/10.1007/978-3-030-13758-8_7

Denote by $\eta^1, \ldots, \eta^n, \zeta^1, \ldots, \zeta^n$ the $2n$-generators of the real Grassmann algebra $\bigwedge_{2n}$. For its complexification $\bigwedge_{2n} \otimes_{\mathbb{R}} \mathbb{C}$ we use the basis

$$\theta^\alpha = \eta^\alpha + \mathrm{i}\zeta^\alpha, \qquad \bar{\theta}^\alpha = \eta^\alpha - \mathrm{i}\zeta^\alpha.$$

The elements of degree k in $\bigwedge_{2n} \otimes_{\mathbb{R}} \mathbb{C}$ can be written as a linear combination of products containing p of the θ^α and q of the $\bar{\theta}^\alpha$, where $k = p + q$. The complex Grassmann algebra $\bigwedge_n^{\mathbb{C}}$ with n generators embeds into $\bigwedge_{2n} \otimes_{\mathbb{R}} \mathbb{C}$ by mapping the n generators to $\theta^1, \ldots, \theta^n$. That is, $\bigwedge_n^{\mathbb{C}}$ contains all the elements of $\bigwedge_{2n} \otimes_{\mathbb{R}} \mathbb{C}$ which do not have a factor of $\bar{\theta}^\alpha$.

Definition 7.1.1 (Complex Superdomains) The linear complex supermanifold $\mathbb{C}^{m|n}$ is defined to be the topological space $\mathbb{C}^m$ together with the sheaf $\mathcal{O}_{\mathbb{C}^{m|n}} = \mathcal{H}_{\mathbb{C}^m} \otimes_{\mathbb{C}} \bigwedge_n^{\mathbb{C}}$. The ringed space that is given by restriction to an open domain $U \subseteq \mathbb{C}^m$ is called a complex superdomain.

A morphism of locally ringed spaces between $\mathbb{C}^{m|n}$ and $\mathbb{C}^{p|q}$ is called a superholomorphic map.

As in the case of real superdomains, any superholomorphic map between complex superdomains can be expressed in coordinates, see Theorem 3.1.3.

Let $z^1, \ldots, z^m, \theta^1, \ldots, \theta^n$ be the standard complex coordinate system of $\mathbb{C}^{m|n}$. Of course, the complex coordinates may be decomposed in pairs of real coordinates as follows: $z^a = x^a + \mathrm{i}y^a$, $\theta^\alpha = \eta^\alpha + \mathrm{i}\zeta^\alpha$. This gives us the underlying real superdomain of dimension $2m|2n$, with coordinates $x^a, y^a, \eta^\alpha, \zeta^\alpha$. The question how the sheaves $\mathcal{O}_{\mathbb{C}^{m|n}}$ and $\mathcal{O}_{\mathbb{R}^{2m|2n}}$ are related is answered by the following lemma:

Lemma 7.1.2 (Cauchy Riemann Equations) *Let $f \in \mathcal{O}_{\mathbb{C}^{m|n}}$ be a superholomorphic function. Then there exist two real functions $u, v \in \mathcal{O}_{\mathbb{R}^{2m|2n}}$ such that $f = u + \mathrm{i}v$ in $\mathcal{O}_{\mathbb{R}^{2m|2n}} \otimes_{\mathbb{R}} \mathbb{C}$ and the following differential equations are fulfilled:*

$$\frac{\partial u}{\partial x^a} = \frac{\partial v}{\partial y^a} \qquad \frac{\partial u}{\partial \eta^\alpha} = \frac{\partial v}{\partial \zeta^\alpha}$$
$$\frac{\partial u}{\partial y^a} = -\frac{\partial v}{\partial x^a} \qquad \frac{\partial u}{\partial \zeta^\alpha} = -\frac{\partial v}{\partial \eta^\alpha}$$

The other way round, let $u, v \in \mathcal{O}_{\mathbb{R}^{2m|2n}}$ be two functions fulfilling the above equations. Then the function $f = u + \mathrm{i}v \in \mathcal{O}_{\mathbb{R}^{2m|2n}} \otimes_{\mathbb{R}} \mathbb{C}$ is in $\mathcal{O}_{\mathbb{C}^{m|n}}$. Consequently, $\mathcal{O}_{\mathbb{C}^{m|n}}$ is a subring of $\mathcal{O}_{\mathbb{R}^{2m|2n}} \otimes_{\mathbb{R}} \mathbb{C}$.

Proof Any $f \in \mathcal{O}_{\mathbb{C}^{m|n}}$ can be developed in a polynomial in the odd coordinates θ^α where every coefficient lies in $\mathcal{H}_{\mathbb{C}^m}$. Then f can be decomposed into the real and imaginary part as it is a finite sum and product of decomposable functions. Each of the coefficients is a holomorphic function, the usual Cauchy–Riemann equations hold for them. This gives the equations for the derivatives with respect to the even coordinates. For the odd derivatives we remember that we can express every function f as

$$f = g + \theta^\alpha h,$$

with $\partial_{\theta^\alpha} g = 0$, and so $\partial_{\eta^\alpha} g = 0$ and $\partial_{\zeta^\alpha} g = 0$. Now decompose θ^α and $h = r + \mathrm{i}s$. Then

$$\begin{aligned} f &= g + (\eta^\alpha + \mathrm{i}\zeta^i)(r + \mathrm{i}s) \\ &= g + (\eta^\alpha r - \zeta^\alpha s) + \mathrm{i}(\eta^\alpha s + \zeta^\alpha r), \end{aligned}$$

from where we can read off the odd Cauchy Riemann equations. The converse follows analogously. □

Let $\partial_{x^a}, \partial_{y^a}, \partial_{\eta^\alpha}, \partial_{\zeta^\alpha}$ be a frame of the real tangent bundle to $\mathbb{C}^{m|n} \simeq \mathbb{R}^{2m|2n}$. Recall that they form a basis for the module $\mathrm{Der}_{\mathbb{R}}\left(\mathcal{O}_{\mathbb{R}^{2m|2n}}\right)$ of $\mathbb{R}$-linear derivations of $\mathcal{O}_{\mathbb{R}^{2m|2n}}$. By scalar extension, we know that the $\mathbb{C}$-linear derivations of $\mathcal{O}_{\mathbb{R}^{2m|2n}} \otimes_{\mathbb{R}} \mathbb{C}$

$$\mathrm{Der}_{\mathbb{C}}\left(\mathcal{O}_{\mathbb{R}^{2m|2n}} \otimes \mathbb{C}\right) = \mathrm{Der}_{\mathbb{R}}\left(\mathcal{O}_{\mathbb{R}^{2m|2n}}\right) \otimes_{\mathbb{R}} \mathbb{C}.$$

are a free $\mathcal{O}_{\mathbb{R}^{2m|2n}} \otimes \mathbb{C}$-module of rank $2m|2n$. The derivations $\partial_{x^a}, \partial_{y^a}, \partial_{\eta^\alpha}, \partial_{\zeta^\alpha}$ form a basis. As an alternative basis we use the following derivations.

$$\begin{aligned} \partial_{z^a} &= \frac{1}{2}(\partial_{x^a} - \mathrm{i}\partial_{y^a}), & \partial_{\theta^\alpha} &= \frac{1}{2}(\partial_{\eta^\alpha} - \mathrm{i}\partial_{\zeta^\alpha}), \\ \partial_{\bar{z}^a} &= \frac{1}{2}(\partial_{x^a} + \mathrm{i}\partial_{y^a}), & \partial_{\bar{\theta}^\alpha} &= \frac{1}{2}(\partial_{\eta^\alpha} + \mathrm{i}\partial_{\zeta^\alpha}). \end{aligned}$$

Notice that $\partial_{z^a} z^b = \delta_a^b$ and $\partial_{\theta^\alpha}\theta^\beta = \delta_\alpha^\beta$. Hence, the Cauchy–Riemann equations can be rewritten as $\partial_{\bar{z}^a} f = 0$ and $\partial_{\bar{\theta}^\alpha} f = 0$. Consequently, one can show that $\mathrm{Der}_{\mathbb{C}}\left(\mathcal{O}_{\mathbb{C}^{m|n}}\right)$ is $\mathcal{O}_{\mathbb{C}^{m|n}}$-linearly generated by ∂_{z^a} and ∂_{θ^α}.

The multiplication with i induces an almost complex structure I on the real tangent bundle. In the coordinate frames the almost complex structure is in standard form, see Example 2.12.2:

$$\begin{aligned} \mathrm{I}\,\partial_{x^a} &= \partial_{y^a} & \mathrm{I}\,\partial_{\eta^\alpha} &= \partial_{\zeta^\alpha} \\ \mathrm{I}\,\partial_{y^a} &= -\partial_{x^a} & \mathrm{I}\,\partial_{\zeta^\alpha} &= -\partial_{\eta^\alpha} \end{aligned} \tag{7.1.3}$$

As described in Sect. 2.12, the almost complex structure I endows $\mathrm{Der}_{\mathbb{R}}\left(\mathcal{O}_{\mathbb{R}^{2m|2n}}\right)$ with the structure of a $\mathcal{O}_{\mathbb{R}^{2m|2n}} \otimes \mathbb{C}$ module via

$$(a + b\mathrm{i})X = aX + b\,\mathrm{I}\,X.$$

On $\mathrm{Der}_{\mathbb{R}}\left(\mathcal{O}_{\mathbb{R}^{2m|2n}}\right)\otimes_{\mathbb{R}}\mathbb{C}$ we have the complexified almost complex structure $\mathrm{I}^{\mathbb{C}}=\mathrm{I}\otimes\mathrm{id}_{\mathbb{C}}$. The complexified tangent sheaf splits into a direct sum

$$\mathrm{Der}_{\mathbb{R}}\left(\mathcal{O}_{\mathbb{R}^{2m|2n}}\right)\otimes_{\mathbb{R}}\mathbb{C}=\mathrm{Der}_{\mathbb{R}}\left(\mathcal{O}_{\mathbb{R}^{2m|2n}}\right)^{1,0}\oplus\mathrm{Der}_{\mathbb{R}}\left(\mathcal{O}_{\mathbb{R}^{2m|2n}}\right)^{0,1}$$

such that

$$\mathrm{I}^{\mathbb{C}}\Big|_{\mathrm{Der}_{\mathbb{R}}\left(\mathcal{O}_{\mathbb{R}^{2m|2n}}\right)^{1,0}}=\mathrm{i}\cdot\mathrm{id},\quad\text{and}\quad\mathrm{I}^{\mathbb{C}}\Big|_{\mathrm{Der}_{\mathbb{R}}\left(\mathcal{O}_{\mathbb{R}^{2m|2n}}\right)^{0,1}}=-\mathrm{i}\cdot\mathrm{id}\,.$$

$\mathrm{Der}_{\mathbb{R}}\left(\mathcal{O}_{\mathbb{R}^{2m|2n}}\right)^{1,0}$ is generated by $\partial_{z^a},\partial_{\theta^\alpha}$ as a $\mathcal{O}_{\mathbb{R}^{2m|2n}}\otimes_{\mathbb{R}}\mathbb{C}$-module, whereas the module $\mathrm{Der}_{\mathbb{R}}\left(\mathcal{O}_{\mathbb{R}^{2m|2n}}\right)^{0,1}$ is generated by $\partial_{\overline{z}^a},\partial_{\overline{\theta}^\alpha}$. The $\mathcal{O}_{\mathbb{R}^{2m|2n}}\otimes_{\mathbb{C}}$-modules $\mathrm{Der}_{\mathbb{R}}\left(\mathcal{O}_{\mathbb{R}^{2m|2n}}\right)$ and $\mathrm{Der}_{\mathbb{R}}\left(\mathcal{O}_{\mathbb{R}^{2m|2n}}\right)^{1,0}$ are isomorphic.

We have seen that on modules, every almost complex structure induces a complex structure. In the case of manifolds this is not true, due to the presence of integrability conditions. The following proposition can be found in Vaintrob (1988).

Proposition 7.1.4 *Let $U\subseteq\mathbb{R}^{2m|2n}$ be a real superdomain with an almost complex structure* I *such that for all vector fields X and Y the Nijenhuis-tensor vanishes:*

$$0=N_{\mathrm{I}}(X,Y)=[\mathrm{I}\,X,\mathrm{I}\,Y]-\mathrm{I}[\mathrm{I}\,X,Y]-\mathrm{I}[X,\mathrm{I}\,Y]-[X,Y]$$

Then there exist coordinates X^A on U such that the almost complex structure I is in standard form with respect to the coordinate frames ∂_{X^A}.

The idea of proof is to use the classical theorem of integrability of almost complex structures for the even directions (see, for example Kobayashi and Nomizu 1996, Appendix 8, and references therein), and to solve the corresponding equations for the odd directions algebraically. It is easy to see that N_{I} vanishes if I is in the standard form (7.1.3). Consequently, an almost complex structure on $\mathbb{R}^{2m|2n}$ induces a complex structure $\mathbb{C}^{m|n}\simeq\mathbb{R}^{2m|2n}$ such that multiplication by i on the tangent bundle is given by I if and only if the Nijenhuis-tensor vanishes.

Proposition 7.1.5 *Let $U\subseteq\mathbb{C}^{m|n}$ and $V\subseteq\mathbb{C}^{p|q}$ two complex superdomains. Let us denote by $U^{\mathbb{R}}$ and $V^{\mathbb{R}}$ the corresponding real superdomains and by I^U and I^V the induced almost complex structures on the real superdomains. A smooth map $f\colon U^{\mathbb{R}}\to V^{\mathbb{R}}$ is holomorphic if and only if*

$$df\left(\mathrm{I}^U\,X\right)=f^*\,\mathrm{I}^V\,df X \tag{7.1.6}$$

for all vector fields X on $U^{\mathbb{R}}$. Consequently, the tangent map df of a holomorphic map is a complex linear map.

Proof Let $z^a=x^a+\mathrm{i}y^a$ and $\theta^\alpha=\eta^\alpha+\mathrm{i}\zeta^\alpha$ be the standard coordinates on U and $t^b=u^b+\mathrm{i}v^b$ and $\mu^\beta=\kappa^\beta+\mathrm{i}\lambda^\beta$ be the standard coordinates on V. Expressing

Tf in coordinates, it can be checked that Eq. (7.1.6) amounts to the holomorphicity of $f^{\#}t^b$ and $f^{\#}\mu^\beta$ by Lemma 7.1.2. □

7.2 Smooth Families of Complex Supermanifolds

Using the results of Sect. 7.1, we can give the definition of a complex supermanifold and obtain the theorem of integrability of almost complex structures:

Definition 7.2.1 A complex supermanifold M of dimension $m|n$ is a locally ringed space $(\|M\|, \mathcal{O}_M)$ which is locally isomorphic to $\mathbb{C}^{m|n} = (\mathbb{C}^m, \mathcal{H}_{\mathbb{C}^m} \otimes_{\mathbb{C}} \bigwedge_n^{\mathbb{C}})$.

Examples of complex supermanifolds can be constructed similarly to real supermanifolds.

Example 7.2.2 (split Complex Supermanifolds) Let $|M|$ be a complex manifold of dimension m over a base B and $E \to |M|$ a holomorphic vector bundle of rank n. The sheaf of holomorphic sections of $\bigwedge E$ endows $|M|$ with the structure of a complex supermanifold M over B of relative dimension $m|n$, see Example 3.2.3.

As in the real case, split complex supermanifolds come with a holomorphic embedding $i\colon |M| \to M$ and a holomorphic projection $p\colon M \to |M|$ such that $p \circ i = \mathrm{id}_{|M|}$. However, in contrast to real supermanifolds, not every complex supermanifold is isomorphic to a split complex supermanifold. In general it is not even projected, see Green (1982), Donagi and Witten (2015).

Example 7.2.3 The super Lie group $\mathrm{GL}_{\mathbb{C}}(m|n)$ and its complex linear subgroups such as $\mathrm{U}(m|n)$ carry naturally the structure of a complex supermanifold.

Example 7.2.4 Complex projective superspaces $\mathbb{P}_{\mathbb{C}}^{m|n}$ can be constructed just as in the real case, compare Example 3.2.5 and Example 3.4.2. In particular the $1|1$-dimensional complex superspace can be covered by two coordinate charts with coordinates (z_1, θ_1) and (z_2, θ_2) such that

$$z_2 = \frac{1}{z_1}, \qquad \theta_2 = \frac{\theta_1}{z_1}.$$

The local theory described in Sect. 7.1 generalizes directly to see that every complex supermanifold is a real supermanifold with almost complex structure. Conversely, only integrable almost complex structures equip a real supermanifold with the structure of a complex supermanifold:

Theorem 7.2.5 (Newlander-Nirenberg Theorem for Supermanifolds, see Vaintrob (1988)) *Let M be a smooth supermanifold of dimension $2m|2n$ over $\mathbb{R}^{0|0}$ equipped with an almost complex structure* I*. M can be equipped with the structure of an $m|n$-dimensional complex supermanifold such that on TM multiplication by* i *coincides with* I *if and only if the Nijenhuis-tensor N_{I} vanishes.*

The theory of complex supermanifolds as presented here can, for example, be found in Deligne and Morgan (1999, §4.6), Carmeli et al. (2011, Chapter 4.8). However, for the applications in Part II, it is crucial to also construct non-trivial smooth families of complex supermanifolds. As we will be very much interested in the interplay between real and complex supermanifolds we should use the same notion of family on them. The notion of a family of real supermanifolds with almost complex structure is already defined for arbitrary base B. The corresponding notion of a smooth family of complex supermanifolds is maybe a bit surprising from the viewpoint of complex geometry, where one is rather used to work with holomorphic families. For results in Part II it is however necessary to work with smooth families. Consequently, we use the following definition:

Definition 7.2.6 Let B be a real supermanifold. A smooth family of complex supermanifolds of dimension $m|n$ over B is a family of real supermanifolds of dimension $2m|2n$ over B together with an almost complex structure I such that $N_{\mathrm{I}} = 0$.

A holomorphic map between smooth families of complex manifolds is a smooth map $f: M \to N$ of families of real manifolds such that $f^* \mathrm{I}^M df = df\, \mathrm{I}^N$.

Proposition 7.2.7 (Local Structure of Smooth Families of Complex Supermanifolds) *Let $b: M \to B$ be a smooth family of complex supermanifolds. M can be covered by relative complex coordinate charts, that is, open submanifolds $M \supseteq U = V \times W$ such that*

- *$W \subseteq B$ and the map $b|_U : U = V \times W \to W$ is given by the projection on the second factor,*
- *V is a real superdomain, that is, $V \subseteq \mathbb{R}^{2m|2n}$,*
- *on V there are coordinates $(x^a, y^a, \eta^\alpha, \zeta^\alpha)$ such that the almost complex structure is in the standard form* (7.1.3).

Consequently, a smooth family of complex supermanifolds can be covered by complex relative coordinate charts such that all coordinate changes are smooth families of holomorphic coordinate changes.

Proof Every family of real supermanifolds can be covered by real relative coordinate charts. A relative version of Proposition 7.1.4 then assures that the coordinates can be chosen in a way such that the almost complex structure is in standard form, see McHugh (1989). □

Corollary 7.2.8 *There is a well-defined sheaf of holomorphic functions $\mathcal{H}_M$ that is a subsheaf of $\mathcal{O}_M \otimes_{\mathbb{R}} \mathbb{C}$. A function $f \in \mathcal{O}_M \otimes_{\mathbb{R}} \mathbb{C}$ is holomorphic if in any relative complex coordinate system $z^a = x^a + \mathrm{i}y^a$, $\theta^\alpha = \eta^\alpha + \mathrm{i}\zeta^\alpha$ the equalities*

$$\partial_{\overline{z}^a} = 0 \qquad\qquad \partial_{\overline{\theta}^\alpha} = 0$$

hold.

In the following we will again assume that any complex manifold is implicitly a smooth family of complex manifolds. Furthermore, we will allow arbitrary base change without notice.

Let us now turn to vector bundles and in particular the tangent bundles of smooth families of complex manifolds. For simplicity will define them via their sheafs of sections, even though they could be defined as manifolds as well. An adapted version of Sect. 4.2 applies here.

Definition 7.2.9 (Complex Vector Bundles on Real Supermanifolds) Let M be a real supermanifold. A complex vector bundle on M is a locally free module over $\mathcal{O}_M \otimes \mathbb{C}$. As explained in Sect. 2.12, a locally free module over $\mathcal{O}_M \otimes_{\mathbb{R}} \mathbb{C}$ of rank $m|n$ is isomorphic to a locally free module of rank $2m|2n$ over $\mathcal{O}_M$ together with an almost complex structure.

Definition 7.2.10 (Holomorphic Vector Bundle) A holomorphic vector bundle on the complex supermanifold M is a locally free module over $\mathcal{H}_M$.

Any holomorphic vector bundle E over a complex supermanifold induces a complex vector bundle $E \otimes_{\mathcal{H}_M} (\mathcal{O}_M \otimes_{\mathbb{R}} \mathbb{C})$ on the underlying real manifold.

Example 7.2.11 Let M be a smooth family of complex manifolds of dimension $m|n$ over B. The holomorphic vector fields $\mathrm{Der}_{\mathcal{O}_B}(\mathcal{H}_M)$ form a locally free module of rank $m|n$ over $\mathcal{H}_M$. The associated real vector bundle is the real tangent bundle to M. This was essentially proven in Sect. 7.1.

7.3 Real and Complex Commutators

This section is more technical in nature. For future reference, we will deal with the comparison of commutators of complex linear derivations with real linear derivations. To be more precise, for a family of supermanifolds $M \to B$ we are looking at

$$\mathrm{Der}_{\mathcal{O}_B \otimes_{\mathbb{R}} \mathbb{C}}(\mathcal{O}_M \otimes \mathbb{C}) = \mathrm{Der}_{\mathcal{O}_B}(\mathcal{O}_M) \otimes_{\mathbb{R}} \mathbb{C}$$

seen as an $\mathcal{O}_M$-module. Any element $X \in \mathrm{Der}_{\mathcal{O}_B}(\mathcal{O}_M) \otimes_{\mathbb{R}} \mathbb{C}$ may be written as

$$X = \frac{1}{2}(X_1 \otimes 1 - X_2 \otimes \mathrm{i})$$

for some $X_1, X_2 \in \mathrm{Der}_{\mathcal{O}_B}(\mathcal{O}_M)$. Consequently, there is also a well-defined complex conjugate $\overline{X}$. We will use the shorthand $X = \frac{1}{2}(X_1 - \mathrm{i}X_2)$ and $\overline{X} = \frac{1}{2}(X_1 + \mathrm{i}X_2)$. By straightforward calculation and polarization we obtain:

Lemma 7.3.1 *Let* $X = \frac{1}{2}(X_1 - \mathrm{i}X_2)$, $Y = \frac{1}{2}(Y_1 - \mathrm{i}Y_2)$ *be two complex vector fields. Then*

$$\begin{aligned} 4[X,Y] &= [X_1,Y_1] - [X_2,Y_2] - \mathrm{i}\left([X_1,Y_2] + [X_2,Y_1]\right), \\ 4[X,\overline{Y}] &= [X_1,Y_1] + [X_2,Y_2] + \mathrm{i}\left([X_1,Y_2] - [X_2,Y_1]\right), \end{aligned} \tag{7.3.2}$$

and

$$\begin{aligned} [X_1,Y_1] &= [X,Y] + [X,\overline{Y}] + [\overline{X},\overline{Y}] + [\overline{X},Y], \\ [X_2,Y_1] &= \mathrm{i}\left([X,Y] + [X,\overline{Y}] - [\overline{X},\overline{Y}] - [\overline{X},Y]\right), \\ [X_1,Y_2] &= \mathrm{i}\left([X,Y] - [X,\overline{Y}] - [\overline{X},\overline{Y}] + [\overline{X},Y]\right), \\ [X_2,Y_2] &= -[X,Y] + [X,\overline{Y}] - [\overline{X},\overline{Y}] + [\overline{X},Y]. \end{aligned} \tag{7.3.3}$$

Let us now suppose that we have an almost complex structure I on M. As $\mathrm{Der}_{\mathcal{O}_B}\,\mathcal{O}_M$ is locally free, one may choose a local basis e_{A_1}, e_{A_2}, such that I is in standard form:

$$\mathrm{I}\,e_{A_1} = e_{A_2} \qquad\qquad \mathrm{I}\,e_{A_2} = -e_{A_1}$$

Consequently, the vectors

$$e_A = \frac{1}{2}\left(e_{A_1} - \mathrm{i}e_{A_2}\right) \qquad\qquad e_{\overline{A}} = \frac{1}{2}\left(e_{A_1} + \mathrm{i}e_{A_2}\right)$$

form a basis of $\mathrm{Der}_{\mathcal{O}_B}(\mathcal{O}_M) \otimes_{\mathbb{R}} \mathbb{C}$ as a $\mathcal{O}_M \otimes_{\mathbb{R}} \mathbb{C}$. Notice that the vectors e_A form a basis of $\mathrm{Der}_{\mathcal{O}_B}(\mathcal{O}_M)^{1,0}$, whereas the vectors $e_{\overline{A}}$ form a basis of $\mathrm{Der}_{\mathcal{O}_B}(\mathcal{O}_M)^{0,1}$.

Lemma 7.3.4 *Let* $e_A = \frac{1}{2}\left(e_{A_1} - \mathrm{i}e_{A_2}\right)$ *be a complex frame and* t_{AB}^{C} *the complex structure constants; that is*

$$[e_A, e_B] = t_{AB}^{C} e_C + t_{AB}^{\overline{C}} e_{\overline{C}}.$$

Let $d_{A_iB_j}^{C_k}$ *be the corresponding real structure constants. That is,*

$$[e_{A_i}, e_{B_j}] = d_{A_iB_j}^{C_1} e_{C_1} + d_{A_iB_j}^{C_2} e_{C_2}.$$

The real parts of the complex structure constants are given by:

$$\begin{aligned} 4\,\mathrm{Re}\,t_{AB}^{C} &= d_{A_1B_1}^{C_1} - d_{A_2B_2}^{C_1} + d_{A_1B_2}^{C_2} + d_{A_2B_1}^{C_2}, \\ 4\,\mathrm{Re}\,t_{AB}^{\overline{C}} &= d_{A_1B_1}^{C_1} - d_{A_2B_2}^{C_1} - d_{A_1B_2}^{C_2} - d_{A_2B_1}^{C_2}, \\ 4\,\mathrm{Re}\,t_{A\overline{B}}^{C} &= d_{A_1B_1}^{C_1} + d_{A_2B_2}^{C_1} - d_{A_1B_2}^{C_2} + d_{A_2B_1}^{C_2}, \\ 4\,\mathrm{Re}\,t_{A\overline{B}}^{\overline{C}} &= d_{A_1B_1}^{C_1} + d_{A_2B_2}^{C_1} + d_{A_1B_2}^{C_2} - d_{A_2B_1}^{C_2}. \end{aligned}$$

The imaginary parts of the complex structure constants are given by:

$$
\begin{aligned}
4\,\mathrm{Im}\, t_{AB}^{C} &= d_{A_1B_1}^{C_2} - d_{A_2B_2}^{C_2} - d_{A_1B_2}^{C_1} - d_{A_2B_1}^{C_1},\\
4\,\mathrm{Im}\, t_{AB}^{\overline{C}} &= -d_{A_1B_1}^{C_2} + d_{A_2B_2}^{C_2} - d_{A_1B_2}^{C_1} - d_{A_2B_1}^{C_1},\\
4\,\mathrm{Im}\, t_{A\overline{B}}^{C} &= d_{A_1B_1}^{C_2} + d_{A_2B_2}^{C_2} + d_{A_1B_2}^{C_1} - d_{A_2B_1}^{C_1},\\
4\,\mathrm{Im}\, t_{A\overline{B}}^{\overline{C}} &= -d_{A_1B_1}^{C_2} - d_{A_2B_2}^{C_2} + d_{A_1B_2}^{C_1} - d_{A_2B_1}^{C_1}.
\end{aligned}
$$

Conversely, the real structure constants can be expressed as functions of the complex structure coefficients as follows:

$$
\begin{aligned}
d_{A_1B_1}^{C_1} &= \mathrm{Re}\left(t_{AB}^{C} + t_{A\overline{B}}^{C} + t_{A\overline{B}}^{\overline{C}} + t_{AB}^{\overline{C}}\right)\\
d_{A_1B_1}^{C_2} &= \mathrm{Im}\left(t_{AB}^{C} + t_{A\overline{B}}^{C} + t_{A\overline{B}}^{\overline{C}} + t_{AB}^{\overline{C}}\right)\\
d_{A_2B_1}^{C_1} &= -\,\mathrm{Im}\left(t_{AB}^{C} + t_{A\overline{B}}^{C} - t_{A\overline{B}}^{\overline{C}} - t_{AB}^{\overline{C}}\right)\\
d_{A_2B_1}^{C_2} &= \mathrm{Re}\left(t_{AB}^{C} + t_{A\overline{B}}^{C} - t_{A\overline{B}}^{\overline{C}} - t_{AB}^{\overline{C}}\right)\\
d_{A_1B_2}^{C_1} &= -\,\mathrm{Im}\left(t_{AB}^{C} - t_{A\overline{B}}^{C} - t_{A\overline{B}}^{\overline{C}} + t_{AB}^{\overline{C}}\right)\\
d_{A_1B_2}^{C_2} &= \mathrm{Re}\left(t_{AB}^{C} - t_{A\overline{B}}^{C} - t_{A\overline{B}}^{\overline{C}} + t_{AB}^{\overline{C}}\right)\\
d_{A_2B_2}^{C_1} &= \mathrm{Re}\left(-t_{AB}^{C} + t_{A\overline{B}}^{C} - t_{A\overline{B}}^{\overline{C}} + t_{AB}^{\overline{C}}\right)\\
d_{A_2B_2}^{C_2} &= \mathrm{Im}\left(-t_{AB}^{C} + t_{A\overline{B}}^{C} - t_{A\overline{B}}^{\overline{C}} + t_{AB}^{\overline{C}}\right)
\end{aligned}
$$

Proof From Eq. (7.3.2) we get:

$$
\begin{aligned}
&2\left(\mathrm{Re}\, t_{AB}^{C} + \mathrm{Re}\, t_{AB}^{\overline{C}}\right) e_{C_1} + 2\left(\mathrm{Im}\, t_{AB}^{C} - \mathrm{Im}\, t_{AB}^{\overline{C}}\right) e_{C_2}\\
&\qquad + 2\mathrm{i}\left(\left(\mathrm{Im}\, t_{AB}^{C} + \mathrm{Im}\, t_{AB}^{\overline{C}}\right) e_{C_1} - \left(\mathrm{Re}\, t_{AB}^{C} - \mathrm{Re}\, t_{AB}^{\overline{C}}\right) e_{C_2}\right)\\
&= 4t_{AB}^{C} e_C + 4t_{AB}^{\overline{C}} e_{\overline{C}}\\
&= [e_{A_1}, e_{B_1}] - [e_{A_2}, e_{B_2}] - \mathrm{i}\left([e_{A_1}, e_{B_2}] + [e_{A_2}, e_{B_1}]\right)\\
&= \left(d_{A_1B_1}^{C_1} - d_{A_2B_2}^{C_1}\right) e_{C_1} + \left(d_{A_1B_1}^{C_2} - d_{A_2B_2}^{C_2}\right) e_{C_2}\\
&\qquad - \mathrm{i}\left(\left(d_{A_1B_2}^{C_1} + d_{A_2B_1}^{C_1}\right) e_{C_1} + \left(d_{A_1B_2}^{C_2} + d_{A_2B_1}^{C_2}\right) e_{C_2}\right)
\end{aligned}
$$

whereas from Eq. (7.3.3) we get that

$$\begin{aligned} d^{C_1}_{A_1B_1} e_{C_1} + d^{C_2}_{A_1B_1} e_{C_2} &= [e_{A_1}, e_{B_1}] \\ &= t^{C}_{AB} e_C + t^{\overline{C}}_{AB} e_{\overline{C}} + t^{C}_{A\overline{B}} e_C + t^{\overline{C}}_{A\overline{B}} e_{\overline{C}} + t^{C}_{\overline{A}B} e_C + t^{\overline{C}}_{\overline{A}B} e_{\overline{C}} \\ &\quad + t^{C}_{\overline{AB}} e_C + t^{\overline{C}}_{\overline{AB}} e_{\overline{C}} \\ &= \operatorname{Re}\left(\left(t^{C}_{AB} + t^{C}_{A\overline{B}} + t^{C}_{\overline{A}B} + t^{C}_{\overline{AB}}\right)(e_{C_1} - \mathrm{i} e_{C_2})\right) \\ &= \operatorname{Re}\left(t^{C}_{AB} + t^{C}_{A\overline{B}} + t^{C}_{\overline{A}B} + t^{C}_{\overline{AB}}\right) e_{C_1} \\ &\quad + \operatorname{Im}\left(t^{C}_{AB} + t^{C}_{A\overline{B}} + t^{C}_{\overline{A}B} + t^{C}_{\overline{AB}}\right) e_{C_2} \end{aligned}$$

The others follow analogously. □

Lemma 7.3.5 *The complexified Nijenhuis tensor* $N^{\mathbb{C}}_{\mathrm{I}}$, *defined by complex linear extension to* $\mathrm{Der}_{\mathcal{O}_B}(\mathcal{O}_M) \otimes_{\mathbb{R}} \mathbb{C}$ *is equivalent to*

$$N^{\mathbb{C}}_{\mathrm{I}}(X, Y) = [\mathrm{I}^{\mathbb{C}} X, \mathrm{I}^{\mathbb{C}} Y] - \mathrm{I}^{\mathbb{C}}[\mathrm{I}^{\mathbb{C}} X, Y] - \mathrm{I}^{\mathbb{C}}[X, \mathrm{I}^{\mathbb{C}} Y] - [X, Y].$$

In addition to the symmetries described in Remark 6.10.5 it holds that

$$N^{\mathbb{C}}_{\mathrm{I}}(\overline{X}, \overline{Y}) = \overline{N^{\mathbb{C}}_{\mathrm{I}}(X, Y)}.$$

Proof

$$\begin{aligned} N^{\mathbb{C}}_{\mathrm{I}}(\overline{X}, \overline{Y}) &= [\mathrm{I}\,\overline{X}, \mathrm{I}\,\overline{Y}] - \mathrm{I}[\mathrm{I}\,\overline{X}, \overline{Y}] - \mathrm{I}[\overline{X}, \mathrm{I}\,\overline{Y}] - [\overline{X}, \overline{Y}] \\ &= \overline{[\mathrm{I}\,X, \mathrm{I}\,Y]} - \overline{\mathrm{I}[\mathrm{I}\,X, Y]} - \overline{\mathrm{I}[X, \mathrm{I}\,Y]} - \overline{[X, Y]} \\ &= \overline{N^{\mathbb{C}}_{\mathrm{I}}(X, Y)} \end{aligned}$$

□

Chapter 8
Integration

This chapter explains the theory of integration for families $M \to B$ of supermanifolds. The integral over a compactly supported Berezinian form on M takes values in $\mathcal{O}_B$.

In the first section we treat Lie-derivatives of Berezinian forms. They appear in a version of Stokes theorem and are used in proofs.

The second section defines two different concepts of orientations for supermanifolds. For purposes of integration theory a purely topological definition of fiberwise orientation is sufficient. In contrast, to define Riemannian volume forms, an orientation of the odd direction is necessary.

The third section defines the local theory of integration, which is then generalized to manifolds in the fourth section. Integration in the odd directions is given by an algebraic procedure that is locally equivalent to a derivation.

The theory of integration for families of supermanifolds that we describe here, has been sketched in Deligne and Morgan (1999). The proofs are adapted from the classical paper Leites (1980), describing the integration theory for trivial families of supermanifolds. In addition to the discussion of orientation, often neglected in other sources, the main new result here is that for a Berezinian form b on M and an underlying even manifold $i : |M| \to M$ there is a top form $|b|$ on $|M|$ such that the integrals over b and $|b|$ coincide.

E. Keßler, *Supergeometry, Super Riemann Surfaces and the Superconformal Action Functional*, Lecture Notes in Mathematics 2230,
https://doi.org/10.1007/978-3-030-13758-8_8

8.1 The Berezinian Line Bundle

Definition 8.1.1 Let $E \to M$ be a vector bundle of rank $p|q$ and P its frame bundle. Let $V = \mathbb{R}^{1|0}$ if q is even and $V = \mathbb{R}^{0|1}$ if q is odd. The Berezinian bundle of E is defined as

$$\operatorname{Ber} E = P \times_{\operatorname{Ber}} V.$$

Here Ber is seen as the representation $\operatorname{Ber}\colon \operatorname{GL}(p|q) \to \operatorname{GL}(V)$ given by the Berezinian.

For any (ordered) frame F_A of E there is an associated frame

$$[F_\bullet] = [F_1 \cdots F_{p+q}]$$

of $\operatorname{Ber} E$. A linear transformation $\tilde{F}_A = {}_A L^B F_B$ yields the transformation

$$[\tilde{F}_\bullet] = \operatorname{Ber} L[F_\bullet].$$

Proposition 8.1.2 (Leites 1980, Lemma 2.4.6) *There is a unique action of the Lie algebra of vector fields $\Gamma(TM)$ on the Berezinian forms $\Gamma\left(\operatorname{Ber} T^\vee M\right)$ such that for all vector fields Y, Berezinian forms b, all functions f and any relative local coordinates X^A it holds that*

$$\begin{aligned} L_Y fb &= (Yf)\, b + (-1)^{p(Y)p(f)} f L_Y b, \\ L_{fY} b &= (-1)^{p(f)p(Y)} L_Y\,(fb), \\ L_{\partial_{X^A}} [dX^\bullet] &= 0. \end{aligned}$$

Proof Uniqueness and existence is easy, as in a coordinate system on can write $Y = Y^A \partial_{X^A}$ and $b = f[dX^\bullet]$. Thus

$$L_Y b = L_{Y^A \partial_{X^A}} (f[dX^\bullet]) = L_{\partial_{X^A}} \left(Y^A f[dX^\bullet]\right) = \partial_{X^A} \left(Y^A f\right) [dX^\bullet].$$

It is straightforward to check that this action is indeed an action of the Lie algebra of vector fields, that is,

$$L_X L_Y b - (-1)^{p(X)p(Y)} L_Y L_X b = L_{[X,Y]} b. \qquad \square$$

Remark 8.1.3 It is appropriate to call this action of the vector fields on the Berezinian forms a Lie derivative, as it holds in a coordinate system X^A for the flow Υ_t of an even vector field $Y = Y^A \partial_{X^A}$ that

$$\begin{aligned}\left.\frac{d}{dt}\right|_{t=0} \Upsilon_t^*[dX^\bullet]\operatorname{Ber} d\Upsilon_t &= [dX^\bullet]\operatorname{Ber} d\Upsilon_0 \operatorname{sTr}\left((d\Upsilon_0)^{-1} \left.\frac{d}{dt}\right|_{t=0} d\Upsilon_t\right)\\ &= [dX^\bullet]\left(\partial_{X^A} Y^A\right)\end{aligned}$$

which coincides with the Lie derivative from Proposition 8.1.2.

Definition 8.1.4 Let b be a Berezinian form which is a generator of $\Gamma\left(\operatorname{Ber} T^\vee M\right)$. That is any other Berezinian form is a multiple of b. The divergence of a vector field X on M with respect to b is defined by

$$(\operatorname{div}_b X)\, b = L_X b$$

in analogy to the classical case.

8.2 Orientation and Riemannian Volume Forms

Recall that in "classical" differential geometry, a manifold is called oriented if it possesses an atlas of coordinate charts, such that the determinant of the Jacobian of every coordinate change is positive. We want to generalize the concept of orientation to families of supermanifolds in a way suitable for integration theory.

Definition 8.2.1 A superdiffeomorphism $\Phi\colon \mathbb{R}^{m|n} \times B \to \mathbb{R}^{m|n} \times B$ of superdomains over B is called orientation preserving if the determinant of the Jacobian of the corresponding reduced diffeomorphism $\Phi_{red}\colon \mathbb{R}^m \times B_{red} \to \mathbb{R}^m \times B_{red}$ is everywhere positive.

Definition 8.2.2 (Orientation) Let M be a supermanifold over B. M is called oriented if it is equipped with an atlas of relative coordinate charts, such that every coordinate change is orientation preserving.

For supermanifolds M over $\mathbb{R}^{0|0}$, an orientation is the same as an orientation of M_{red}. Thus the concept of orientation given here does not depend on the odd directions. While this concept of orientation is sufficient for integration theory, we also need the following stronger concept that we will call Shander orientation, see Shander (1988):

Definition 8.2.3 (Shander Orientation) A superdiffeomorphism $\Phi\colon \mathbb{R}^{m|n} \times B \to \mathbb{R}^{m|n} \times B$ of supermanifolds over B is called Shander orientation preserving if both diagonal blocks of the reduction of the differential $d\Phi$ have everywhere positive determinant. A supermanifold M over B is called Shander oriented if it is equipped with an atlas of relative coordinate charts, such that every coordinate change is Shander orientation preserving.

Example 8.2.4 Every complex supermanifold is Shander oriented, as a holomorphic coordinate change is orientation preserving, compare Shander (1988, Proposition 5). Every Shander oriented supermanifold is oriented.

Remark 8.2.5 Notice that neither the existence of an orientation nor the existence of a Shander orientation on the supermanifold M does imply that the total space of M_{red} is orientable. The product of $\mathbb{R}$ with the Möbius strip as a family over the Möbius strip is still orientable.

We are now going to define the volume form associated to a metric g on TM. To this end we need to define oriented frames of TM.

Definition 8.2.6 Let M be an oriented family of supermanifolds, $X^A = (x^a, \eta^\alpha)$ local oriented coordinates and F_A a local frame of TM over the same open subset. Denote by ${F_A}^B$ the frame coefficients of F_A, that is

$$F_A = {F_A}^B \partial_{X^B}.$$

The frame F_A is called oriented if the reduction of ${F_a}^b$ has positive determinant. The frame F_A is called Shander oriented if both the reductions of ${F_a}^b$ and ${F_\alpha}^\beta$ have positive determinant.

Definition 8.2.7 Let F_A be a g-orthonormal, Shander oriented frame on M and F^A its dual basis. We define the Riemannian volume form $[dvol_g] \in \Gamma\left(\operatorname{Ber} T^\vee M\right)$ by

$$[dvol_g] = [F^\bullet].$$

Remark that this definition is independent of the chosen orthonormal oriented frame, as the Berezinian of the change of frame is $+1$. Furthermore, $[dvol_g]$ is a generator for $\Gamma\left(\operatorname{Ber} T^\vee M\right)$. Consequently, the divergence of a vector field X with respect to the metric g is defined to be

$$\operatorname{div}_g X = \operatorname{div}_{[dvol_g]} X.$$

8.3 Integrals on $\mathbb{R}^{m|n}$

In this section let $U = \mathbb{R}^{m|n} \times \mathbb{R}^{p|q} \to \mathbb{R}^{p|q}$ be an oriented, trivial family of superdomains and b be a fiberwise compact section of its Berezinian bundle. The goal of this section is to define the integral

$$\int_{U/\mathbb{R}^{p|q}} b \in \mathcal{O}_{\mathbb{R}^{p|q}}$$

as an even $\mathcal{O}_{\mathbb{R}^{p|q}}$-linear functional on fiberwise compactly supported sections of $\mathrm{Ber}\, T^\vee U$ and to study its basic properties.

Definition 8.3.1 Let $\pi : M \to B$ be a family of supermanifolds. A subset $U \subseteq \|M\|$ is called fiberwise compact if for any $b \in \|B\|$ the set

$$\pi^{-1} b \cap U \subseteq \|M\|$$

is compact.

We define the sheaf of sections with fiberwise compact support of the vector bundle $E \to M$ to be

$$\Gamma_c(E) = \{s \in \Gamma(E) \mid \mathrm{supp}(s) \text{ fiberwise compact}\}.$$

Let us first consider the integral in the case $n = 0$. In this case the relative tangent bundle is purely even dimensional and thus its Berezinian coincides with its determinant. Let x^a be oriented coordinates on $\mathbb{R}^m$ and l^b, λ^β coordinates on $\mathbb{R}^{p|q}$. Any fiberwise compact section $b \in \Gamma_c\left(\mathrm{Ber}\, T^\vee U\right)$ can then be written as

$$b = \sum_{\underline{\alpha}} \lambda^{\underline{\alpha}}\, {}_{\underline{\alpha}}b\, dx^1 \cdots dx^m$$

for odd multiindices $\underline{\alpha}$. Here the functions ${}_{\underline{\alpha}}b$ are usual smooth functions on $\mathbb{R}^m \times \mathbb{R}^p$. Then by the $\mathcal{O}_{\mathbb{R}^{p|q}}$-linearity it is clear that

$$\int_{\mathbb{R}^{m|0} \times \mathbb{R}^{p|q}/\mathbb{R}^{p|q}} b = \sum_{\underline{\alpha}} \lambda^{\underline{\alpha}} \int_{\mathbb{R}^m \times \mathbb{R}^p/\mathbb{R}^p} {}_{\underline{\alpha}}b\, dx^1 \cdots dx^m$$

where

$$\int_{\mathbb{R}^m \times \mathbb{R}^p/\mathbb{R}^p} {}_{\underline{\alpha}}b\, dx^1 \cdots dx^m$$

is the usual, purely even, fiberwise integral. This is independent of the choice of (oriented) coordinates on $\mathbb{R}^{p|q}$ as by definition the integral is $\mathcal{O}_{\mathbb{R}^{p|q}}$-linear. Its independence from the choice of coordinates on $\mathbb{R}^m$ is given by the classical argument. Furthermore the value of the integral is an element of $\mathcal{O}_{\mathbb{R}^{p|q}}$, as all classical integrals are finite, as b has fiberwise compact support.

Definition 8.3.2 Let $X^A = (x^a, \eta^\alpha)$ be oriented, relative coordinates on U. Then the Berezinian form b can be written

$$b = \left({}_0b + \eta^\mu\, {}_\mu b + \cdots + \eta^n \cdots \eta^1\, {}_{n\ldots 1}b\right)[dX^\bullet].$$

Define the integral of b over U relative to $\mathbb{R}^{p|q}$ by

$$\int_{U/\mathbb{R}^{p|q}}^{X^A} b = \int_{\mathbb{R}^m \times \mathbb{R}^{p|q}/\mathbb{R}^{p|q}} {}_{n\ldots 1}b\, dx^1 \cdots dx^m,$$

using the case $n = 0$ we have treated before. We keep track of the coordinates X^A used until we have shown the independence of the coordinates in Proposition 8.3.5.

Remark 8.3.3 Changing the orientation of the domain of integration leads to a flip of sign in the result of the integration as in the classical case. It is obvious that one does not need to give an orientation on the odd part as any reordering of the odd coordinates leaves the sign of the integral invariant. As an example in the case of two odd dimensions:

$$\int_{\mathbb{R}^{m|2}} \eta^2\eta^1 f(x)[dx\, d\eta^1\, d\eta^2] = \int_{\mathbb{R}^{m|2}} \eta^1\eta^2 f(x)[dx\, d\eta^2\, d\eta^1]$$

In order to keep track of the orientation and to allow to generalize the integral to the non-orientable case, it is proposed in Deligne and Morgan (1999) to twist the sheaf of Berezinian forms with the orientation sheaf of the underlying manifold. However, we will only need to integrate over oriented supermanifolds.

Lemma 8.3.4 *For all oriented, relative coordinates X^A, every vector field Y and every $b \in \Gamma_c\left(\operatorname{Ber} T^\vee U\right)$ on U it holds that*

$$\int_{U/\mathbb{R}^{p|q}}^{X^A} L_Y b = 0.$$

Proof By the additivity of the Lie derivative of Berezinian forms, we can assume $Y = Z\partial_{X^C}$ for some function Z and some index C. Then it follows

$$\begin{aligned} \int_{U/\mathbb{R}^{p|q}}^{X^A} L_Y b &= \int_{U/\mathbb{R}^{p|q}}^{X^A} (-1)^{p(C)p(Z)} L_{\partial_{X^C}} (Zb) \\ &= \int_{U/\mathbb{R}^{p|q}}^{X^A} \left(\partial_{X^C} f\right) [dX^\bullet] \end{aligned}$$

where f is the function given by

$$(-1)^{p(C)p(Z)} Zb = f[dX^\bullet].$$

Suppose now that C is an odd index and $\partial_{X^C} = \partial_{\eta^\gamma}$. In this case, the function $\partial_{X^C} f$ does not have a term proportional to $\eta^n \cdots \eta^1$ and the integral is zero. On the other hand if C is an even index and $\partial_{X^C} = \partial_{x^c}$ for some c, then the integral reduces to

$$\int_{\mathbb{R}^m \times \mathbb{R}^{p|q}/\mathbb{R}^{p|q}} (\partial_{x^c}\, {}_{n\ldots 1}f)\, dx^1 \cdots dx^m = 0.$$

For the last step it is essential that b and consequently f have fiberwise compact support. □

Proposition 8.3.5 *The definition of the integral is independent of the coordinates used.*

Proof In given coordinates X^A the Berezinian form b can be decomposed into a top degree part b_1 and a part of lower degree b_0 that is

$$b_1 = \eta^n \cdots \eta^1 g(x)[dX^\bullet]$$

and $b_0 = b - b_1$ does not have a top degree part. Clearly b_0 is of the form $L_Q\tilde{b}$ for some vector field Q and Berezinian form $\tilde{b}$. Thus the integral over b_0 vanishes independently of the coordinate system. It remains to treat b_1. Any change of coordinates can be written as the composite of a coordinate change

$$x^a = y^a \qquad\qquad \eta^\alpha = q^\alpha + \theta^\alpha$$

where q^α does not depend on η and a coordinate change

$$x^a = f^a(y) + \theta \cdots \qquad\qquad \eta^\alpha = \theta^\mu\, {}_\mu f^\alpha(y) + \theta^2 \cdots$$

Let us start with the first type of coordinate change. In this case

$$\operatorname{Ber}\frac{\partial X^A}{\partial Y^B} = 1$$

and consequently the integral over

$$b_1 = \eta^n \cdots \eta^1 g(x)[dX^\bullet] = \left(\theta^n \cdots \theta^1 g(y) + \text{lower order terms}\right)[dY^\bullet]$$

is invariant. For the second type of coordinate change it holds that

$$\operatorname{Ber}\frac{\partial X^A}{\partial Y^B} = \det\frac{\partial x^a}{\partial y^b}\left(\det\frac{\partial \eta^\alpha}{\partial \theta^\beta}\right)^{-1} + \theta \ldots$$

and

$$\begin{aligned} b_1 &= \eta^n \cdots \eta^1 g(x)[dX^\bullet] \\ &= \theta^n \cdots \theta^1 \left(\det{}_\mu f^\alpha(y)\right) g(x(y)) \left(\det \frac{\partial x^a}{\partial y^b}\right) \left(\det \frac{\partial \eta^\alpha}{\partial \theta^\beta}\right)^{-1} [dY^\bullet] \\ &= \theta^n \cdots \theta^1 \left(\det \frac{\partial x^a}{\partial y^b}\right) g(x(y))[dY^\bullet] \end{aligned}$$

Thus, the result follows from the classical theorem on the change of variables:

$$\begin{aligned} \int_{U/\mathbb{R}^{p|q}}^{X^A} b_1 &= \int_{\mathbb{R}^m/\mathbb{R}^{p|q}} g(x)\, dx^1 \cdots dx^m \\ &= \int_{\mathbb{R}^m/\mathbb{R}^{p|q}} g(x(y)) \left(\det \frac{\partial x^a}{\partial y^b}\right) dy^1 \cdots dy^m = \int_{U/\mathbb{R}^{p|q}}^{Y^A} b_1 \qquad \square \end{aligned}$$

Lemma 8.3.6 (Integral and Base Change) *Any map $f\colon \mathbb{R}^{r|s} \to \mathbb{R}^{p|q}$ induces a base change $g\colon f^*U \to U$:*

$$\begin{array}{ccc} f^*U = \mathbb{R}^{m|n} \times \mathbb{R}^{r|s} & \xrightarrow{g=\mathrm{id}_{\mathbb{R}^{m|n}} \times f} & U = \mathbb{R}^{m|n} \times \mathbb{R}^{p|q} \\ \downarrow & & \downarrow \\ \mathbb{R}^{r|s} & \xrightarrow{\quad f \quad} & \mathbb{R}^{p|q} \end{array}$$

For any Berezinian form $b \in \Gamma_c\left(\operatorname{Ber} T^\vee U\right)$ it holds that

$$f^* \int_{U/\mathbb{R}^{p|q}} b = \int_{f^*U/\mathbb{R}^{r|s}} g^* b.$$

Remark 8.3.7 It might be tempting to loosen the condition of fiberwise compactness of b, for example imposing decay properties on b, so that the integral still is finite. However, fiberwise compactness is required to prove that the integral is independent of the coordinates, see Proposition 8.3.5 and Lemma 8.3.4. The following example, adapted from Rothstein (1987, Section 3), shows that fiberwise compactness cannot be replaced by decay properties naïvely.

Let x, η^1, η^2 be coordinates on $\mathbb{R}^{1|2}$. Then

$$\int_{\mathbb{R}^{1|2}} e^{-x^2} [dx\, d\eta^1\, d\eta^2] = 0.$$

However after the change of coordinates

$$x = y - \frac{1}{2}\theta^2\theta^1 e^{y^2} f(y), \qquad \eta^\alpha = \theta^\alpha,$$

where $f\colon \mathbb{R} \to \mathbb{R}$ is an arbitrary smooth function, one finds

$$\int_{\mathbb{R}^{1|2}} e^{-x^2}[dx\, d\eta^1\, d\eta^2] = \int_{\mathbb{R}^{1|2}} \left(e^{-y^2} + \theta^2\theta^1 y f(y)\right)[dy\, d\theta^1\, d\theta^2] = \int_{\mathbb{R}} y f(y)\, dy.$$

Hence, the integral is dependent on the choice of coordinates despite the decay of the integrand.

For approaches to integration with non-compact support and with boundaries on supermanifolds in the case of $B = \mathbb{R}^{0|0}$, see Rothstein (1987) and Alldridge et al. (2012). Notice that fiberwise compact support is slightly more general than compact support, as the base may be non-compact.

8.4 Integrals on Supermanifolds and Stokes Theorem

The integral on $\mathbb{R}^{m|n}$ defined before extends to oriented supermanifolds in just the same way, as m-dimensional integrals on $\mathbb{R}^m$ extend to integration of top forms on m-dimensional manifolds.

Definition 8.4.1 Let $M \to B$ be a family of supermanifolds, (U_α, ϕ_α) a finite covering of M by charts $\phi_\alpha\colon \mathbb{R}^{m|n} \times \mathbb{R}^{p|q} \to U_\alpha$ and f_α a partition of unity subordinate to U_α. The integral of a Berezinian form $b \in \Gamma_c\left(T^\vee M\right)$ with fiberwise compact support relative to B is defined as

$$\int_{M/B} b = \sum_\alpha \int_{\mathbb{R}^{m|n} \times \mathbb{R}^{p|q}/\mathbb{R}^{p|q}} \phi_\alpha^*\left(f_\alpha b\right) \in \mathcal{O}_B.$$

The integrals on the superdomains are well-defined as $\phi_\alpha^*\left(f_\alpha b\right)$ has compact support. As b has fiberwise compact support, the sum is finite on sufficiently small open sets on B. The argument for independence from the choice of covering and partition of unity is the same as for the classical integral, which can be found, for example, in Lück (2005).

The following two propositions are immediate global variants of Lemma 8.3.4 and Proposition 8.3.5.

Proposition 8.4.2 *Let b be a compactly supported Berezinian form on $M \to B$ and Y a vector field on M. Then*

$$\int_{M/B} L_Y b = 0.$$

Proposition 8.4.3 *Let* $\Xi \in \mathrm{Diff}(M)$ *and* $b \in \Gamma_c\left(\mathrm{Ber}\, T^\vee M\right)$. *For*

$$b_\Xi = \left(\mathrm{Ber}\,(d\Xi)^\vee\right) \Xi^* b \in \Gamma_c\left(\mathrm{Ber}\, T^\vee M\right)$$

it holds

$$\int_{M/B} b_\Xi = \int_{M/B} b.$$

Proposition 8.4.4 *Let* M *be an oriented supermanifold of dimension* $m|n$, $i\colon |M| \to M$ *an embedding of the underlying even manifold and* b *a fiberwise compact Berezinian form on* M. *Define a top form* $|b|$ *on* $|M|$ *by setting*

$$|f[dX^\bullet]| = \left(i^* \partial_{\eta^1} \cdots \partial_{\eta^n} f\right) dy^1 \cdots dy^m$$

in any local coordinates $X^A = (x^a, \eta^\alpha)$ *on* M *and* y^a *on* $|M|$ *such that* $i^\# x^a = y^a$ *and* $i^\# \eta^\alpha = 0$. *Then*

$$\int_{M/B} b = \int_{|M|/B} |b|.$$

Proof Without loss of generality we may assume that M is a trivial family of superdomains over $\mathbb{R}^{p|q}$. In that case the equality of the integrals over b and $|b|$ is by definition of the integral over Berezinian forms. It remains to show that $|b|$ transforms as a top form under coordinate changes that preserve the condition $i^\# \eta^\alpha = 0$. Those coordinate changes are precisely the coordinate changes of the second type in the proof of Proposition 8.3.5. The correct transformation behaviour was also proven there. □

Remark 8.4.5 The Berezinian forms are not part of the de Rham sequence. One can replace the de Rham sequence by the sequence of integral forms for questions arising when integrating over submanifolds. See Deligne and Morgan (1999), Witten (2012) and references therein.

Part II
Super Riemann Surfaces

Chapter 9
Super Riemann Surfaces and Reductions of the Structure Group

This chapter describes super Riemann surfaces and the additional structure of an adapted metric with the help of reductions of the structure group.

The most common definition of super Riemann surfaces, to be found, for example, in Friedan (1986); LeBrun and Rothstein (1988), is an algebro-geometric one. We will give this definition and some of its consequences in Sect. 9.1.

However for the larger part of this book we will work with an alternative description of super Riemann surfaces in terms of reductions of the structure group. This description, to be found in Giddings and Nelson (1988), is presented in Sect. 9.2. Furthermore, different descriptions of the integrability conditions will be derived for later reference.

Section 9.3 studies local deformations of super Riemann surfaces in the language of reductions of the structure group.

Uniformization of super Riemann surfaces is the starting point for a super Teichmüller theory and is described in Sect. 9.4.

In Sect. 9.5, we introduce the concept of a metric compatible with the super Riemann surface structure. Such a metric yields a further reduction of the structure group to $\mathrm{U}(1)$ and will be used for the definition of metric and gravitino on $|M|$ in Chap. 11.

9.1 Definition and Algebraic Properties of Super Riemann Surfaces

In contrast to most of the supergeometric concepts encountered so far, there is not a unique, straightforward generalization of Riemann surfaces to supergeometry. For example, Sachse (2009, Chapter 4) lists several families of possible generalizations

E. Keßler, *Supergeometry, Super Riemann Surfaces and the Superconformal Action Functional*, Lecture Notes in Mathematics 2230,
https://doi.org/10.1007/978-3-030-13758-8_9

of Riemann surfaces to supergeometry. Here, we will work exclusively with super Riemann surfaces, a particular generalization of Riemann surfaces:

Definition 9.1.1 (See, for Example, LeBrun and Rothstein (1988)) A super Riemann surface M is a smooth family of complex supermanifolds of complex dimension $1|1$ together with a holomorphic subbundle $\mathcal{D} \subseteq TM$ of complex rank $0|1$ such that the Lie bracket induces an isomorphism

$$\frac{1}{2}[\cdot,\cdot]\colon \mathcal{D}\otimes_{\mathbb{C}}\mathcal{D} \to {}^{TM}\!/_{\mathcal{D}}.$$

This condition is called the complete non-integrability of $\mathcal{D}$, reflecting the notion of integrability of distributions of Frobenius theorem.

A holomorphic map $M \to M'$ over B is called superconformal if it preserves the distribution $\mathcal{D}$. The set of superconformal automorphisms of a super Riemann surface M over B is denoted by $\mathrm{SC}_B(M)$.

The non-integrability condition is crucial to the results of Part II. Hence the results of Part II most probably do not generalize to the other generalizations of Riemann surfaces.

Example 9.1.2 Let (z,θ) be the standard complex coordinates on $\mathbb{C}^{1|1}$. The standard super Riemann surface structure on $\mathbb{C}^{1|1}$ is given by the line bundle $\mathcal{D} \subseteq TM$ that is generated by $D = \partial_\theta + \theta\partial_z$. Indeed, the distribution $\mathcal{D}$ is completely non-integrable, because

$$[D, D] = 2\partial_z.$$

The Example 9.1.2 is generic, because every super Riemann surface is locally equivalent to the standard super Riemann surface on $\mathbb{C}^{1|1}$:

Lemma 9.1.3 (See LeBrun and Rothstein (1988, Lemma 1.2)) *Let M be a super Riemann surface over B. Around every point of $\|M\|$ there is an relative coordinate neighbourhood with adapted holomorphic coordinates (z,θ) such that $\mathcal{D}$ is generated by $\partial_\theta + \theta\partial_z$.*

Proof Let (u,η) be relative holomorphic coordinates and suppose that the distribution $\mathcal{D}$ is generated locally by the vector field $D = a\partial_\eta + b\partial_u$ for some even holomorphic function $a = a(u,\eta)$ and odd holomorphic function $b = b(u,\eta)$. The distribution $\mathcal{D}$ is locally free and the vector fields ∂_u and ∂_η form a local frame for TM. Consequently, the function a has to be invertible and without loss of generality we may assume $a = 1$. Due the complete non-integrability condition, the remaining coefficient of $[D, D] = 2\left(\partial_\eta b\right)\partial_u$ has to be invertible.

Under the coordinate change $z = f(u) + \eta\zeta(u)$, $\theta = \eta$ we obtain

$$D = \partial_\eta + b\partial_u = \partial_\theta + \left(b\frac{\partial z}{\partial u} + \frac{\partial z}{\partial \eta}\right)\partial_z.$$

Writing $b(u,\eta) = {}_0b(u) + \eta\, {}_\eta b(u)$ equation $b\frac{\partial z}{\partial u} + \frac{\partial z}{\partial \eta} = \eta$ decomposes in

$$ {}_0b(u) f'(u) + \zeta(u) = 0, \qquad {}_0b(u)\zeta'(u) + {}_\eta b(u) f'(u) = 1. $$

This set of differential equations is solved by $\zeta(u) = -\, {}_0b(u)\, {}_\eta b(u)^{-1}$ and $f(u)$ an anti-derivative of $\left({}_\eta b - {}_0b\, {}_0b'\right) {}_\eta b(u)^{-2}$. □

Definition 9.1.4 We call holomorphic coordinates (z, θ) on a super Riemann surface superconformal coordinates if the distribution $\mathcal{D}$ is locally generated by $\partial_\theta + \theta\partial_z$.

Super Riemann surfaces are completely determined by an atlas of superconformal coordinates such that all coordinate changes preserve the line bundle $\mathcal{D}$. The most general holomorphic coordinate change is given by

$$ \tilde{z} = f(z) + \theta\zeta(z), \qquad \tilde{\theta} = \xi(z) + \theta g(z), \tag{9.1.5} $$

where f, g are even holomorphic functions and ζ, ξ are odd holomorphic functions. Recall that we are working over a base B, and hence the odd functions ζ and ξ can be non-zero. The differential of the coordinate change is

$$ \begin{pmatrix} \partial_z \\ \partial_\theta \end{pmatrix} = \begin{pmatrix} f' + \theta\zeta' & \xi' + \theta g' \\ \zeta & g \end{pmatrix} \begin{pmatrix} \partial_{\tilde{z}} \\ \partial_{\tilde{\theta}} \end{pmatrix} $$

and hence D transforms according to

$$ \begin{aligned} D &= \left(D\tilde{\theta}\right)\tilde{D} + \left(D\tilde{z} - \tilde{\theta} D\tilde{\theta}\right)\partial_z \\ &= \left(g + \theta\xi'\right)\tilde{D} + \left(\zeta + \theta f' - (\xi + g\theta)\left(g + \theta\xi'\right)\right)\partial_{\tilde{z}}. \end{aligned} $$

The coefficient of $\partial_{\tilde{z}}$ must vanish because D and $\tilde{D}$ should define the same line bundle $\mathcal{D}$. So we get the constraints

$$ \zeta = g\xi, \qquad f' = g^2 - \xi\xi'. $$

We call a holomorphic coordinate change that fulfils those constraints a superconformal coordinate change. Superconformal coordinate changes map superconformal coordinates to superconformal coordinates. The derivation of the most general form of a superconformal coordinate change given here can be found in Crane and Rabin (1988).

Under a superconformal coordinate change the base frame ∂_z, D transforms as follows:

$$ \begin{pmatrix} \partial_z \\ D \end{pmatrix} = \begin{pmatrix} g^2 + 2\theta g\xi' & \xi' + \theta g' \\ 0 & g + \theta\xi' \end{pmatrix} \begin{pmatrix} \partial_{\tilde{z}} \\ \tilde{D} \end{pmatrix} \tag{9.1.6} $$

Remark that this base change matrix is completely determined by $g(z) + \theta\xi'$ as $g^2+2\theta g\xi' = \left(g(z) + \theta\xi'\right)^2$ and $\xi'(z)+\theta g'(z) = D(g(z)+\theta\xi')$. The particular form of the superconformal coordinate changes gives rise to a correspondence between super Riemann surfaces and Riemann surfaces together with a square root of the canonical line bundle. Indeed, if ξ and ζ vanish, the glueing function g for the line bundle $\mathcal{D}$ is the square root of the glueing function f' for the tangent bundle. This yields:

Proposition 9.1.7 (See, for Example, Sachse (2009, Proposition 4.2.2)) *There exists a bijection between the set of super Riemann surfaces over the point $\mathbb{R}^{0|0}$ and the set of pairs (M, S), where M is Riemann surface over a point and S is a spinor bundle on M.*

More generally, all trivial families of super Riemann surfaces are completely determined by a complex structure (or a conformal class of metrics) and a spinor bundle on the reduced space. We will see in Chap. 11 that Proposition 9.1.7 does not extend to all non-trivial families of super Riemann surfaces. In general a super Riemann surface is determined by the metric g, a spinor bundle S and the gravitino χ on the underlying even manifold $|M|$, see Theorem 11.3.5.

Example 9.1.8 (Simply Connected Super Riemann Surfaces) The simply connected supermanifolds $\mathbb{C}^{1|1}$, $S\mathbb{H}$ and $\mathbb{P}^{1|1}_{\mathbb{C}}$ posses canonical super Riemann surface structures because their reduced manifolds have unique spin structures. Indeed, the super Riemann surface structure on $\mathbb{C}^{1|1}$ has been studied in Example 9.1.2 and Lemma 9.1.3. The supermanifold $S\mathbb{H}$ is the open sub-manifold of $\mathbb{C}^{1|1}$ where the absolute value of the even complex coordinate is positive. The super Riemann surface structure on $S\mathbb{H}$ is the one induced from $\mathbb{C}^{1|1}$.

The super Riemann surface structure on $\mathbb{P}^{1|1}_{\mathbb{C}}$ is given in the coordinates (z_1, θ_1) and (z_2, θ_2) from Example 7.2.4 by $D_1 = \partial_{\theta_1} + \theta_1\partial_{z_1}$ and $D_2 = \partial_{\theta_2} - \theta_2\partial_{z_2}$. That is, (z_1, θ_1) and $(-z_2, \theta_2)$ are superconformal coordinates of $\mathbb{P}^{1|1}_{\mathbb{C}}$.

In Sect. 9.4, we will prove that families of super Riemann surfaces with simply connected fibers are trivial families, where the fiber is either $\mathbb{C}^{1|1}$, $S\mathbb{H}$ or $\mathbb{P}^{1|1}_{\mathbb{C}}$.

In the following Sect. 9.2, we will look at super Riemann surfaces as a particular integrable geometric structure. As a preparation, we need real coordinates and frames on a super Riemann surface. Let (x^a, η^α) be real coordinates obtained as real and imaginary part of some superconformal coordinates (z, θ) on M, that is

$$z = x^1 + \mathrm{i}x^2, \qquad \theta = \eta^3 + \mathrm{i}\eta^4.$$

Let furthermore D_α, $\alpha = 3, 4$ be the real and imaginary part of D:

$$D = \partial_\theta + \theta\partial_z = \frac{1}{2}\left(D_3 - \mathrm{i}D_4\right).$$

Then

$$D_\alpha = \partial_{\eta^\alpha} + \eta^\mu \Gamma^k_{\mu\alpha} \partial_{x^k},$$
$$[D_\alpha, D_\beta] = 2\Gamma^k_{\alpha\beta} \partial_{x^k},$$

where Γ is the real basis expression of the complex squaring $D \otimes D = 2\partial_z$, compare Sect. A.1.

9.2 The Theorem of Giddings and Nelson

It is proven in Giddings and Nelson (1988) that any super Riemann surface is a real 2|2-dimensional supermanifold together with an integrable reduction of the structure group of the frame bundle. We will refer to this fact as the Theorem of Giddings and Nelson. The precise statement can be found in Theorem 9.2.3. This theorem is central to our treatment here. We will give a rather detailed presentation of it. The proof presented here is is based on the theory of complex supermanifolds which is developed in Chap. 7 and emphasizes the geometric meaning of the different integrability conditions.

Definition 9.2.1 We denote by

$$\mathrm{Tr}_{\mathbb{C}}(1|1) = \left\{ \begin{pmatrix} A & B \\ 0 & C \end{pmatrix} \right\} \subset \mathrm{GL}_{\mathbb{C}}(1|1) \subset \mathrm{GL}_{\mathbb{R}}(2|2)$$

the group of upper triangular complex matrices. Furthermore we need its subgroup

$$\mathrm{SCL} = \left\{ \begin{pmatrix} A^2 & B \\ 0 & A \end{pmatrix} \right\} \subset \mathrm{Tr}_{\mathbb{C}}(1|1) \subset \mathrm{GL}_{\mathbb{C}}(1|1).$$

Lemma 9.2.2 *Every super Riemann surface M allows for a reduction of the structure group of its tangent bundle TM to* $\mathrm{Tr}_{\mathbb{C}}(1|1)$ *and* SCL.

Proof We need to show that one can find an open cover of M and a trivializing frame over each open set, such that the change of frame is in $\mathrm{Tr}_{\mathbb{C}}(1|1)$ or SCL. This property is fulfilled by the superconformal frame ∂_{x^a}, D_α, as seen in Eq. (9.1.6). □

Any reduction of the structure group of the frame bundle of a 2|2-dimensional supermanifold to $\mathrm{Tr}_{\mathbb{C}}(1|1)$ induces an almost complex structure I and a distribution $\mathcal{D} \subset TM$ on M. It is however not true that every such reduction of the structure group gives M the structure of a super Riemann surface. The following additional conditions, referred to as integrability conditions, need to be satisfied:

- The almost complex structure I defined by this reduction is integrable.
- The real $0|2$-dimensional distribution $\mathcal{D}$ defined by the reduction is a holomorphic $0|1$-dimensional subbundle with respect to the complex structure that is defined by I.
- $\mathcal{D}$ and $[\mathcal{D}, \mathcal{D}]$ together generate TM.

The integrability conditions will be restated in terms of commutators of the $\operatorname{Tr}_{\mathbb{C}}(1|1)$-frames in Theorem 9.2.3. We fix the following notation: Let $F_A = (F_a, F_\alpha)$ be a $\operatorname{Tr}_{\mathbb{C}}(1|1)$-frame, where $a = 1, 2$ and $\alpha = 3, 4$. In any $\operatorname{Tr}_{\mathbb{C}}(1|1)$-frame the almost complex structure I is in standard form, see Example 2.12.2, and the distribution $\mathcal{D}$ is locally generated by F_α. The structure functions of a $\operatorname{Tr}_{\mathbb{C}}(1|1)$-frame are denoted by d_{AB}^C:

$$[F_A, F_B] = d_{AB}^C F_C$$

Following the notation in Giddings and Nelson (1988), we will denote the corresponding frames in $TM \otimes \mathbb{C}$ by

$$F_z = \frac{1}{2}(F_1 - \mathrm{i}F_2) \quad F_{\overline{z}} = \frac{1}{2}(F_1 + \mathrm{i}F_2)$$
$$F_+ = \frac{1}{2}(F_3 - \mathrm{i}F_4) \quad F_- = \frac{1}{2}(F_3 + \mathrm{i}F_4)$$

and their structure functions equally by d_{AB}^C, where the indices run over $z, \overline{z}, +, -$. Notice that for the complex commutators it holds $\overline{d_{AB}^C} = d_{\overline{AB}}^{\overline{C}}$ and that the complex commutators are completely determined by the real commutators as explained in Sect. 7.3.

Theorem 9.2.3 (Theorem of Giddings and Nelson, Complex Version, See Giddings and Nelson (1988)) *Let M be a $2|2$-dimensional real family of supermanifolds with a reduction of the structure group to* $\operatorname{Tr}_{\mathbb{C}}(1|1)$. *Then M is a super Riemann surface, if and only if for any (complex)* $\operatorname{Tr}_{\mathbb{C}}(1|1)$*-frame the following holds:*

- *Integrability of the almost complex structure:*

$$d_{z+}^{\overline{z}} = d_{z+}^{-} = d_{++}^{\overline{z}} = d_{++}^{-} = 0 \tag{9.2.4}$$

This is equivalent to

$$d_{\overline{z}-}^{z} = d_{\overline{z}-}^{+} = d_{--}^{z} = d_{--}^{+} = 0.$$

- *Holomorphicity of $\mathcal{D}$:*

$$d_{+-}^{z} = d_{+\overline{z}}^{z} = 0 \tag{9.2.5}$$

Equivalently,

$$d^{\bar{z}}_{+-} = d^{\bar{z}}_{-z} = 0.$$

- *Complete non-integrability:*

$$d^{z}_{++} \textit{ invertible}$$

Equivalently,

$$d^{\bar{z}}_{--} \textit{ invertible}.$$

Furthermore, if M possesses a reduction of the structure group to SCL, $d^{z}_{++} = 2$ *can be assumed.*

The original proof shows the integrability of the almost complex and superconformal structure by a fixed-point argument. We will rephrase the integrability conditions using the Newlander–Nirenberg Theorem as explained in Sect. 7.2.

Proof of Theoem 9.2.3 To show the integrability of the almost complex structure, it suffices to verify that the Nijenhuis tensor vanishes. For this we remark that I acts on F_z and F_+ as i and on $F_{\bar{z}}$ and F_- as $-$i. By the symmetries of the Nijenhuis tensor it suffices to check equations

$$\begin{aligned} 0 = N_{\mathrm{I}}^{\mathbb{C}}(F_z, F_+) &= [\mathrm{I}\,F_z, \mathrm{I}\,F_+] - \mathrm{I}[\mathrm{I}\,F_z, F_+] - \mathrm{I}[F_z, \mathrm{I}\,F_+] - [F_z, F_+] \\ &= -2[F_z, F_+] - 2\mathrm{i}\,\mathrm{I}[F_z, F_+] = 4d^{\bar{z}}_{z+}F_{\bar{z}} + 4d^{-}_{+z}F_-, \\ 0 = N_{\mathrm{I}}^{\mathbb{C}}(F_+, F_+) &= [\mathrm{I}\,F_+, \mathrm{I}\,F_+] - \mathrm{I}[\mathrm{I}\,F_+, F_+] - \mathrm{I}[F_+, \mathrm{I}\,F_+] - [F_+, F_+] \\ &= -2[F_+, F_+] - 2\mathrm{i}\,\mathrm{I}[F_+, F_+] = 4d^{\bar{z}}_{++}F_{\bar{z}} + 4d^{-}_{++}F_-. \end{aligned}$$

This yields (9.2.4).

For the holomorphicity of $\mathcal{D}$, we suppose that there is a local holomorphic section $s = s^+F_+ + s^-F_-$ of $\mathcal{D}$ being an infinitesimal automorphism of the almost complex structure I, that is:

$$\begin{aligned} 0 &= \mathrm{I}[s, F_A] - [s, \mathrm{I}\,F_A] = \mathrm{I}[s^+F_+ + s^-F_-, F_A] - [s^+F_+ + s^-F_-, \mathrm{I}\,F_A] \\ &= s^+\left(\mathrm{I}[F_+, F_A] - [F_+, \mathrm{I}\,F_A]\right) + s^-\left(\mathrm{I}[F_-, F_A] - [F_-, \mathrm{I}\,F_A]\right) \\ &\quad - \left(F_A s^+ - \mathrm{I}\,F_A s^+\right)F_+ - \left(F_A s^- - \mathrm{I}\,F_A s^-\right)F_- \end{aligned}$$

This can only be zero, if the coefficients of F_z and $F_{\bar{z}}$ coming from the terms linear in s^α vanish. The terms for $A = z, +$ vanish as a consequence of the integrability of the almost complex structure. It remains

$$I[F_+, F_{\bar{z}}] - [F_+, I F_{\bar{z}}] = 2id^z_{+\bar{z}}F_z + 2id^+_{+\bar{z}}F_+,$$
$$I[F_+, F_-] - [F_+, I F_-] = 2id^z_{+-}F_z + 2id^+_{+-}F_+,$$

which gives (9.2.5).

Complete non-integrability is obviously equivalent to the fact that d^z_{++} is invertible. In case of a SCL-reduction, the structure coefficient d^z_{++} is invariant under frame change and the frames F_z and $F_{\bar{z}}$ may be rescaled such that $d^z_{++} = 2$.   □

Proposition 9.2.6 (Theorem of Giddings and Nelson, Real Version) *Let M be a 2|2-dimensional real family of supermanifolds with a reduction of the structure group to SCL. M is a super Riemann surface, if for any SCL-frame the following holds true:*

- *Integrability of the complex structure:*

$$0 = \left(d^1_{24} - d^1_{13}\right) + \left(d^2_{23} + d^2_{14}\right) \quad 0 = \left(d^2_{24} - d^2_{13}\right) - \left(d^1_{23} + d^1_{14}\right)$$
$$0 = \left(d^3_{24} - d^3_{13}\right) + \left(d^4_{23} + d^4_{14}\right) \quad 0 = \left(d^4_{24} - d^4_{13}\right) - \left(d^3_{23} + d^3_{14}\right)$$
$$0 = \left(d^1_{44} - d^1_{33}\right) + 2d^2_{34} \quad\quad 0 = \left(d^2_{44} - d^2_{33}\right) - 2d^1_{34} \tag{9.2.7}$$
$$0 = \left(d^3_{44} - d^3_{33}\right) + 2d^4_{34} \quad\quad 0 = \left(d^4_{44} - d^4_{33}\right) - 2d^3_{34}$$

- *To check that $\mathcal{D}$ is holomorphic it suffices to verify:*

$$d^1_{31} - d^2_{32} = 0 \quad d^2_{31} + d^1_{32} = 0$$
$$d^1_{33} - d^2_{34} = 0 \quad d^2_{33} + d^1_{34} = 0$$

Together with (9.2.7) this is equivalent to:

$$d^1_{41} - d^2_{42} = 0 \quad d^2_{41} + d^1_{42} = 0$$
$$d^1_{43} - d^2_{44} = 0 \quad d^2_{43} + d^1_{44} = 0$$

- *Complete non-integrability:*

$$d^c_{\alpha\beta} = 2\Gamma^c_{\alpha\beta}$$

Proof With the help of Sect. 7.3, it can be verified that the conditions on the structure functions of Theorem 9.2.3 coincide with the conditions given here.   □

Lemma 9.2.8 *In order to satisfy the integrability conditions of Theorem 9.2.3 it is sufficient to verify that for any* $\mathrm{Tr}_{\mathbb{C}}(1|1)$*-frame it holds:*

$$d^{\bar{z}}_{++} = d^{-}_{++} = d^{z}_{+-} = 0$$
$$d^{z}_{++} \text{ invertible}$$

Proof We need to show $d^{\bar{z}}_{z+} = d^{-}_{z+} = d^{z}_{+\bar{z}} = 0$. For this consider the Jacobi identity

$$\begin{aligned} 0 &= [F_+, [F_+, F_+]] = [F_+, d^{z}_{++}F_z + d^{+}_{++}F_+] \\ &= \left(F_+ d^{z}_{++}\right) F_z + d^{z}_{++} d^{A}_{+z} F_A + \left(F_+ d^{+}_{++}\right) F_+ - d^{+}_{++} d^{A}_{++} F_A. \end{aligned} \tag{9.2.9}$$

Its $\bar{z}$ and $-$-components are

$$0 = d^{z}_{++} d^{\bar{z}}_{+z}, \qquad 0 = d^{z}_{++} d^{-}_{+z}.$$

To show $d^{z}_{+\bar{z}} = 0$ we use

$$\begin{aligned} 0 &= [F_-, [F_+, F_+]] + 2[F_+, [F_-, F_+]] = [F_-, d^{A}_{++}F_A] + 2[F_+, d^{B}_{-+}F_B] \\ &= \left(F_- d^{A}_{++}\right) F_A + (-1)^{p(A)} d^{A}_{++}[F_-, F_A] + 2\left(F_+ d^{B}_{-+}\right) F_B \\ &\quad + (-1)^{p(B)} 2 d^{B}_{-+}[F_+, F_B] \\ &= \Big(\left(F_- d^{A}_{++}\right) + 2\left(F_+ d^{A}_{-+}\right) + d^{z}_{++} d^{A}_{-z} - d^{+}_{++} d^{A}_{-+} - 2d^{+}_{-+} d^{A}_{++} \\ &\quad -2d^{-}_{-+} d^{A}_{+-}\Big) F_A \end{aligned} \tag{9.2.10}$$

whose $\bar{z}$-coefficient gives the result. □

Lemma 9.2.11 *The remaining non-zero commutators are completely determined by* d^{z}_{++}*,* d^{+}_{++} *and* d^{+}_{+-}*. In particular, in the case* $d^{z}_{++} = 2$ *they fulfil the following equations:*

$$\begin{aligned} d^{z}_{z+} &= -d^{+}_{++} & d^{+}_{\bar{z}+} &= F_- d^{+}_{+-} - \frac{1}{2}\overline{d^{+}_{++}} d^{+}_{+-} \\ d^{+}_{z+} &= \frac{1}{2} F_+ d^{+}_{++} & d^{-}_{\bar{z}+} &= F_- \overline{d^{+}_{+-}} + \frac{1}{2}\left(F_+ \overline{d^{+}_{++}} - \overline{d^{+}_{++} d^{+}_{+-}}\right) \\ d^{\bar{z}}_{\bar{z}+} &= 2\overline{d^{+}_{+-}} & d^{z}_{z\bar{z}} &= -2d^{+}_{\bar{z}+} = \overline{d^{+}_{++}} d^{+}_{+-} - 2F_- d^{+}_{+-} \end{aligned}$$

Proof The terms d^{z}_{z+} and d^{+}_{z+} are determined from the z- and $+$-coefficient of (9.2.9), whereas $d^{\bar{z}}_{\bar{z}+}$, $d^{+}_{\bar{z}+}$ and $d^{-}_{\bar{z}+}$ are determined by the conjugates of the

z-, $+$- and $-$-coefficient of (9.2.10) respectively. The terms $d_{\bar{z}z}^{A}$ are determined by the A-coefficient of

$$\begin{aligned}
0 &= [F_{\bar{z}}, [F_+, F_+]] + 2[F_+, [F_+, F_{\bar{z}}]] = [F_{\bar{z}}, d_{++}^{A} F_A] + 2[F_+, d_{+\bar{z}}^{A} F_A] \\
&= \left(F_{\bar{z}} d_{++}^{A}\right) F_A + d_{++}^{A}[F_{\bar{z}}, F_A] + 2\left(F_+ d_{+\bar{z}}^{A}\right) F_A \\
&\quad + 2(-1)^{p(A)+1} d_{+\bar{z}}^{A}[F_+, F_A] \\
&= \left(\left(F_{\bar{z}} d_{++}^{A}\right) + 2\left(F_+ d_{+\bar{z}}^{A}\right) + d_{++}^{z} d_{\bar{z}z}^{A} + d_{++}^{+} d_{\bar{z}+}^{A}\right. \\
&\quad \left. + 2\left(-d_{+\bar{z}}^{\bar{z}} d_{+\bar{z}}^{A} + d_{+\bar{z}}^{+} d_{++}^{A} + d_{+\bar{z}}^{-} d_{+-}^{A}\right)\right) F_A
\end{aligned}$$

□

9.3 Infinitesimal Deformations

By the Theorem of Giddings and Nelson (Theorem 9.2.3) the super Riemann surface is locally determined by an integrable $\operatorname{Tr}_{\mathbb{C}}(1|1)$-frame F_A. In this section, we are studying under which conditions an infinitesimal change δF_A of F_A fulfils the integrability conditions up to first order.

As a preparation we need the following lemma:

Lemma 9.3.1 *Denote the structure functions of the frame F_A by d_{AB}^{C}, and let $F'_A = L_A{}^B F_B$ be an invertible transformation with inverse $F_A = M_A{}^B F_B$. The structure functions of F'_A are given by*

$$\begin{aligned}
[F'_A, F'_B] = &\left(L_A{}^C \left(F_C L_B{}^E\right) - (-1)^{p(A)p(B)} L_B{}^D \left(F_D L_A{}^E\right)\right. \\
&\left. + (-1)^{p(C)(p(B)+p(D))} L_A{}^C L_B{}^D d_{CD}^{E}\right) M_E{}^G F'_G.
\end{aligned}$$

If furthermore $L_A{}^B$ and $M_A{}^B$ are formal power series in t such that

$$L_A{}^B = \delta_A^B + H_A{}^B t + \sum_{n>1} \left(L_A{}^B\right)_n t^n,$$

then the above commutator is also a formal power series in t whose zero and first order term are:

$$\begin{aligned}
[F'_A, F'_B] = d_{AB}^{G} F'_G + &\left(-d_{AB}^{E} H_E{}^G + \left(F_A H_B{}^G\right) - (-1)^{p(A)p(B)} \left(F_B H_A{}^G\right)\right. \\
&\left. + H_A{}^C d_{CB}^{G} + (-1)^{p(A)(p(B)+p(D))} H_B{}^D d_{AD}^{G}\right) t F'_G + \ldots
\end{aligned}$$

Proof The commutator of F'_A and F'_B can be calculated as follows:

$$\begin{aligned}
[F'_A, F'_B] &= [L_A{}^C F_C, L_B{}^D F_D] \\
&= L_A{}^C \left(F_C L_B{}^D\right) F_D - (-1)^{p(A)p(B)} L_B{}^D \left(F_D L_A{}^C\right) F_C \\
&\quad + (-1)^{p(C)(p(B)+p(D))} L_A{}^C L_B{}^D d_{CD}^E F_E \\
&= \Big(L_A{}^C \left(F_C L_B{}^E\right) - (-1)^{p(A)p(B)} L_B{}^D \left(F_D L_A{}^E\right) \\
&\qquad + (-1)^{p(C)(p(B)+p(D))} L_A{}^C L_B{}^D d_{CD}^E\Big) M_E{}^G F'_G
\end{aligned}$$

In case $L_A{}^B$ is a formal power series, its inverse starts by

$$M_A{}^B = \delta_A^B - H_A{}^B t + \sum_{n>1} \left(M_A{}^B\right)_n t^n,$$

and hence the zero order term of $[F'_A, F'_B]$ is $d_{AB}^G F'_G$ and the first order term is

$$\begin{aligned}
&- d_{AB}^E H_E{}^G + \left(F_A H_B{}^G\right) - (-1)^{p(A)p(B)} \left(F_B H_A{}^G\right) \\
&+ H_A{}^C d_{CB}^G + (-1)^{p(A)(p(B)+p(D))} H_B{}^D d_{AD}^G.
\end{aligned}$$

□

Proposition 9.3.2 *Let $H_A{}^B$ be a first order deformation of the complex frame F_A on $\mathbb{R}^{2|2}$ that is integrable in the sense of the Theorem of Giddings and Nelson (Theorem 9.2.3). The deformation $H_A{}^B$ is integrable up to first order if and only if the following equations hold:*

$$0 = -d_{++}^z H_z{}^{\overline{z}} - \left(2d_{\overline{z}+}^{\overline{z}} + d_{++}^+\right) H_+{}^{\overline{z}} + 2F_+ H_+{}^{\overline{z}} \tag{9.3.3a}$$

$$0 = -d_{++}^z H_z{}^- + \left(2d_{-+}^- - d_{++}^+\right) H_+{}^- + 2H_+{}^{\overline{z}} d_{\overline{z}+}^- + 2F_+ H_+{}^- \tag{9.3.3b}$$

$$\begin{aligned}
0 = {} & H_-{}^+ d_{++}^z - \left(d_{+-}^+ + d_{z-}^z\right) H_+{}^z + \left(d_{+z}^z - d_{+-}^-\right) H_-{}^z \\
& + F_+ H_-{}^z + F_- H_+{}^z
\end{aligned} \tag{9.3.3c}$$

Equations (9.3.3) are solvable for $H_z{}^{\overline{z}}$, $H_z{}^-$ and $H_-{}^+$ as functions of $H_+{}^z$ and $H_+{}^{\overline{z}}$ because d_{++}^z is invertible. In order to be an infinitesimal deformation of an SCL-frame, the deformation $H_A{}^B$ has to satisfy also

$$0 = -2H_z{}^z + 4H_+{}^+ - \left(2d_{z+}^z + d_{++}^+\right) H_+{}^z + 2\left(F_+ H_+{}^z\right). \tag{9.3.4}$$

Proof By Lemma 9.2.8 it suffices to check that the first order perturbations of $d^{\bar{z}}_{++}$, d^{-}_{++} and d^{z}_{+-} are zero. The condition that d^{z}_{++} is invertible is an open condition, so that for small $H_A{}^B$ it is always fulfilled. To calculate the first order perturbations we use Lemma 9.3.1:

$$
\begin{aligned}
0 &= -d^E_{++}H_E{}^{\bar{z}} + 2F_+H_+{}^{\bar{z}} + H_+{}^C d^{\bar{z}}_{C+} - (-1)^{p(D)} H_+{}^D d^{\bar{z}}_{+D}\\
&= -d^E_{++}H_E{}^{\bar{z}} + 2F_+H_+{}^{\bar{z}} + 2H_+{}^C d^{\bar{z}}_{C+}\\
&= -d^z_{++}H_z{}^{\bar{z}} - d^+_{++}H_+{}^{\bar{z}} + 2F_+H_+{}^{\bar{z}} + 2H_+{}^{\bar{z}} d^{\bar{z}}_{\bar{z}+}\\
&= -d^z_{++}H_z{}^{\bar{z}} - \left(2d^{\bar{z}}_{\bar{z}+} + d^+_{++}\right) H_+{}^{\bar{z}} + 2F_+H_+{}^{\bar{z}}
\end{aligned}
$$

$$
\begin{aligned}
0 &= -d^E_{++}H_E{}^- + 2F_+H_+{}^- + H_+{}^C d^-_{C+} - (-1)^{p(D)} H_+{}^D d^-_{+D}\\
&= -d^E_{++}H_E{}^- + 2F_+H_+{}^- + 2H_+{}^C d^-_{C+}\\
&= -d^z_{++}H_z{}^- - d^+_{++}H_+{}^- + 2F_+H_+{}^- + 2H_+{}^{\bar{z}} d^-_{\bar{z}+} + 2H_+{}^- d^-_{-+}\\
&= -d^z_{++}H_z{}^- + \left(2d^-_{-+} - d^+_{++}\right) H_+{}^- + 2H_+{}^{\bar{z}} d^-_{\bar{z}+} + 2F_+H_+{}^-
\end{aligned}
$$

$$
\begin{aligned}
0 &= -d^E_{+-}H_E{}^z + F_+H_-{}^z + F_-H_+{}^z + H_+{}^C d^z_{C-} - (-1)^{p(D)} H_-{}^D d^z_{+D}\\
&= -d^+_{+-}H_+{}^z - d^-_{+-}H_-{}^z + F_+H_-{}^z + F_-H_+{}^z + H_+{}^z d^z_{z-}\\
&\quad - H_-{}^z d^z_{+z} + H_-{}^+ d^z_{++}\\
&= -\left(d^+_{+-} + d^z_{z-}\right) H_+{}^z + \left(d^z_{+z} - d^-_{+-}\right) H_-{}^z + H_-{}^+ d^z_{++}\\
&\quad + F_+H_-{}^z + F_-H_+{}^z
\end{aligned}
$$

Equation (9.3.4) comes from the fact in a SCL-reduction of the structure group the frames F_+ and F_z are not completely independent due to the condition of complete non-integrability. Hence, additionally the change of d^z_{++} needs to be zero:

$$
\begin{aligned}
0 &= -d^E_{++}H_E{}^z + 2\left(F_+H_+{}^z\right) + H_+{}^C d^z_{C+} - (-1)^{p(D)} H_+{}^D d^z_{+D}\\
&= -d^E_{++}H_E{}^z + 2\left(F_+H_+{}^z\right) + 2H_+{}^C d^z_{C+}\\
&= -d^z_{++}H_z{}^z - d^+_{++}H_+{}^z + 2\left(F_+H_+{}^z\right) + 2H_+{}^z d^z_{z+} + 2H_+{}^+ d^z_{++}\\
&= -2H_z{}^z + 4H_+{}^+ - \left(2d^z_{z+} + d^+_{++}\right) H_+{}^z + 2\left(F_+H_+{}^z\right)
\end{aligned}
$$

□

Proposition 9.3.2 shows that the only true deformations are given by $H_+{}^z$ and $H_+{}^{\bar{z}}$. The deformations $H_z{}^z$, $H_z{}^+$ and $H_+{}^+$, which do not appear in the constraints, do not deform the $\mathrm{Tr}_{\mathbb{C}}(1|1)$-structure.

Any vector field X on $\mathbb{R}^{2|2}$ induces an infinitesimal deformation of F_A given by $[X, F_A] = H_A{}^B F_B$. The integrable deformations $H_A{}^B$ that are not induced by vector fields constitute tangent vectors to the moduli space of super Riemann surfaces. We will not pursue this issue here but refer instead to Sect. 11.5 and Sachse (2009, Chapter 8.1).

Remark 9.3.5 In the case of the superconformal frames, that is, $F_z = \partial_z$ and $F_+ = D$ for some superconformal coordinates (z, θ), Eqs. (9.3.3), simplify considerably to

$$H_z{}^{\overline{z}} = DH_{\overline{z}}{}^{+}, \qquad H_z{}^{-} = DH_+{}^{-}, \qquad H_-{}^{+} = -\frac{1}{2}\left(DH_-{}^{z} + \overline{D}H_+{}^{z}\right),$$

and Eq. (9.3.4) simplifies to

$$H_z{}^{z} = 2H_+{}^{+} + DH_+{}^{z}.$$

In this form Eqs. (9.3.3) and (9.3.4) can be found in D'Hoker and Phong (1988, Equation (3.22)). However, the derivation in D'Hoker and Phong (1988) is based on covariant derivatives and torsion constraints, which we do not need here.

9.4 Uniformization

Recall that the universal cover of a Riemann surface Σ is either $\mathbb{P}^1_{\mathbb{C}}$, the complex plane $\mathbb{C}$, or the upper half-plane $\mathbb{H}$ in $\mathbb{C}$. If the genus of the surface Σ is zero, it is biholomorphic to $\mathbb{P}^1_{\mathbb{C}}$. Surfaces of genus one are tori, that is a quotient of $\mathbb{C}$ by a lattice. Higher genus surfaces are quotients of $\mathbb{H}$ by the properly discontinuos action of a discrete subgroup of the Möbius transformations fixing the boundary of $\mathbb{H}$. This is the statement of the uniformization theorem, see for example Jost (2006, Theorem 4.4.1).

Similar to the case of Riemann surfaces, we have already seen in Proposition 9.1.7 that $\mathbb{P}^{1|1}_{\mathbb{C}}$, $\mathbb{C}^{1|1}$ and $S\mathbb{H}$ are the only simply connected super Riemann surfaces over $B = \mathbb{R}^{0|0}$. A generalization of this fact for more general bases B was given in Crane and Rabin (1988):

Theorem 9.4.1 (Crane and Rabin (1988)) *Let $B = \mathbb{R}^{0|N}$ for $N \in \mathbb{N}$ and $M \to B$ a family of super Riemann surfaces over B such that $\|M\|$ is simply connected. Then M is isomorphic to one of the trivial families of super Riemann surfaces $\mathbb{C}^{1|1} \times B \to B$, $\mathbb{P}^{1|1}_{\mathbb{C}} \times B \to B$ or $S\mathbb{H} \times B \to B$.*

We restrict to the case of $B = \mathbb{R}^{0|N}$ for the remainder of this section. That is, we will not consider bases with even dimensions to keep the topological questions simple. Before giving the proof of Theorem 9.4.1 at the end of this section let us first look at some consequences.

The universal cover $\widetilde{M}$ of a super Riemann surface M over $B = \mathbb{R}^{0|N}$ is a super Riemann surface over B because the super Riemann surface structure can be lifted locally. Assume that $\|M\|$ has genus p. In the case $p = 0$, both M and $\widetilde{M}$ are isomorphic to $\mathbb{P}^{1|1}_{\mathbb{C}} \times B \to B$. In the case $p = 1$, the universal cover $\widetilde{M}$ must be $\mathbb{C}^{1|1} \times B \to B$, whereas in the case $p > 1$ the universal cover $\widetilde{M}$ is given by $S\mathbb{H} \times B \to B$. The group of deck transformations is a subgroup of the superconformal automorphisms of $\widetilde{M}$. Consequently the set of all super Riemann surfaces can be characterized as the set of group homomorphisms from $\pi_1(\|M\|) \to \mathrm{SC}_B\left(\widetilde{M}\right)$ which act properly discontinuosly on $\|\widetilde{M}\|$ up to conjugations in $\mathrm{SC}_B\left(\widetilde{M}\right)$:

$$\mathrm{Hom}\left(\pi_1(\|M\|), \mathrm{SC}_B\left(\widetilde{M}\right)\right)\big/_{\mathrm{SC}_B\left(\widetilde{M}\right)}.$$

This space is called super Teichmüller space in Hodgkin (1987a) and shown to be a complex supermanifold of dimension $3p-3|2p-2$ for $p > 1$. Super elliptic curves, that is super Riemann surfaces with $p = 1$, enjoy particular properties studied in Rabin (1995).

Example 9.4.2 (Superconformal Automorphisms of $\mathbb{C}^{1|1}$, $\mathbb{P}^{1|1}_{\mathbb{C}}$ and $S\mathbb{H}$) Recall from Sect. 9.1 that a superconformal coordinate change is given by

$$\tilde{z} = f(z) + \theta\zeta(z), \qquad \tilde{\theta} = \xi(z) + \theta g(z),$$

such that

$$f' = g^2 - \xi\xi', \qquad \zeta = g\xi.$$

Here f and g are even functions, ζ and ξ are odd functions depending on z and implicitly on λ the odd coordinate on $B = \mathbb{R}^{0|N}$. Consequently they posses decompositions of the form

$$f = \sum_{\underline{\nu}\text{ even}} \lambda^{\underline{\nu}}{}_{\underline{\nu}}f(z), \qquad \zeta = \sum_{\underline{\nu}\text{ odd}} \lambda^{\underline{\nu}}{}_{\underline{\nu}}\zeta(z),$$

$$\xi = \sum_{\underline{\nu}\text{ odd}} \lambda^{\underline{\nu}}{}_{\underline{\nu}}\xi(z), \qquad g = \sum_{\underline{\nu}\text{ even}} \lambda^{\underline{\nu}}{}_{\underline{\nu}}g(z),$$

for holomorphic functions ${}_{\underline{\nu}}f(z)$, ${}_{\underline{\nu}}\zeta(z)$, ${}_{\underline{\nu}}\xi(z)$ and ${}_{\underline{\nu}}g(z)$. We require this coordinate change to be an automorphism of $\mathbb{C}^{1|1}$, hence ${}_0f(z) = az + b$ for some complex numbers $a \neq 0$ and b. Consequently, ${}_0g(z)$ is constant and invertible.

The case of $S\mathbb{H}$ works similarly. The condition that the coordinate change be an automorphism of $S\mathbb{H}$, implies that ${}_0f(z)$ must be a Möbius transformation of the form

$${}_0f(z) = \frac{az+b}{cz+d},$$

where $a, b, c, d \in \mathcal{O}_B$ and $ad - bc = 1$. Consequently, ${}_0g(z) = \pm\frac{1}{cz+d}$ and g is invertible.

For $\mathbb{P}^{1|1}_{\mathbb{C}}$, we know that its reduction must be a complex Möbius transformation. Working in the coordinates (z_1, θ_1) and (z_2, θ_2) introduced in Example 7.2.4, the lowest order term ${}_0f(z_1)$ must be a Möbius transformation, with coefficients from $\mathcal{O}_{B_{red}} \otimes \mathbb{C}$. Furthermore all terms in the expansions of f, g, ξ and ζ need to be extendable to infinity. Let us express this extension to infinity in coordinates. Up to a Möbius transformation, we may assume that

$$\tilde{z}_1 = z_1 + f(z_1) + \theta_1\zeta(z_1), \qquad \tilde{\theta}_1 = \xi(z_1) + \theta_1\left(1 + g(z_1)\right),$$

where f and g are even nilpotent and ζ and ξ are odd. This yields for $z_2 \neq 0$:

$$\begin{aligned}
\tilde{z}_2 &= z_2\left(1 + \sum_{n>0}\left(-z_2 f\left(\frac{1}{z_2}\right)\right)^n - \theta_2\zeta\left(\frac{1}{z_2}\right)\right)\\
\tilde{\theta}_2 &= \left(z_2\xi\left(\frac{1}{z_2}\right) + \theta_2\left(1 + g\left(\frac{1}{z_2}\right)\right)\right)\\
&\quad\cdot\left(1 + \sum_{n>0}\left(-z_2 f\left(\frac{1}{z_2}\right)\right)^n - \theta_2\zeta\left(\frac{1}{z_2}\right)\right)\\
&= z_2\xi\left(\frac{1}{z_2}\right)\left(1 + \sum_{n>0}\left(-z_2 f\left(\frac{1}{z_2}\right)\right)^n\right) + \theta_2\left(\left(1 + g\left(\frac{1}{z_2}\right)\right)\right.\\
&\quad\left.\cdot\left(1 + \sum_{n>0}\left(-z_2 f\left(\frac{1}{z_2}\right)\right)^n\right) + z_2\xi\left(\frac{1}{z_2}\right)\zeta\left(\frac{1}{z_2}\right)\right)
\end{aligned}$$

The expressions for $\tilde{z}_2$ and $\tilde{\theta}_2$ need to have a limit for $z_2 = 0$. By expanding the functions f, g, ζ and ξ with respect to the odd generators of the base one obtains the allowed pole order at $z_2 = 0$. For example, the functions ${}_\nu\zeta\left(\frac{1}{z_2}\right)$ and ${}_\nu\xi\left(\frac{1}{z_2}\right)$ are allowed poles of order one at $z_2 = 0$, the function ${}_{\underline{\nu}}f\left(\frac{1}{z_2}\right)$ may have a pole of order two for $|\underline{\nu}| = 2$, whereas ${}_{\underline{\nu}}g\left(\frac{1}{z_2}\right)$ has a pole of order one for $|\underline{\nu}| = 2$ such that

$${}_{\underline{\nu}}g\left(\frac{1}{z_2}\right) + z_2\left({}_{\underline{\nu}}f\left(\frac{1}{z_2}\right) + \sum_{\alpha+\beta=\underline{\nu}} {}_\alpha\xi\left(\frac{1}{z_2}\right){}_\beta\zeta\left(\frac{1}{z_2}\right)\right)$$

is well-defined for $z_2 = 0$.

Example 9.4.3 ($\mathrm{Sp}_{\mathcal{O}_B\otimes\mathbb{C}}(2|1)$ *as a Subgroup of* $\mathrm{SC}_B\left(\mathbb{P}^{1|1}_{B\otimes\mathbb{C}}\right)$*, following Manin (1991, Chapter 2.1))* Recall from Example 3.4.2 that a linear transformation of the projective coordinates $[Z^1 : Z^2 : \Theta]$ yields an automorphism of $\mathbb{P}^{1|1}_{B\otimes\mathbb{C}}$. We will now show that the subgroup $\mathrm{Sp}_{\mathcal{O}_B \otimes \mathbb{C}}(2|1) \subset \mathrm{GL}_{\mathcal{O}_B \otimes \mathbb{C}}(2|1)$ induces automorphisms which preserve the superconformal structure of $\mathbb{P}^{1|1}_{B \otimes \mathbb{C}}$. Let the linear transformation L be given by the matrix ${}_A L^B$:

$$\left(\tilde{Z}^1\ \tilde{Z}^2\ \tilde{\Theta}\right) = \left(Z^1\ Z^2\ \Theta\right) \begin{pmatrix} a & c & \gamma \\ b & d & \delta \\ \alpha & \beta & e \end{pmatrix}.$$

Here a, b, c, d and e are even sections of $\mathcal{O}_B \otimes \mathbb{C}$, whereas α, β, γ and δ are odd sections of $\mathcal{O}_B \otimes \mathbb{C}$. The induced coordinate transformations on U_1 are

$$\tilde{z}_1 = \frac{az_1 + b + \theta_1 \alpha}{cz_1 + d + \theta_1 \beta}, \qquad \tilde{\theta}_1 = \frac{\gamma z_1 + \delta + \theta_1 e}{cz_1 + d + \theta_1 \beta}.$$

The linear transformation L is a superconformal transformation if and only if $D\tilde{z}_1 - \tilde{\theta}_1 D\tilde{\theta}_1 = 0$, or equivalently after multiplication by $(cz_1 + d + \theta_1\beta)^2$:

$$\begin{aligned} 0 = &\ (\alpha + \theta_1 a)(cz_1 + d + \theta_1 \beta) - (az_1 + b + \theta_1 \alpha)(\beta + \theta_1 c) \\ &- (\gamma z_1 + \delta + \theta_1 e)(e + \theta_1 \gamma) \\ = &\ \alpha d - b\beta - \delta e + (\alpha c - a\beta - \gamma e) z_1 \\ &+ \theta_1 \left(ad - bc - 2\alpha\beta - e^2 - \gamma\delta\right). \end{aligned} \tag{9.4.4}$$

The equality is fulfilled if $L \in \mathrm{Sp}_{\mathcal{O}_B \otimes \mathbb{C}}(2|1)$, compare Example 2.10.11. The requirement that L also induces a superconformal transformation on the coordinate patch U_2 does not lead to additional conditions on L. Hence $\mathrm{Sp}_{\mathcal{O}_B \otimes \mathbb{C}}(2|1)$ can be seen as a subgroup of the superconformal automorphisms of $\mathbb{P}^{1|1}_{B \otimes \mathbb{C}}$.

At first it might seem that the requirement for $L \in \mathrm{GL}_{\mathcal{O}_B \otimes \mathbb{C}}(2|1)$ to be superconformal is a little weaker than the requirement $L \in \mathrm{Sp}_{\mathcal{O}_B \otimes \mathbb{C}}(2|1)$, because Eq. (9.4.4) requires only

$$ad - bc - \gamma\delta = e^2 + 2\alpha\beta,$$

but not necessarily $e^2 + 2\alpha\beta = 1$. The set of matrices M such that there is an invertible λ and

$$\begin{aligned} ad - bc - \gamma\delta &= \lambda^2, & a\beta - c\alpha + e\gamma &= 0, \\ e^2 + 2\alpha\beta &= \lambda^2, & b\beta - d\alpha + e\delta &= 0, \end{aligned}$$

form the subgroup of matrices that preserve the standard symplectic structure up to rescaling, that is ${}_A M^B\, {}_B b_C\, {}^C M_D = \lambda^2\, {}_A b_D$. However every such conformal symplectic isomorphism M can be written as $M = \lambda L$ for $L \in \mathrm{Sp}_{\mathcal{O}_B \otimes \mathbb{C}}(2|1)$.

The linear transformation L and its multiples λL induce the same transformation on $\mathbb{P}^{1|1}_{B\otimes\mathbb{C}}$.

Let us now rewrite the coordinate transformation induced by L using the relations from Example 2.10.11:

$$\begin{aligned}
\tilde{z}_1 &= \frac{az_1 + b + \theta_1\alpha}{cz_1 + d + \theta_1\beta} = \frac{az_1 + b + \theta_1\alpha}{cz_1 + d}\left(1 - \theta_1\frac{\beta}{cz_1 + d}\right)\\
&= \frac{az_1 + b}{cz_1 + d} + \theta_1\frac{(c\alpha - a\beta)\, z_1 + d\alpha - b\beta}{(cz_1 + d)^2} = \frac{az_1 + b}{cz_1 + d} + \theta_1\frac{e\,(\gamma z_1 + \delta)}{(cz_1 + d)^2}\\
&= \frac{az_1 + b}{cz_1 + d} \pm \theta_1\frac{\gamma z_1 + \delta}{(cz_1 + d)^2}\\
\tilde{\theta}_1 &= \frac{\gamma z_1 + \delta + \theta_1 e}{cz_1 + d + \theta_1\beta} = \frac{\gamma z_1 + \delta + \theta_1 e}{cz_1 + d}\left(1 - \theta_1\frac{\beta}{cz_1 + d}\right)\\
&= \frac{\gamma z_1 + \delta}{cz_1 + d} + \theta_1\frac{e^2\,(cz_1 + d) + e\,(\gamma z_1 + \delta)\,\beta}{e(cz_1 + d)^2}\\
&= \frac{\gamma z_1 + \delta}{cz_1 + d} + \theta_1\frac{\left(e^2 + \alpha\beta\right)(cz_1 + d)}{e(cz_1 + d)^2} = \frac{\gamma z_1 + \delta}{cz_1 + d} + \theta_1\frac{1 - \gamma\delta}{e\,(cz_1 + d)}\\
&= \frac{\gamma z_1 + \delta}{cz_1 + d} \pm \theta_1\frac{1}{cz_1 + d}
\end{aligned}$$

The sign of the terms proportional to θ_1 corresponds to the choice of sign in $e = \pm\,(1 - \gamma\delta)$. It can be absorbed by replacing a, b, c, d, γ and δ by their negative. Consequently, any superconformal automorphism induced by $\mathrm{Sp}_{\mathcal{O}_B\otimes\mathbb{C}}(2|1)$ is in fact given by a tuple $(a, b, c, d, \gamma, \delta)$ such that $ad - bc - \gamma\delta = 1$. Very similar formulas for superconformal automorphisms induced by $\mathrm{Sp}_{\mathcal{O}_B\otimes\mathbb{C}}(2|1)$ are given, for instance, in Crane and Rabin (1988, Equation (2.12)); Hodgkin (1987b, Equation (9)).

Notice that $\mathrm{Sp}_{\mathcal{O}_B\otimes\mathbb{C}}(2|1)$ is a proper subgroup of $\mathrm{SC}_B\left(\mathbb{P}^{1|1}_{B\otimes\mathbb{C}}\right)$. Indeed, the following example of a superconformal automorphism of $\mathbb{P}^{1|1}_{B\otimes\mathbb{C}}$ is not induced by $\mathrm{Sp}_{\mathcal{O}_B\otimes\mathbb{C}}(2|1)$:

$$\tilde{z}_1 = z_1 + \lambda_1\lambda_2 z_1^2 + \theta_1\,(\lambda_1 - \lambda_2)\, z_1, \qquad \tilde{\theta}_1 = (\lambda_1 - \lambda_2)\, z_1 + \theta_1.$$

Here λ_1 and λ_2 are odd generators on B. Notice that the expressions for $\tilde{z}_1$ and $\tilde{\theta}_1$ extend to infinity as explained in Example 9.4.2.

Example 9.4.5 ($\mathrm{Sp}_{\mathcal{O}_B}(2|1)$ as a Subgroup of $\mathrm{SC}\,(S\mathbb{H})$) The group $\mathrm{Sp}_{\mathcal{O}_B}(2|1)$ induces superconformal automorphisms of $S\mathbb{H}$ in the same way as detailed for $\mathbb{P}^{1|1}_{B\otimes\mathbb{C}}$ in Example 9.4.3. In this case it is easier to find examples of elements of $\mathrm{SC}\,(S\mathbb{H})$, which do not arise as Möbius transformations. The following example is given in Crane and Rabin (1988):

$$\tilde{z}_1 = z_1 + \theta_1 \eta z_1^n \qquad \tilde{\theta} = \theta_1 + \eta z_1^n$$

Here, η is an odd element of $\mathcal{O}_B \otimes \mathbb{C}$ and $n > 1$.

The paper Hodgkin (1987b) gives an argument that for purposes of Teichmüller theory it is sufficient to consider $\mathrm{Sp}_{\mathcal{O}_B}(2|1)$ instead of $\mathrm{SC}_B(S\mathbb{H})$. More specifically,

$$\begin{aligned}&\mathrm{Hom}\left(\pi_1\left(\|M\|\right), \mathrm{SC}_B\left(S\mathbb{H}\right)\right)\big/_{\mathrm{SC}_B\left(S\mathbb{H}\right)}\\ &= \mathrm{Hom}\left(\pi_1\left(\|M\|\right), \mathrm{Sp}_{\mathcal{O}_B}(2|1)\right)\big/_{\mathrm{Sp}_{\mathcal{O}_B}(2|1)}.\end{aligned}$$

Based on this result, explicit coordinates on super Teichmüller space are constructed in Penner and Zeitlin (2015).

With this preparation and motivation we do now turn to the proof of Theorem 9.4.1. The presentation follows mostly Crane and Rabin (1988).

Proof of Theorem 9.4.1 Let $M \to B = \mathbb{R}^{0|N}$ be a simply connected family of super Riemann surfaces. The Riemann surface $M_{red} \to B_{red} = pt$ is simply connected and hence one of the three simply connected Riemann surfaces $\mathbb{P}^1_{\mathbb{C}}$, $\mathbb{C}$ and $S\mathbb{H}$. The super Riemann surface structure on M is determined by an open cover $\{U_\alpha\}$ of M, superconformal coordinates $(z_\alpha, \theta_\alpha)$ on U_α and superconformal coordinate changes

$$z_\beta = f_{\beta\alpha}(z_\alpha) + \theta_\alpha \zeta_{\beta\alpha}(z), \qquad \theta_\beta = \xi_{\beta\alpha}(z_\alpha) + \theta_\alpha g_{\beta\alpha}(z_\alpha).$$

Recall that $f'_{\beta\alpha} = g^2_{\beta\alpha} - \xi_{\beta\alpha}\xi'_{\beta\alpha}$ and $\zeta_{\beta\alpha} = g_{\beta\alpha}\xi_{\beta\alpha}$. Since simply connected Riemann surfaces possess a unique spin structure, we can regard $g_{\beta\alpha}$ and $\zeta_{\beta\alpha}$ as functions of $f_{\beta\alpha}$ and $\xi_{\beta\alpha}$. The composition of two coordinate changes induces a corresponding transformation on the functions f and ξ. For example, the cocycle condition of the glueing is given by

$$\begin{aligned} f_{\gamma\alpha}(z_\alpha) &= f_{\gamma\beta}\left(f_{\beta\alpha}(z_\alpha)\right) + \xi_{\beta\alpha}(z_\alpha)\xi_{\gamma\beta}(f_{\beta\alpha}(z_\alpha))\sqrt{f'_{\gamma\beta}\left(f_{\beta\alpha}(z_\alpha)\right)}, \qquad (9.4.6)\\ \xi_{\gamma\alpha}(z_\alpha) &= \xi_{\gamma\beta}\left(f_{\beta\alpha}(z_\alpha)\right)\\ &\quad + \xi_{\beta\alpha}(z_\alpha)\sqrt{f'_{\gamma\beta}\left(f_{\beta\alpha}(z_\alpha)\right) - \xi_{\gamma\beta}(f_{\beta\alpha}(z_\alpha))\xi'_{\gamma\beta}(f_{\beta\alpha}(z_\alpha))}. \end{aligned}$$

The functions $f_{\beta\alpha}$, $\xi_{\beta\alpha}$, $\zeta_{\beta\alpha}$ and $g_{\beta\alpha}$ posses an expansion in the coordinates λ^μ of B as in Example 9.4.2. In this proof we will show that we can actually set the dependence of all the patching functions $f_{\alpha\beta}$ and $\xi_{\alpha\beta}$ on λ^μ simultaneously to zero by applying coordinate changes on the open sets U_α. This shows that any simply connected family of super Riemann surfaces is actually a trivial family, and hence either $\mathbb{P}^{1|1}_B \times B \to B$, $\mathbb{C}^{1|1} \times B \to B$ or $S\mathbb{H} \times B \to B$.

Let now $\underline{\nu}$ be a $\mathbb{Z}_2$-multiindex of the lowest order of λ^μ appearing in any of the $\xi_{\alpha\beta}$. The coefficient of $\lambda^{\underline{\nu}}$ in Eq. (9.4.6) is given by

$$ {}_{\underline{\nu}}\xi_{\gamma\alpha}(z_\alpha) = {}_{\underline{\nu}}\xi_{\gamma\beta}\left({}_0f_{\beta\alpha}(z_\alpha)\right) + {}_{\underline{\nu}}\xi_{\beta\alpha}(z_\alpha)\sqrt{{}_0f'_{\gamma\beta}\left({}_0f_{\beta\alpha}(z_\alpha)\right)} \,. \tag{9.4.7} $$

The term ${}_0f_{\beta\alpha}(z_\alpha)$ is a coordinate change of the reduced manifold M_{red}. Let $S = i^*_{red}\mathcal{D}$ be the spinor bundle induced by $i_{red}\colon M_{red} \to M$. The term $\sqrt{{}_0f'_{\gamma\beta}}$ is the change of trivialization $S_\alpha \to S_\beta$ of the complex line bundle S. Hence, Eq. (9.4.7) yields that the collection of ${}_{\underline{\nu}}\xi_{\alpha\beta}$ form a cocycle in the Čech-cohomology $H^1(M_{red}, S)$ and

$$ {}_{\underline{\nu}}\xi_{\beta\alpha} : U_\alpha \cap U_\beta \to S_\beta. $$

We can assume without loss of generality that $\{U_\alpha\}$ is a good cover in the sense that Čech cohomology on M can be calculated directly in the covering $\{U_\alpha\}$. Since $H^1(M_{red}, S)$ vanishes in all three cases $M_{red} = \mathbb{P}^1_{\mathbb{C}}$, $M = \mathbb{C}$ and $M = \mathbb{H}$ there are local sections s_α of S_α over U_α, such that

$$ {}_{\underline{\nu}}\xi_{\beta\alpha} = {}_{\underline{\nu}}s_\beta - {}_{\underline{\nu}}s_\alpha $$

for all indices α, β. The coordinate changes

$$ \widetilde{z}_\alpha = z_\alpha - \theta_\alpha\lambda^{\underline{\nu}}\,{}_{\underline{\nu}}s_\alpha \qquad\qquad \widetilde{\theta}_\alpha = -\lambda^{\underline{\nu}}\,{}_{\underline{\nu}}s_\alpha + \theta_\alpha $$

yield a covering of M by coordinates $(\widetilde{z}_\alpha, \widetilde{\theta}_\alpha)$ such that the coordinate changes are determined by $\widetilde{f}_{\beta\alpha} = f_{\alpha\beta}$ and $\widetilde{\xi}_{\alpha\beta} = \xi_{\alpha\beta} - \lambda^{\underline{\nu}}\,{}_{\underline{\nu}}s_\alpha$ up to higher orders of λ^μ. Consequently, we can remove all terms of lowest order in λ^μ in the $\xi_{\alpha\beta}$.

Similarly, for the lowest order terms in $f_{\alpha\beta}$ can be removed. Indeed, if $\underline{\nu}$ is now a $\mathbb{Z}_2$-multiindex of the lowest order of λ^μ appearing in any of the $f_{\beta\alpha}$ and $\xi_{\beta\alpha}$ the coefficient of $\lambda^{\underline{\nu}}$ in Eq. (9.4.6) is given by

$$ {}_{\underline{\nu}}f_{\gamma\alpha}(z_\alpha) = {}_{\underline{\nu}}f_{\gamma\beta}\left({}_0f_{\beta\alpha}(z_\alpha)\right) + {}_0f'_{\gamma\beta}\left({}_{\underline{\nu}}f_{\beta\alpha}(z_\alpha)\right). $$

The term ${}_0f'_{\gamma\beta}$ is a change of trivialization of TM_{red}. Hence the collection of ${}_{\underline{\nu}}f_{\beta\alpha}$ form a cocycle in $H^1(M_{red}, TM)$. By the triviality of this cohomology group in the simply connected cases, there are local tangent fields v_α over U_α such that

$$ {}_{\underline{\nu}}f_{\beta\alpha} = {}_{\underline{\nu}}v_\beta - {}_{\underline{\nu}}v_\alpha. $$

Consequently the coordinate changes

$$ \widetilde{z}_\alpha = z_\alpha - \lambda^{\underline{\nu}}\,{}_{\underline{\nu}}v_\alpha \qquad\qquad \widetilde{\theta}_\alpha = \theta_\alpha\left(1 - \frac{1}{2}\lambda^{\underline{\nu}}\,{}_{\underline{\nu}}v'_\alpha\right) $$

yield a covering of M by superconformal coordinates $(\widetilde{z}_\alpha, \widetilde{\theta}_\alpha)$ with patching functions $\widetilde{f}_{\beta\alpha} = f_{\beta\alpha} - \lambda^{\underline{\nu}}{}_{\nu} f_{\beta\alpha}$ and $\widetilde{\xi}_{\beta\alpha} = \xi_{\beta\alpha}$ up to higher orders of λ^μ.

Therefore, by an iterative procedure, all nilpotent terms can be removed from the patching of the super Riemann surfaces. Since trivial families of super Riemann surfaces are classified in Proposition 9.1.7, this completes the proof. □

In addition to the classification of super Riemann surfaces as quotients of $S\mathbb{H}$, there are several other approaches super Teichmüller theory. In particular in LeBrun and Rothstein (1988); Sachse (2009, Theorem 8.4.4) it has been shown that there is a semi-universal family $\mathcal{E} \to \mathcal{ST}_p$ of super Riemann surfaces of genus p. That is any family $M \to B$ of super Riemann surfaces can be obtained in a non-unique way as a pullback of $\mathcal{E}$ along a map $B \to \mathcal{ST}_p$. The base manifold $\mathcal{ST}_p$ is a supermanifold over $\mathbb{R}^{0|0}$ of real dimension $6p - 6|4p - 4$ and also possesses a complex structure. Proposition 9.1.7 proves that the points of $|\mathcal{ST}_p|$, that is maps $\mathbb{R}^{0|0} \to \mathcal{ST}_p$, are in one to one correspondence to Riemann surfaces with a chosen spinor bundle. Hence, the study of non-trivial families of super Riemann surfaces provides an understanding of the super structure of $\mathcal{ST}_p$.

9.5 Metrics on Super Riemann Surfaces

In this section we discuss metrics which are compatible with the super Riemann surface structure. That is they are compatible with the almost complex structure and the isomorphism $\mathcal{D} \otimes_{\mathbb{C}} \mathcal{D} \simeq {}^{TM}\!/_{\mathcal{D}}$.

As a first step we consider the linear algebra case. Following the notation of Sect. 2.12 we let R be a superalgebra over $\mathbb{R}$, $S = R \otimes_{\mathbb{R}} \mathbb{C}$ and T be a free module over S of dimension $1|1$. We assume, furthermore, that all invertible elements of S posses a square root. A superconformal structure on T is given by an odd submodule $D \subseteq T$ and an isomorphism $\Gamma\colon D \otimes_S D \to {}^{T}\!/_{D}$. Mirroring the notation on TM and $\mathcal{D}$ we will denote by f_z, f_+ a basis that fulfils

1. f_+ generates D,
2. $\Gamma(f_+ \otimes f_+) = f_z$ in ${}^{T}\!/_{D}$.

Any two of those bases are related by a matrix in SCL.

Definition 9.5.1 An even, positive hermitian form on T is called compatible with the superconformal structure if the induced hermitian forms on $D \otimes_S D$ and ${}^{T}\!/_{D} \simeq D^\perp$ coincide.

Lemma 9.5.2 *Let f_+, f_z be a superconformal basis on T with superconformal structure D. A hermitian form h on T is compatible with the superconformal structure, if*

$$ {}_z h_z = -{}_+h_+{}^2 - \frac{\overline{{}_z h_+}\, {}_z h_+}{{}_+ h_+} $$

where ${}_A h_B$ are the entries of the matrix representation of h. Notice that ${}_z h_z$ is real, $\overline{{}_z h_+ = {}_+ h_z}$ and ${}_+ h_+$ is imaginary.

Proof We are first going to construct a superconformal basis F_z, F_+ of T such that $\langle F_+, h, F_+\rangle = \mathrm{i}$ and $\langle F_z, h, F_+\rangle = 0$. To this end let

$$F_z = \frac{\mathrm{i}}{{}_+h_+} f_z + \frac{\mathrm{i}}{{}_+h_+} \frac{\overline{{}_z h_+}}{{}_+h_+} f_+,$$

$$F_+ = \sqrt{\frac{\mathrm{i}}{{}_+h_+}} f_+.$$

It holds that

$$\langle F_+, h, F_+\rangle = \frac{\mathrm{i}}{{}_+h_+} \langle f_+, h, f_+\rangle = \mathrm{i},$$

$$\langle F_z, h, F_+\rangle = \frac{\mathrm{i}}{{}_+h_+}\sqrt{\frac{\mathrm{i}}{{}_+h_+}} \langle f_z, h, f_+\rangle - \sqrt{\frac{\mathrm{i}}{{}_+h_+}} \frac{\mathrm{i}}{{}_+h_+} \frac{{}_z h_+}{{}_+h_+} \langle f_+, h, f_+\rangle = 0.$$

For h to be compatible with the superconformal structure the hermitian forms on $D \otimes D$ and ${}^{T}\!/_{D} \cong D^\perp$ need to coincide. That is $F_+ \otimes F_+$ and F_z need to have the same length.

$$\langle F_+ \otimes F_+, h \otimes h, F_+ \otimes F_+\rangle = -\langle F_+, h, F_+\rangle \langle F_+, h, F_+\rangle = 1$$

$$\begin{aligned} 1 &= \langle F_z, h, F_z\rangle \\ &= -\frac{1}{{}_+h_+{}^2} \langle f_z, h, f_z\rangle + \frac{\mathrm{i}}{{}_+h_+} \frac{\mathrm{i}}{{}_+h_+} \frac{\overline{{}_z h_+}}{{}_+h_+} \langle f_z, h, f_+\rangle \\ &\quad - \frac{\mathrm{i}}{{}_+h_+} \frac{\mathrm{i}}{{}_+h_+} \frac{{}_z h_+}{{}_+h_+} \langle f_+, h, f_z\rangle + \frac{\mathrm{i}}{{}_+h_+} \frac{{}_z h_+}{{}_+h_+} \frac{\mathrm{i}}{{}_+h_+} \frac{\overline{{}_z h_+}}{{}_+h_+} \langle f_+, h, f_+\rangle \\ &= -\frac{1}{{}_+h_+{}^2} \left({}_z h_z + \frac{\overline{{}_z h_+}\, {}_z h_+}{{}_+h_+} \right) \end{aligned}$$

□

Proposition 9.5.3 *Let T be equipped with a superconformal structure D. The following are equivalent:*

i) a hermitian form h compatible with the superconformal structure,
ii) a hermitian form $h|_D$ on D and a splitting of the short exact sequence

$$0 \to D \to T \overset{\curvearrowleft}{\to} {}^{T}\!/_{D} \to 0.$$

That is a direct sum decomposition $T = D \oplus D^\perp$.

Proof Any hermitian form h restricts to a hermitian form $h|_D$ on D and gives a direct sum decomposition $T = D \oplus D^{\perp}$. The hermitian form h that is compatible with the superconformal structure can be reconstructed from $h|_D$ and $T = D \oplus D^{\perp}$ because the hermitian form on $D^{\perp}$ is determined by $h|_D$. □

Definition 9.5.4 We call two hermitian forms h and $\tilde{h}$ related by a *superconformal rescaling* if the hermitian forms $h|_D$ and $\tilde{h}|_D$ are related by an invertible scalar $\lambda \in R$:

$$h|_D = \lambda \tilde{h}|_D$$

We call the two hermitian forms $h|_D$ and $\tilde{h}|_D$ related by a *superconformal change of splitting* if they differ only in the associated splitting of the short exact sequence. We call the two hermitian forms $h|_D$ and $\tilde{h}|_D$ related by a *superconformal transformation* if they differ by the composition of a superconformal rescaling and a superconformal change of splitting.

Consequently, any two hermitian forms which are compatible with the superconformal structure are related by a superconformal transformation. In particular every hermitian form which is compatible with the superconformal structure is related by a superconformal transformation to the standard hermitian form that is given in the basis f_z, f_+ by the matrix

$$\begin{pmatrix} 1 & 0 \\ 0 & \mathrm{i} \end{pmatrix}.$$

Every element of SCL can be written as

$$\begin{pmatrix} A^2 & B \\ 0 & A \end{pmatrix} = \begin{pmatrix} U^2 & 0 \\ 0 & U \end{pmatrix} \begin{pmatrix} V^2 & 0 \\ 0 & V \end{pmatrix} \begin{pmatrix} 1 & Q \\ 0 & 1 \end{pmatrix}$$

with $U \in \mathrm{U}(1)$, $V \in R$ and $Q \in S$. The matrix

$$\begin{pmatrix} V^2 & 0 \\ 0 & V \end{pmatrix}$$

acts as a superconformal rescaling, whereas the matrix

$$\begin{pmatrix} 1 & Q \\ 0 & 1 \end{pmatrix}$$

acts as a superconformal change of splitting. In particular every element of SCL is a composition of a U(1)-transformation with a superconformal one. Notice that the decomposition of elements of G into a U(1)-transformation, a superconformal rescaling and a superconformal change of splitting is unique up to the sign of U and V. In the case that R possesses a subgroup of positive elements R_+ (with multiplication), for example when R is a ring of functions on a supermanifold, V can be assumed to be positive, and the decomposition is unique. In this case the group SCL is given as a semidirect product $\mathrm{SCL} = (\mathrm{U}(1) \times R_+) \ltimes R$.

Let us now consider T as a free module over R of dimension $2|2$. The complex module structure induces an almost complex structure I on T, which is given by the multiplication with i. The submodule D is now of dimension $0|2$ and there is a surjective bilinear map $D \otimes_R D \to {}^T\!/_D$ that commutes with I. The adapted bases to this situation are even vectors f_1, f_2 and odd vectors f_3, f_4 such that

1. the almost complex structure I has the standard form with respect to this basis, that is,

$$\begin{aligned} \mathrm{I}\, f_1 &= f_2, & \mathrm{I}\, f_2 &= -f_1, \\ \mathrm{I}\, f_3 &= f_4, & \mathrm{I}\, f_4 &= -f_3. \end{aligned}$$

2. D is generated by f_3 and f_4.
3. The coefficients of $\Gamma\left(f_\alpha \otimes f_\beta\right) = \Gamma_{\alpha\beta}^k f_k$ are constant such that $\Gamma_{33}^1 = 1$, see also Sect. A.1.

Again, any two adapted frames are related by a matrix in SCL, where the complex number i acts on T via I.

Proposition 9.5.5 *Any hermitian form which is compatible with the superconformal structure induces a positive symmetric R-bilinear form b on T such that $b(\mathrm{I}t, \mathrm{I}t') = b(t, t')$. With respect to the adapted basis f_A the bilinear form b is given by a matrix of the form*

$$\begin{pmatrix} {}_1b_1 & 0 & {}_1b_3 & {}_1b_4 \\ 0 & {}_2b_2 & {}_2b_3 & {}_2b_4 \\ {}_3b_1 & {}_3b_2 & 0 & {}_3b_4 \\ {}_4b_1 & {}_4b_2 & {}_4b_3 & 0 \end{pmatrix},$$

where ${}_Ab_B = (-1)^{p(A)p(B)}\, {}_Bb_A$, ${}_1b_1 = {}_2b_2$, ${}_1b_3 = {}_2b_4$, ${}_1b_4 = -\,{}_2b_3$ and

$${}_1b_1 = {}_3b_4{}^2 - 2\frac{{}_1b_3 \cdot {}_1b_4}{{}_3b_4}.$$

Thus, the bilinear form b is completely determined by the three scalars ${}_1b_3$, ${}_1b_4$ and ${}_3b_4$, which all are in R. These scalars are related to the matrix entries of the hermitian form h by

$$\begin{aligned} {}_zh_z &= {}_1b_1, \\ {}_zh_+ &= {}_1b_3 + \mathrm{i}\,{}_1b_4, \\ {}_+h_+ &= \mathrm{i}\,{}_3b_4. \end{aligned}$$

A set of orthonormal frames is given by

$$F_a = \frac{1}{{}_3b_4} f_a - \frac{1}{{}_3b_4} {}_a b_\beta {}^\beta b^\gamma f_\gamma,$$

$$F_\alpha = \frac{1}{\sqrt{{}_3b_4}} f_\alpha,$$

where ${}^\beta b^\gamma$ *is the inverse to* ${}_\alpha b_\beta$*, that is* ${}_\alpha b_\beta {}^\beta b^\gamma = \delta_\alpha^\gamma$*. The frames* F_a *and* F_α *are the real and imaginary part of the complex frames* F_z *and* F_+*, that is*

$$F_z = \frac{1}{2}(F_1 - \mathrm{i}F_2), \qquad F_+ = \frac{1}{2}(F_3 - \mathrm{i}F_4).$$

Identifying ${}^T\!/_D$ with $D^\perp$, we can introduce a bilinear map $\gamma\colon D^\perp \otimes D \to D$ by setting $b(\Gamma(t,t'),s) = -b(\gamma(s)t,t')$ for all t, $t' \in D$ and $s \in D^\perp$. The following is then immediate from the description of the superconformal transformations on hermitian forms compatible with the superconformal structure:

Proposition 9.5.6 *Let b be a metric compatible with the superconformal structure on T and denote the orthogonal complement of D with respect to b by* $D^\perp$*. For any other metric* $\tilde{b}$ *compatible with the superconformal structure on T there exists a* $\lambda \in R$ *and* $l \in D$ *such that*

$$\begin{aligned}
\tilde{b}(t,t') &= \lambda^2 b(t,t'),\\
\tilde{b}(s,t) &= -\lambda^2 b(\gamma(s)l,t),\\
\tilde{b}(s,s') &= \lambda^4 b(s,s') + \lambda^2 b(\gamma(s)l,\gamma(s')l),
\end{aligned}$$

for all s, $s' \in D^\perp$ *and* t, $t' \in D$*. Let* F_A *be an orthonormal frame with respect to b and* $l = l^\mu F_\mu$*. A* $\tilde{b}$*-orthonormal frame is then given by*

$$\tilde{F}_a = \frac{1}{\lambda^2}\left(F_a + l^\mu \gamma_{a\mu}{}^\nu F_\nu\right), \qquad \tilde{F}_\alpha = \frac{1}{\lambda} F_\alpha.$$

Denote by p and $\tilde{p}$ *the orthogonal projection* $T \to D$ *induced by b and* $\tilde{b}$ *respectively. Then* $p(t) - \tilde{p}(t) = \gamma(p(t))l$ *for all* $t \in T$*.*

Let us now come back to super Riemann surfaces.

Definition 9.5.7 A positive hermitian form h on a super Riemann surface M is called compatible with the super Riemann surface structure or superconformal, if the induced hermitian forms on $\mathcal{D} \otimes_{\mathbb{C}} \mathcal{D}$ and ${}^{TM}\!/_{\mathcal{D}}$ coincide.

Any such hermitian form gives rise to a reduction of the structure group of M to U(1). The tangent bundle is associated to the U(1)-principal bundle via the representation

$$\begin{aligned} \rho\colon \mathrm{U}(1) &\to \mathrm{GL}_{\mathcal{O}_B\otimes\mathbb{C}}(1|1) \subset \mathrm{GL}_{\mathcal{O}_B}(2|2) \\ U &\mapsto \begin{pmatrix} U^2 & 0 \\ 0 & U \end{pmatrix} \end{aligned}.$$

Every other hermitian form $\tilde{h}$ on M, which is compatible with the super Riemann surface structure, is related to h by a combination of a superconformal rescaling and a superconformal change of splitting. Equivalently, a reduction of the structure group to U(1) can also be given by a metric m compatible with the super Riemann surface structure, compare Proposition 9.5.5. Such a metric may also be called a superconformal metric.

Example 9.5.8 (Hermitian Form on $\mathbb{C}^{1|1}$) Let (z, θ) be the standard coordinates on $\mathbb{C}^{1|1}$ and (∂_z, D) the corresponding superconformal frame. The standard hermitian form h on the super Riemann surface $\mathbb{C}^{1|1}$ is given by

$$h(\partial_z, \partial_z) = 1, \qquad h(\partial_z, D) = 0, \qquad h(D, D) = \mathrm{i}.$$

With respect to the dual frames $(dz + \theta\, d\theta, d\theta)$ the standard hermitian form h is given by

$$h = \overline{(dz + \theta\, d\theta)} \otimes (dz + \theta\, d\theta) + \mathrm{i}\overline{d\theta} \otimes d\theta.$$

The most general hermitian form is

$$\begin{aligned} h = {}& \lambda^4\, \overline{(dz + \theta\, d\theta)} \otimes (dz + \theta\, d\theta) + \tau\, \overline{(dz + \theta\, d\theta)} \otimes d\theta \\ &+ \overline{\tau}\overline{d\theta} \otimes (dz + \theta\, d\theta) + \lambda^2 \mathrm{i}\overline{d\theta} \otimes d\theta \end{aligned}$$

for an even real function λ and an odd complex function τ on $\mathbb{C}^{1|1}$.

Proposition 9.5.9 *On every super Riemann surfaces exists a hermitian form compatible with the super Riemann surface structure.*

Proof Let $\{U_\alpha\}$ be a cover of M by relative coordinate charts and $\{\varphi_\alpha\}$ a partition of unity subordinate to $\{U_\alpha\}$. On any open set U_α a hermitian form h_α on $\mathcal{D}$ can be chosen, for example the standard hermitian form on $\mathcal{D}$ with respect to the frame D of some superconformal coordinates. The hermitian form

$$h|_{\mathcal{D}} = \sum_\alpha \varphi_\alpha h_\alpha$$

is a hermitian form on $\mathcal{D}$.

Similarly, on every open set U_α we can choose a splitting $\iota_\alpha \colon {}^{TU_\alpha}\!/_{\mathcal{D}} \to TU_\alpha$, for example given by $\iota_\alpha(\partial_z) = \partial_z$ in some superconformal coordinates. The map

$$\iota = \sum_\alpha \varphi_\alpha p_\alpha \colon {}^{TM}\!/_{\mathcal{D}} \to TM$$

is a splitting of the short exact sequence

$$0 \longrightarrow \mathcal{D} \longrightarrow TM \overset{\iota}{\underset{\longleftarrow}{\longrightarrow}} {}^{TM}\!/_{\mathcal{D}} \longrightarrow 0.$$

By Proposition 9.5.3, p and $h|_{\mathcal{D}}$ characterize a hermitian form h on M that is compatible with the super Riemann surface structure. □

Example 9.5.10 (Hyperbolic Hermitian Form on $S\mathbb{H}$) Let (∂_z, D) be the standard superconformal frames on $S\mathbb{H}$ with respect to the standard relative coordinates (z, θ). The hyperbolic hermitian form on the super Riemann surface $S\mathbb{H}$ is given by the orthonormal frames

$$F_+ = \sqrt{\operatorname{Im} z + \frac{1}{2}\mathrm{i}\theta\overline{\theta}}\, D, \quad F_z = \left(\operatorname{Im} z + \frac{1}{2}\mathrm{i}\theta\overline{\theta}\right)\partial_z + \frac{\mathrm{i}}{2}\left(\overline{\theta} - \theta\right) D. \tag{9.5.11}$$

Notice that all coefficient functions in (9.5.11) are real. The dual frames of (9.5.11) are

$$\begin{aligned} F^z &= \frac{1}{\operatorname{Im} z + \frac{1}{2}\mathrm{i}\theta\overline{\theta}}\left(dz + \theta\, d\theta\right), \\ F^+ &= \frac{1}{\sqrt{\operatorname{Im} z + \frac{1}{2}\mathrm{i}\theta\overline{\theta}}}\, d\theta + \frac{\mathrm{i}\left(\theta - \overline{\theta}\right)}{2\left(\sqrt{\operatorname{Im} z + \frac{1}{2}\mathrm{i}\theta\overline{\theta}}\right)^3}\left(dz + \theta\, d\theta\right). \end{aligned}$$

In this form the hyperbolic hermitian form on $S\mathbb{H}$ can be found in Crane and Rabin (1988, Equation (3.6)). While the hyperbolic hermitian form h on $S\mathbb{H}$ is clearly a nilpotent extension of the hyperbolic metric on $\mathbb{H}$, its main motivation is given by the following Lemma 9.5.12.

Lemma 9.5.12 *The hyperbolic hermitian form h on $S\mathbb{H}$ is invariant under the action of* $\operatorname{Sp}_{\mathcal{O}_B}(2|1)$ *on $S\mathbb{H}$. Consequently, the hyperbolic hermitian form h on induces a standard hermitian form on all super Riemann surface of genus $p > 1$.*

Proof Let (z, θ) be the hyperbolic superconformal coordinates on $S\mathbb{H}$ and let $f\colon S\mathbb{H} \to S\mathbb{H}$ be a superconformal automorphism induced by an element of $\mathrm{Sp}_{\mathcal{O}_B}(2|1)$ as in Example 9.4.3. Denote by (u, η) the image of (z, θ) under f:

$$f^{\#}u = \frac{az+b}{cz+d} \pm \theta \frac{\gamma z + \delta}{(cz+d)^2}, \qquad f^{\#}\eta = \frac{\gamma z+\delta}{cz+d} \pm \theta \frac{1}{cz+d}.$$

Recall that $df^{\vee}$ acts by

$$df^{\vee}\begin{pmatrix} du + \eta\, d\eta \\ d\eta \end{pmatrix} = \begin{pmatrix} A^2 & 0 \\ DA & A \end{pmatrix}\begin{pmatrix} dz + \theta\, d\theta \\ d\theta \end{pmatrix},$$

where $A = \frac{1}{cz+d}\left(\pm 1 + \theta \frac{d\gamma - \delta c}{cz+d}\right)$. Abbreviating $\lambda = \frac{1}{\sqrt{\operatorname{Im} u + \frac{1}{2}\mathrm{i}\eta\overline{\eta}}}$ and $D_\eta = \partial_\eta + \eta\partial_u$, notice that λ and $D_\eta\lambda$ are real:

$$\begin{aligned} D_\eta\lambda &= -\frac{1}{2}\frac{1}{\sqrt{\operatorname{Im} u + \frac{1}{2}\mathrm{i}\eta\overline{\eta}}^{\,3}} D_\eta\left(\frac{\mathrm{i}}{2}\left(\overline{u} - u + \eta\overline{\eta}\right)\right) \\ &= \frac{\mathrm{i}\left(\eta - \overline{\eta}\right)}{4\left(\sqrt{\operatorname{Im} u + \frac{1}{2}\mathrm{i}\eta\overline{\eta}}\right)^3}. \end{aligned}$$

We obtain

$$\begin{aligned} h &= F^{\overline{u}} \otimes F^{u} + F^{-} \otimes F^{+} \\ &= \lambda^4 \left(d\overline{u} + \overline{\eta}\, d\overline{\eta}\right) \otimes \left(du + \eta\, d\eta\right) \\ &\quad + \left(\lambda\, d\overline{\eta} + 2\overline{D_\eta}\lambda\left(d\overline{u} + \overline{\eta}\, d\overline{\eta}\right)\right) \otimes \left(\lambda\, d\eta + 2D_\eta\lambda\left(du + \eta\, d\eta\right)\right) \\ &= \lambda^4 \left(d\overline{u} + \overline{\eta}\, d\overline{\eta}\right) \otimes \left(du + \eta\, d\eta\right) - 2\lambda D_\eta\lambda\, d\overline{\eta} \otimes \left(du + \eta\, d\eta\right) \\ &\quad + 2\lambda\overline{D_\eta}\lambda\left(d\overline{u} + \overline{\eta}\, d\overline{\eta}\right) \otimes d\eta + \lambda^2\, d\overline{\eta} \otimes d\eta, \end{aligned}$$

and hence

$$\begin{aligned} h_f &= f^{\#}\lambda^4\overline{A}^2 A^2 \left(d\overline{z} + \overline{\theta}\, d\overline{\theta}\right) \otimes \left(dz + \theta\, d\theta\right) \\ &\quad - 2f^{\#}\left(\lambda D_\eta\lambda\right)\left(\overline{DA}\left(d\overline{z} + \overline{\theta}\, d\overline{\theta}\right) + \overline{A}\, d\overline{\theta}\right) \otimes A^2\left(dz + \theta\, d\theta\right) \\ &\quad + 2f^{\#}\left(\lambda\overline{D_\eta}\lambda\right)\overline{A}^2\left(d\overline{z} + \overline{\theta}\, d\overline{\theta}\right) \otimes \left(DA\left(dz + \theta\, d\theta\right) + A\, d\theta\right) \\ &\quad + f^{\#}\lambda^2\left(\overline{DA}\left(d\overline{z} + \overline{\theta}\, d\overline{\theta}\right) + \overline{A}\, d\overline{\theta}\right) \otimes \left(DA\left(dz + \theta\, d\theta\right) + A\, d\theta\right) \end{aligned}$$

$$= \Big(f^{\#}\lambda^4\overline{A}^2A^2 - 2f^{\#}\left(\lambda D_\eta\lambda\right)A^2\overline{DA} + 2f^{\#}\left(\lambda\overline{D_\eta}\lambda\right)\overline{A}^2DA$$
$$+ f^{\#}\lambda^2\overline{DA}DA\Big)\left(d\overline{z} + \overline{\theta}\,d\theta\right)\otimes\left(dz + \theta\,d\theta\right)$$
$$- \Big(2f^{\#}\left(\lambda D_\eta\lambda\right)\overline{A}A^2 + f^{\#}\lambda^2\overline{A}DA\Big)\,d\overline{\theta}\otimes\left(dz + \theta\,d\theta\right)$$
$$+ \Big(2f^{\#}\left(\lambda\overline{D_\eta}\lambda\right)\overline{A}^2A + f^{\#}\lambda^2A\overline{DA}\Big)\left(d\overline{z} + \overline{\theta}\,d\overline{\theta}\right)\otimes d\theta$$
$$+ f^{\#}\lambda^2\overline{A}A\,d\overline{\theta}\otimes d\theta.$$

In order to prove $h_f = h$ it remains to show

$$\frac{1}{\operatorname{Im} z + \frac{1}{2}\mathrm{i}\theta\overline{\theta}} = f^{\#}\lambda^2\overline{A}A \tag{9.5.13}$$

$$\frac{\mathrm{i}\left(\theta - \overline{\theta}\right)}{2\left(\operatorname{Im} z + \frac{1}{2}\mathrm{i}\theta\overline{\theta}\right)^2} = 2f^{\#}\left(\lambda D_\eta\lambda\right)\overline{A}A^2 + f^{\#}\lambda^2\overline{A}DA \tag{9.5.14}$$

$$-\frac{\mathrm{i}}{2\left(\operatorname{Im} z + \frac{1}{2}\mathrm{i}\theta\overline{\theta}\right)^2} = -\frac{\mathrm{i}}{2}f^{\#}\lambda^4A^2\overline{A}^2 + 2f^{\#}\left(\lambda\overline{D_\eta}\lambda\right)\overline{A}^2DA$$
$$- 2f^{\#}\left(\lambda D_\eta\lambda\right)A^2\overline{DA} + f^{\#}\lambda^2\overline{DA}DA \tag{9.5.15}$$

Since both sides of (9.5.13) are invertible, we show its inverse instead. The product of

$$f^{\#}\lambda^{-2} = f^{\#}\left(\operatorname{Im} u + \frac{1}{2}\mathrm{i}\eta\overline{\eta}\right) = \frac{\mathrm{i}}{2}f^{\#}\left(\overline{u} - u + \eta\overline{\eta}\right)$$
$$= \frac{\mathrm{i}}{2}\left(\frac{a\overline{z}+b}{c\overline{z}+d} \pm \overline{\theta}\frac{\gamma\overline{z}+\delta}{(c\overline{z}+d)^2} - \frac{az+b}{cz+d} \mp \theta\frac{\gamma z+\delta}{(cz+d)^2}\right.$$
$$\left. + \left(\frac{\gamma z+\delta}{cz+d} \pm \theta\frac{1}{cz+d}\right)\left(\frac{\gamma\overline{z}+\delta}{c\overline{z}+d} \pm \overline{\theta}\frac{1}{c\overline{z}+d}\right)\right)$$
$$= \frac{\mathrm{i}}{2|cz+d|^4}\Big(\big((a\overline{z}+b)(cz+d) - (az+b)(c\overline{z}+d)$$
$$+ (\gamma z+\delta)(\gamma\overline{z}+\delta)\big)|cz+d|^2$$
$$\pm\overline{\theta}\left((\gamma\overline{z}+\delta)(cz+d)^2 - (\gamma z+\delta)|cz+d|^2\right)$$
$$\pm\theta\left(-(\gamma z+\delta)(c\overline{z}+d)^2 + (\gamma\overline{z}+\delta)|cz+d|^2\right) + \theta\overline{\theta}|cz+d|^2\Big)$$

$$= \frac{\mathrm{i}}{2|cz+d|^4}\Big((\overline{z}-z)\,|cz+d|^2 \pm \overline{\theta}\,(\gamma d - \delta c)\,(\overline{z}-z)\,(cz+d)$$
$$\pm\, \theta\,(\gamma d - c\delta)\,(\overline{z}-z)\,(c\overline{z}+d) + \theta\overline{\theta}|cz+d|^2\Big)$$

and

$$\overline{A}^{-1}A^{-1} = |cz+d|^2\left(\pm 1 - \overline{\theta}\frac{d\gamma-\delta c}{c\overline{z}+d}\right)\left(\pm 1 - \theta\frac{d\gamma-\delta c}{cz+d}\right)$$
$$= |cz+d|^2\left(1 \mp \overline{\theta}\frac{d\gamma-\delta c}{c\overline{z}+d} \mp \theta\frac{d\gamma-\delta c}{cz+d} - \theta\overline{\theta}\frac{\gamma\delta c d}{|cz+d|^2}\right).$$

gives the desired result. Equation (9.5.14) follows from (9.5.13) by differentiation. Indeed,

$$\frac{\mathrm{i}\,(\theta-\overline{\theta})}{2\left(\operatorname{Im} z + \frac{1}{2}\mathrm{i}\theta\overline{\theta}\right)^2} = D\left(\frac{1}{\operatorname{Im} z + \frac{1}{2}\mathrm{i}\theta\overline{\theta}}\right) = D\left(f^{\#}\lambda^2\overline{A}A\right)$$
$$= 2Af^{\#}\left(\lambda D_\eta\lambda\right)\overline{A}A + f^{\#}\lambda^2\overline{A}DA.$$

Finally, (9.5.15) follows from a second order derivative of (9.5.13):

$$-\frac{\mathrm{i}}{2\left(\operatorname{Im} z + \frac{1}{2}\mathrm{i}\theta\overline{\theta}\right)^2} = \overline{D}D\frac{1}{\operatorname{Im} z + \frac{1}{2}\mathrm{i}\theta\overline{\theta}} = \overline{D}D\left(f^{\#}\lambda^2 A\overline{A}\right)$$
$$= f^{\#}\left(\overline{D_\eta}D_\eta\lambda^2\right)A^2\overline{A}^2 + 2f^{\#}\left(\lambda\overline{D_\eta}\lambda\right)\overline{A}^2 DA$$
$$- 2f^{\#}\left(\lambda D_\eta\lambda\right)A^2\overline{DA} + f^{\#}\lambda^2\overline{DA}DA$$
$$= -\frac{\mathrm{i}}{2}f^{\#}\lambda^4A^2\overline{A}^2 + 2f^{\#}\left(\lambda\overline{D_\eta}\lambda\right)\overline{A}^2 DA$$
$$- 2f^{\#}\left(\lambda D_\eta\lambda\right)A^2\overline{DA} + f^{\#}\lambda^2\overline{DA}DA \qquad \square$$

Remark 9.5.16 Every Riemann surface is determined by a conformal class of metrics, see Sect. A.2. For super Riemann surfaces, a similar statement is the following: The super Riemann surface structure is completely determined by an integrable U(1)-reduction of the structure group. All U(1)-reductions that are related by a superconformal transformation give the same super Riemann surface structure. Notice that by Theorem 9.2.3, if a U(1)-reduction of the structure group is integrable, all U(1)-structures in the same superconformal class are integrable.

However, an arbitrary metric or hermitian form on a 2|2-dimensional manifold M does not determine a super Riemann surface structure, since both do not determine the distribution $\mathcal{D}$. Expressed differently: a general metric reduces the structure group of M to O(2|2), and a general hermitian form reduces the structure group to $\mathrm{U}(1|1)$. However, neither O(2|2) nor $\mathrm{U}(1|1)$ are subgroups of neither $\mathrm{Tr}_{\mathbb{C}}(1|1)$ nor SCL.

Chapter 10
Connections on Super Riemann Surfaces

The goal of this chapter is to study the torsion tensor of connections on the reductions of the frame bundle of a super Riemann surface M to $\mathrm{Tr}_{\mathbb{C}}(1|1)$, SCL and $\mathrm{U}(1)$ respectively. Intuitively, the smaller the structure group, the more of the torsion tensor is determined by the geometry of the frame bundle of M and less by the choice of the connection. The main interest for connections on super Riemann surfaces is that $\mathrm{U}(1)$-connections and their torsion tensors enter in supersymmetry transformations in the following chapters.

However, the study of connections on super Riemann surfaces allows also to compare the theory of super Riemann surfaces to the earlier approaches of supergravity where the geometry is specified by constraints on the torsion of a connection. The following torsion constraints can be found in the literature:

$$T_{ab}{}^{c} = 0 \qquad T_{\alpha\beta}{}^{\gamma} = 0 \qquad T_{\alpha\beta}{}^{c} = 2\Gamma^{c}_{\alpha\beta} \tag{10.0.1}$$

Compare, for instance, Howe (1979, Equation 2.13); D'Hoker and Phong (1988, Equation 3.11a), but note that the definition of connection on supermanifolds does not necessarily coincide with the one we use here.

In Sect. 10.1, we will see that the integrability conditions in the Theorem of Giddings and Nelson (Theorem 9.2.3) can be expressed as conditions on the torsion of any connection of the SCL-principal bundle. It is known, see Giddings and Nelson (1987, 1988) that the torsion constraints (10.0.1) imply integrability of the G-structure. For any integrable G-structure there is a connection fulfilling the torsion constraints, but this connection is not unique.

The case of $\mathrm{U}(1)$-structures on super Riemann surfaces, treated in Sect. 10.2, is different. Not every choice of $\mathrm{U}(1)$-structure on a super Riemann surface allows for a connection fulfilling (10.0.1). Rather, the torsion constraints can be read as a restriction on the class of allowable superconformal metrics.

E. Keßler, *Supergeometry, Super Riemann Surfaces and the Superconformal Action Functional*, Lecture Notes in Mathematics 2230,
https://doi.org/10.1007/978-3-030-13758-8_10

10.1 SCL-Connections and Integrability

Recall that the group $\mathrm{Tr}_{\mathbb{C}}(1|1)$ is given by

$$\mathrm{Tr}_{\mathbb{C}}(1|1) = \left\{ \begin{pmatrix} A & B \\ 0 & C \end{pmatrix} \;\middle|\; A, B, C \in \mathbb{C} \right\} \subseteq \mathrm{GL}_{\mathbb{C}}(1|1) \subseteq \mathrm{GL}_{\mathbb{R}}(2|2)$$

and SCL is the subset of $\mathrm{Tr}_{\mathbb{C}}(1|1)$ where $A = C^2$. Consequently, the Lie algebra of $\mathrm{Tr}_{\mathbb{C}}(1|1)$ is given by upper triangular complex matrices, whereas the Lie algebra $\mathfrak{scl}$ of SCL is given by

$$\mathfrak{scl} = \left\{ \begin{pmatrix} 2A & B \\ 0 & A \end{pmatrix} \;\middle|\; A, B \in \mathbb{C} \right\} \subseteq \mathrm{Mat}_{\mathbb{C}}(1|1) \subseteq \mathrm{Mat}_{\mathbb{R}}(2|2).$$

Any connection on the SCL-principal bundle is given locally by a differential form with values in $\mathfrak{scl}$. We obtain the following properties for the associated covariant derivatives:

Lemma 10.1.1 *Any covariant derivative associated to a connection on the* $\mathrm{Tr}_{\mathbb{C}}(1|1)$*-reduction of the frame bundle has the following properties:*

i) The connection is compatible with the almost complex structure J, that is,

$$\nabla J = 0. \tag{10.1.2}$$

ii) The covariant derivative preserves the distribution $\mathcal{D}$ that is for any vector field X, we have that

$$\nabla_X \mathcal{D} \subseteq \mathcal{D}. \tag{10.1.3}$$

Furthermore, for covariant derivatives associated to connections on the SCL*-reduction of the frame bundle of M it holds in addition*

iii) There is an induced covariant derivative on ${}^{TM}\!/_{\mathcal{D}}$. The induced covariant derivatives on $\mathcal{D} \otimes \mathcal{D}$ and ${}^{TM}\!/_{\mathcal{D}}$ are the same.

Proof Equations (10.1.2) and (10.1.3) directly follow from the fact that the covariant derivative applied to a $\mathrm{Tr}_{\mathbb{C}}(1|1)$-frame is an endomorphism from the Lie algebra of $\mathrm{Tr}_{\mathbb{C}}(1|1)$. For property iii) recall that

$$\begin{aligned} \mathcal{D} \otimes \mathcal{D} &\to {}^{TM}\!/_{\mathcal{D}} \\ F_\beta \otimes F_\gamma &\mapsto \frac{1}{2}[F_\beta, F_\gamma] = \frac{1}{2} d^d_{\beta\gamma} F_d \end{aligned} \tag{10.1.4}$$

is an isomorphism. For SCL-frames it holds that $d^d_{\beta\gamma} = 2\Gamma^d_{\beta\gamma}$, thus constant. Furthermore, the induced connection on $\mathcal{D} \otimes \mathcal{D}$ is given by

$$\begin{aligned}\nabla^{\mathcal{D}\otimes\mathcal{D}}_{F_A} F_\beta \otimes F_\gamma &= \left(\nabla_{F_A} F_\beta\right) \otimes F_\gamma + (-1)^{p(A)} F_\beta \otimes \left(\nabla_{F_A} F_\gamma\right)\\ &= \omega_{A\beta}{}^{\delta} F_\delta \otimes F_\gamma + \omega_{A\gamma}{}^{\delta} F_\beta \otimes F_\delta,\end{aligned}$$

which is mapped under the isomorphism (10.1.4) to

$$\left(\omega_{A\beta}{}^{\delta}\Gamma^k_{\delta\gamma} + \omega_{A\gamma}{}^{\delta}\Gamma^k_{\beta\delta}\right) F_k.$$

On the other hand

$$\nabla^{TM/\mathcal{D}}_{F_A}\left(\Gamma^d_{\beta\gamma} F_d\right) = \Gamma^d_{\beta\gamma}\omega_{Ad}{}^{k} F_k.$$

The equality

$$\omega_{A\beta}{}^{\delta}\Gamma^k_{\delta\gamma} + \omega_{A\gamma}{}^{\delta}\Gamma^k_{\beta\delta} = \Gamma^d_{\beta\gamma}\omega_{Ad}{}^{k}$$

follows because ω is a $\mathfrak{scl}$-valued form. □

Proposition 10.1.5 *For the torsion tensor of any connection on the* $\mathrm{Tr}_{\mathbb{C}}(1|1)$*-reduced frame bundle, it holds*

$$\begin{aligned}&T(\mathrm{I}X, \mathrm{I}Y) - \mathrm{I}T(X, \mathrm{I}Y) - \mathrm{I}T(\mathrm{I}X, Y) - T(X, Y)\\ &\quad= -[\mathrm{I}X, \mathrm{I}Y] + \mathrm{I}[X, \mathrm{I}Y] + \mathrm{I}[\mathrm{I}X, Y] + [X, Y] = -N_{\mathrm{I}}(X, Y) \qquad (10.1.6)\end{aligned}$$

for all vector fields X and Y. Furthermore, with respect to any $\mathrm{Tr}_{\mathbb{C}}(1|1)$*-frame it holds that*

$$\begin{aligned}T^c_{\alpha\beta} &= -d^c_{\alpha\beta},\\ T^1_{\alpha 1} - T^2_{\alpha 2} &= -\left(d^1_{\alpha 1} - d^2_{\alpha 2}\right), \qquad (10.1.7)\\ T^2_{\alpha 1} + T^1_{\alpha 2} &= -\left(d^2_{\alpha 1} + d^1_{\alpha 2}\right).\end{aligned}$$

Proof Equation (10.1.6) holds for all connections compatible with an almost complex structure I, see Proposition 6.10.6. The equations (10.1.7) are a consequence of Eq. (10.1.3). It holds for any $\mathrm{Tr}_{\mathbb{C}}(1|1)$-frames that

$$T(F_\alpha, F_\beta) = \nabla_{F_\alpha} F_\beta + \nabla_{F_\beta} F_\alpha - [F_\alpha, F_\beta] = -d^c_{\alpha\beta} F_c + \left(\omega_{\alpha\beta}{}^{\gamma} + \omega_{\beta\alpha}{}^{\gamma} - d^\gamma_{\alpha\beta}\right) F_\gamma$$

and

$$\begin{aligned}T(F_\alpha, F_b) &= \nabla_{F_\alpha} F_b + \nabla_{F_b} F_\alpha - [F_\alpha, F_b] \\ &= \left(\omega_{\alpha b}{}^c - d^c_{\alpha b}\right) F_c + \left(\omega_{\alpha b}{}^\gamma - \omega_{b\alpha}{}^\gamma - d^\gamma_{\alpha b}\right) F_\gamma.\end{aligned}$$

The claim follows because ω is a form with values in the Lie algebra of $\mathrm{Tr}_{\mathbb{C}}(1|1)$:

$$\begin{aligned}0 &= \omega_{\alpha 1}{}^1 - \omega_{\alpha 2}{}^2 = T^1_{\alpha 1} + d^1_{\alpha 1} - T^2_{\alpha 2} - d^2_{\alpha 2} \\ 0 &= \omega_{\alpha 1}{}^2 + \omega_{\alpha 2}{}^1 = T^2_{\alpha 1} + d^2_{\alpha 1} + T^1_{\alpha 2} + d^1_{\alpha 2}\end{aligned}$$

□

Corollary 10.1.8 *Let P be a* SCL-*reduction of the structure group of a* 2|2-*dimensional supermanifold M. P is integrable, that is, defines a super Riemann surface if for any connection on P the torsion tensor fulfils*

$$\begin{aligned}T(\mathrm{I}X, \mathrm{I}Y) - \mathrm{I}T(X, \mathrm{I}Y) - \mathrm{I}T(\mathrm{I}X, Y) - T(X, Y) &= 0, \\ T^1_{\alpha 1} - T^2_{\alpha 2} &= 0, \\ T^2_{\alpha 1} + T^1_{\alpha 2} &= 0, \\ T_{\alpha\beta}{}^c &= -2\Gamma^c_{\alpha\beta}.\end{aligned} \tag{10.1.9}$$

Here X, Y are vector fields on M and indices refer to an arbitrary SCL-*frame F_A.*

Proof By Proposition 10.1.5 the conditions (10.1.9) are conditions on the commutators of the SCL-frame F_A and hence independent of the connection. The resulting commutator conditions are sufficient to guarantee the integrability of P by Proposition 9.2.6. □

By Lemma 9.2.8 it is sufficient to check the subset of the full integrability conditions involving $d^C_{\alpha\beta}$, the others follow from Jacobi identities. Consequently, the supergravity torsion constraints given in Eq. (10.0.1) are sufficient to guarantee the integrability of the SCL-structure. This fact can be found in Giddings and Nelson (1987). Similarly, in Lott (1990), the supergravity torsion constraints in different dimensions are interpreted as first order flatness of the SCL-structure.

The integrability conditions constitute only a subset of the supergravity torsion constraints. We will now verify that the remaining conditions can indeed be fulfilled in the case of a connection on the SCL-reduction of the structure group. Let a connection on the SCL-principal bundle P be given by the associated covariant derivative ∇ on TM. Any other connection has then a covariant derivative $\tilde{\nabla} = \nabla + A$ for some $A \in \Omega^1(M, \operatorname{End} TM)$. For all vector fields X the endomorphism $A(X)$ has a matrix with respect to any SCL-frame F_A that lies in the

Lie algebra $\mathfrak{scl}$. The torsion tensor of $\tilde{\nabla}$ is given by

$$\tilde{T}_{ab}{}^{c} = T_{ab}{}^{c} + A_{ab}{}^{c} - A_{ba}{}^{c} \qquad \tilde{T}_{ab}{}^{\gamma} = T_{ab}{}^{\gamma} + A_{ab}{}^{\gamma} - A_{ba}{}^{\gamma}$$
$$\tilde{T}_{\alpha b}{}^{c} = T_{\alpha b}{}^{c} + A_{\alpha b}{}^{c} \qquad \tilde{T}_{\alpha b}{}^{\gamma} = T_{\alpha b}{}^{\gamma} + A_{\alpha b}{}^{\gamma} - A_{b\alpha}{}^{\gamma}$$
$$\tilde{T}_{\alpha\beta}{}^{c} = T_{\alpha\beta}{}^{c} \qquad \tilde{T}_{\alpha\beta}{}^{\gamma} = T_{\alpha\beta}{}^{\gamma} + A_{\alpha\beta}{}^{\gamma} + A_{\beta\alpha}{}^{\gamma}$$

Consequently, there is enough freedom to choose a connection that fulfills the supergravity torsion constraints (10.0.1). However such a connection will not be unique, as only the (super) anti-symmetric part of A enters into the torsion.

10.2 U(1)-Connections on Super Riemann Surfaces

Usually the supergravity torsion constraints are considered for U(1)-connections instead of SCL-connections. First, we will treat $U(1)$-connections in a similar way as SCL-connections were treated in Sect. 10.1 and obtain algebraic properties and restrictions to the torsion. In a second step we will consider the change of connection under a change of superconformal metric.

Let $P_{\mathrm{U}(1)}$ be a given reduction of the frame bundle of a $2|2$-dimensional supermanifold that sits inside a SCL-reduction of the frame bundle P as explained in Sect. 9.5.

Lemma 10.2.1 *Let any connection on $P_{\mathrm{U}(1)}$ be given by the associated covariant derivative on TM. Then in addition to the properties i), ii) and iii) of Lemma 10.1.1 we have that*

iv) The covariant derivative is metric. That is,

$$X\left\langle Y, Z\right\rangle = \left\langle \nabla_X Y, Z\right\rangle + (-1)^{p(X)p(Y)}\left\langle Y, \nabla_X Z\right\rangle .$$

v) The covariant derivative preserves $\mathcal{D}^{\perp}$, that is,

$$\nabla_X \mathcal{D}^{\perp} \subseteq \mathcal{D}^{\perp}.$$

Proof The result that the connections on the $U(1)$-reduction of the frame bundle give rise to metric covariant derivatives is standard, see also Proposition 6.9.5. The fact that the bundle $\mathcal{D}^{\perp}$ is preserved under the connection is then an easy consequence of the fact that $\mathcal{D}$ is preserved. □

Proposition 10.2.2 *The torsion tensor of any connection on $P_{\mathrm{U}(1)}$ fulfills in addition to the properties from Proposition 10.1.5*

$$
\begin{aligned}
T_{ab}{}^{\mu} &= -d_{ab}^{\mu} \\
T_{a3}{}^{3} + d_{a3}^{3} &= T_{a4}{}^{4} + d_{a4}^{4} = 0 \\
T_{a3}{}^{4} + T_{a4}{}^{3} &= -\left(d_{a3}^{4} + d_{a4}^{3}\right) \\
T_{\alpha 1}{}^{1} + d_{\alpha 1}^{1} &= T_{\alpha 2}{}^{2} + d_{\alpha_2}^{2} = 0 \\
T_{33}{}^{3} + d_{33}^{3} &= T_{44}{}^{4} + d_{44}^{4} = 0
\end{aligned}
$$

Here again all indices refer to a U(1)*-frame.*

Proof The covariant derivative preserves $\mathcal{D}$ and $\mathcal{D}^{\perp}$. Thus

$$
\begin{aligned}
T(F_a, F_\beta) &= \nabla_{F_a} F_\beta - \nabla_{F_\beta} F_a - [F_a, F_\beta] \\
&= -\left(\omega_{\beta a}{}^{n} + d_{a\beta}^{n}\right) F_n + \left(\omega_{a\beta}{}^{\nu} - d_{a\beta}^{\nu}\right) F_\nu, \\
T(F_a, F_b) &= \nabla_{F_a} F_b - \nabla_{F_b} F_a - [F_a, F_b] \\
&= \left(\omega_{ab}{}^{n} - \omega_{ba}{}^{n} - d_{ab}^{n}\right) F_n - d_{ab}^{\nu} F_\nu.
\end{aligned}
$$

The claim then follows from the fact that ω is a form with values in $\mathfrak{u}(1)$. □

By Property iii) of U(1)-connections, the connection on $\mathcal{D}^{\perp}$ is completely determined by the connection on $\mathcal{D}$. As explained in Sect. 6.4, the space of U(1)-connections on $P_{\mathrm{U}(1)}$ is an affine space over the space of sections of $T^{\vee}M \otimes P_{\mathrm{U}(1)} \times_{\mathrm{ad}} \mathfrak{u}(1) = T^{\vee}M$. Let ∇ be a U(1)-connection with torsion tensor T and $A \in \Omega^1(M)$. The connection $\tilde{\nabla} = \nabla + A\,\mathrm{I}$ has the following torsion tensor:

$$
\begin{aligned}
\tilde{T}_{ab}{}^{c} &= T_{ab}{}^{c} + 2A_a\,\mathrm{I}_b{}^{c} - 2A_b\,\mathrm{I}_a{}^{c} & \tilde{T}_{ab}{}^{\gamma} &= T_{ab}{}^{\gamma} \\
\tilde{T}_{\alpha b}{}^{c} &= T_{\alpha b}{}^{c} + 2A_\alpha\,\mathrm{I}_b{}^{c} & \tilde{T}_{\alpha b}{}^{\gamma} &= T_{\alpha b}{}^{\gamma} - A_b\,\mathrm{I}_\alpha{}^{\gamma} \\
\tilde{T}_{\alpha\beta}{}^{c} &= T_{\alpha\beta}{}^{c} & \tilde{T}_{\alpha\beta}{}^{\gamma} &= T_{\alpha\beta}{}^{\gamma} + A_\alpha\,\mathrm{I}_\beta{}^{\gamma} + A_\beta\,\mathrm{I}_\alpha{}^{\gamma}
\end{aligned}
$$

Consequently, the form A can be determined such that $\tilde{T}_{ab}{}^{c} = 0$. In contrast, the torsion constraint $\tilde{T}_{\alpha\beta}{}^{\gamma} = 0$ can in general not be achieved, as $\tilde{T}_{33}{}^{3} = -d_{33}^{3} \neq 0$ in general. However, for any tensor $T \in \Omega^2(M, TM)$ there is at most one connection ∇ on $P_{\mathrm{U}(1)}$ such that $T^{\nabla} = T$. This follows from Proposition 6.9.6, as any realizable torsion tensor determines the connection uniquely.

Proposition 10.2.3 *On a super Riemann surface with* U(1)*-structure, there exists a unique* U(1)*-connection such that*

$$T_{ab}{}^{c} = 0 \qquad\qquad T_{\alpha\beta}{}^{\gamma}\,\mathrm{I}_{\gamma}{}^{\beta} = 0$$

for all U(1)*-frames* F_A*. We call this connection standard connection to the* U(1)*-structure. Its connection form is given by* $\omega_{AB}{}^{C} = \omega_A J_B{}^{C}$*, where* J *is the image of* $i \in \mathfrak{u}(1)$ *in* End TM*, that is,* $J_B{}^{C} = 2\,\mathrm{I}_b{}^{c} + \mathrm{I}_\beta{}^{\gamma}$ *and*

$$\omega_a = -\frac{1}{2} d_{12}^{b}\delta_{ba}, \qquad\qquad \omega_\alpha = -\frac{1}{3} d_{\alpha\beta}^{\gamma}\,\mathrm{I}_{\gamma}{}^{\beta}.$$

Proof From the discussion above it is clear that the torsion constraints are realizable, hence we only have to show the local expressions for the connection form:

$$\begin{aligned} T_{ab}{}^{c}\,\mathrm{I}_b{}^{c} &= \left(2\omega_a\,\mathrm{I}_b{}^{c} - 2\omega_b\,\mathrm{I}_a{}^{c} - d_{ab}^{c}\right)\mathrm{I}_c{}^{b} = -2\omega_a - d_{12}^{b}\delta_{ba} \\ T_{\alpha\beta}{}^{\gamma}\,\mathrm{I}_\gamma{}^{\beta} &= \left(\omega_\alpha\,\mathrm{I}_\beta{}^{\gamma} + \omega_\beta\,\mathrm{I}_\alpha{}^{\gamma} - d_{\alpha\beta}^{\gamma}\right)\mathrm{I}_\gamma{}^{\beta} = -3\omega_\alpha - d_{\alpha\beta}^{\gamma}\,\mathrm{I}_\gamma{}^{\beta} \end{aligned}$$

□

Example 10.2.4 (Standard Connection in Superconformal Coordinates) Let (z, θ) be superconformal coordinates for a super Riemann surface and (x^a, η^α) be the corresponding real coordinates. For any metric m compatible with the super Riemann surface structure, an orthonormal frame is given by

$$F_\alpha = \frac{1}{\lambda} D_\alpha, \qquad\qquad F_a = \frac{1}{\lambda^2}\left(\partial_{x^a} + l^\mu \gamma_{a\mu}{}^{\alpha} D_\alpha\right).$$

Here $\lambda \in \mathcal{O}_M$ and $l = l^\mu D_\mu$ is an even section of $\mathcal{D}$. The matrices $\gamma_{a\mu}{}^{\alpha}$ are constant and given in Eq. (A.1.2). We have

$$\begin{aligned} [F_\alpha, F_\beta] &= \left[\frac{1}{\lambda} D_\alpha, \frac{1}{\lambda} D_\beta\right] = \frac{2}{\lambda^2}\Gamma_{\alpha\beta}^{c}\partial_{x^c} + \frac{1}{\lambda}\left(D_\alpha \frac{1}{\lambda}\right) D_\beta + \frac{1}{\lambda}\left(D_\beta \frac{1}{\lambda}\right) D_\alpha \\ &= 2\Gamma_{\alpha\beta}^{c} F_c + \left(\left(D_\alpha \frac{1}{\lambda}\right)\delta_\beta^\gamma + \left(D_\beta \frac{1}{\lambda}\right)\delta_\alpha^\gamma - \frac{2}{\lambda}\Gamma_{\alpha\beta}^{c} l^\mu \gamma_{c\mu}{}^{\gamma}\right) F_\gamma. \end{aligned}$$

Notice that the coefficients l^μ are odd and hence, by the symmetries of γ we have $l^\mu \gamma_{a\mu}{}^\alpha \Gamma^c_{\alpha\beta} l^\nu \gamma_{c\nu}{}^\gamma = 0$ and $\varepsilon^{cd} l^\mu \gamma_{c\mu}{}^\gamma l^\nu \gamma_{d\nu}{}^\delta \Gamma^k_{\gamma\delta} = 0$. Consequently

$$\begin{aligned}
[F_a, F_\beta] &= \left[\frac{1}{\lambda^2}\left(\partial_{x^a} + l^\mu \gamma_{a\mu}{}^\alpha D_\alpha\right), \frac{1}{\lambda} D_\beta\right] \\
&= \frac{1}{\lambda^2}\left(\left(\partial_{x^a} + l^\mu \gamma_{a\mu}{}^\alpha D_\alpha\right)\frac{1}{\lambda}\right) D_\beta - \frac{1}{\lambda}\left(D_\beta \frac{1}{\lambda^2}\right)\left(\partial_{x^a} + l^\mu \gamma_{a\mu}{}^\alpha D_\alpha\right) \\
&\quad - \frac{1}{\lambda^3}\left(D_\beta l^\mu\right)\gamma_{a\mu}{}^\alpha D_\alpha + \frac{2}{\lambda^3} l^\mu \gamma_{a\mu}{}^\alpha \Gamma^c_{\alpha\beta} \partial_{x^c} \\
&= 2\left(\frac{1}{\lambda} l^\mu \gamma_{a\mu}{}^\alpha \Gamma^c_{\alpha\beta} - \left(D_\beta \frac{1}{\lambda}\right)\delta^c_a\right) F_c \\
&\quad + \left(\frac{1}{\lambda}\left(\left(\partial_{x^a} + l^\mu \gamma_{a\mu}{}^\alpha D_\alpha\right)\frac{1}{\lambda}\right)\delta^\gamma_\beta - \frac{1}{\lambda^2}\left(D_\beta l^\mu\right)\gamma_{a\mu}{}^\gamma\right) F_\gamma
\end{aligned}$$

and

$$\begin{aligned}
[F_a, F_b] &= \left[\frac{1}{\lambda^2}\left(\partial_{x^a} + l^\mu \gamma_{a\mu}{}^\alpha D_\alpha\right), \frac{1}{\lambda^2}\left(\partial_{x^b} + l^\nu \gamma_{b\nu}{}^\beta D_\beta\right)\right] \\
&= \varepsilon_{ab}\varepsilon^{cd}\frac{1}{\lambda^2}\left(\partial_{x^c} + l^\mu \gamma_{c\mu}{}^\gamma D_\gamma\right)\frac{1}{\lambda^2}\left(\partial_{x^d} + l^\nu \gamma_{d\nu}{}^\delta D_\delta\right) \\
&\quad + \frac{1}{\lambda^4}\left[\partial_{x^a} + l^\mu \gamma_{a\mu}{}^\alpha D_\alpha, \partial_{x^b} + l^\nu \gamma_{b\nu}{}^\beta D_\beta\right] \\
&= \varepsilon_{ab}\varepsilon^{cd}\frac{1}{\lambda^2}\left(\partial_{x^c} + l^\mu \gamma_{c\mu}{}^\gamma D_\gamma\right)\frac{1}{\lambda^2}\left(\partial_{x^d} + l^\nu \gamma_d{}^\delta D_\delta\right) \\
&\quad + \varepsilon_{ab}\varepsilon^{cd}\left(\frac{1}{\lambda^4}\left(\partial_{x^c} + l^\mu \gamma_{c\mu}{}^\alpha D_\alpha\right) l^\nu\right)\gamma_{d\nu}{}^\gamma D_\gamma \\
&= \varepsilon_{ab}\left(\left(\partial_{x^c} + l^\mu \gamma_{c\mu}{}^\gamma D_\gamma\right)\frac{1}{\lambda^2}\right)\varepsilon^{cd} F_d \\
&\quad + \varepsilon_{ab}\frac{1}{\lambda^3}\left(\left(\partial_{x^c} + l^\mu \gamma_{c\mu}{}^\alpha D_\alpha\right) l^\nu\right)\varepsilon^{cd}\gamma_{d\nu}{}^\gamma F_\gamma.
\end{aligned}$$

Hence, by Proposition 10.2.3 we have

$$\begin{aligned}
\omega_a &= \frac{1}{2}\mathrm{I}_a{}^c\left(\partial_{x^c} + l^\mu \gamma_{c\mu}{}^\gamma D_\gamma\right)\frac{1}{\lambda^2}, \\
\omega_\alpha &= -\frac{1}{3}\left(\mathrm{I}_\alpha{}^\beta\left(D_\beta \frac{1}{\lambda}\right) + \frac{4}{\lambda}\delta_{\alpha\mu} l^\mu\right).
\end{aligned}$$

Example 10.2.5 (Standard Connection on $\mathbb{SH}$ with Respect to the Hyperbolic Metric) This is a special case of the previous Example 10.2.4, where

$$\lambda = \frac{1}{\sqrt{x^2 + \eta^3\eta^4}}, \qquad l^\nu = \lambda \left(D_\mu \frac{1}{\lambda} \right) \varepsilon^{\mu\nu}.$$

Notice that $\partial_{x^1}\frac{1}{\lambda} = 0$, $\partial_{x^2}\frac{1}{\lambda} = \frac{1}{2}\lambda$, $D_3\frac{1}{\lambda} = \lambda\eta^4$ and $D_4\frac{1}{\lambda} = 0$. We obtain the following non-zero commutators:

$$\begin{array}{llll} d_{12}^1 = -1 & d_{12}^3 = \eta^4\lambda & d_{14}^2 = 2\eta^4\lambda & d_{24}^1 = -2\eta^4\lambda \\ d_{14}^3 = -1 & d_{23}^3 = \dfrac{1}{2} & d_{24}^4 = -\dfrac{1}{2} & d_{33}^1 = 2 \\ d_{34}^2 = 2 & d_{44}^1 = -2 & d_{34}^4 = -\eta^4\lambda & d_{44}^3 = 2\eta^4\lambda \end{array}$$

Those commutator coefficients which do not arise from the above by (super) anti-symmetry in the lower indices are zero. The standard connection from Proposition 10.2.3 to this U(1)-structure is given by

$$\omega_1 = \frac{1}{2}, \qquad \omega_2 = 0, \qquad \omega_3 = 0, \qquad \omega_4 = -\eta^4\lambda.$$

For the torsion of this connection we obtain:

$$T_{ab}{}^c = T_{\alpha b}{}^c = T_{\alpha\beta}{}^\gamma = 0, \qquad T_{ab}{}^\gamma = -\varepsilon_{ab}\delta^{\gamma 3}\eta^4\lambda,$$

$$T_{\alpha b}{}^\gamma = \frac{1}{2}\left(\mathrm{I}_\alpha{}^\gamma - \gamma_{b\alpha}{}^\gamma\right) \qquad T_{\alpha\beta}{}^c = -2\Gamma_{\alpha\beta}^c,$$

In particular the torsion fulfills the torsion constraints (10.0.1). The curvature tensor R is also proportional to the generator J of $\mathfrak{u}(1)$:

$$R(X, Y)Z = r(X, Y)JZ,$$

where $r(X, Y)$ is the two-form given by

$$r(F_1, F_2) = -d_{12}^A\omega_A = -d_{12}^1\omega_1 = \frac{1}{2}$$

$$r(F_a, F_\beta) = F_a\omega_\beta - F_\beta\omega_a - d_{a\beta}^A\omega_A = 0$$

$$r(F_\alpha, F_\beta) = F_\alpha\omega_\beta + F_\beta\omega_\alpha - d_{\alpha\beta}^A\omega_A = -\delta_{\alpha\beta}$$

The example of the standard connection on the superhyperbolic space motivates to look at generalizations of scalar curvature.

Definition 10.2.6 Let m be a superconformal metric on a super Riemann surface, F_A an m-orthonormal frame and $R(X, Y)Z$ the curvature tensor of a U(1)-connection. We define

$$R|_{\mathcal{D}} = m\left(R(F_\alpha, F_\beta)F_\gamma, F_\delta\right) m^{\delta\alpha} m^{\beta\gamma}, \quad R|_{\mathcal{D}^\perp} = m\left(R(F_a, F_b)F_c, F_d\right) m^{da} m^{bc}.$$

The sum

$$R = R|_{\mathcal{D}} + R|_{\mathcal{D}^\perp} = m\left(R(F_A, F_B)F_C, F_D\right) m^{DA} m^{BC}$$

is a supergeometric generalization of scalar curvature.

Example 10.2.7 (Scalar Curvature of the Standard Connection on $S\mathbb{H}$) From Example 10.2.5 it follows that

$$\begin{aligned} R|_{\mathcal{D}} &= m\left(R(F_\alpha, F_\beta)F_\gamma, F_\delta\right) m^{\delta\alpha} m^{\beta\gamma} = -\delta_{\alpha\beta}\, \mathrm{I}_\gamma^{\ \tau}\, \varepsilon_{\tau\delta}\varepsilon^{\delta\alpha}\varepsilon^{\beta\gamma} = -2, \\ R|_{\mathcal{D}^\perp} &= m\left(R(F_a, F_b)F_c, F_d\right) m^{da} m^{bc} = \frac{1}{2}\varepsilon_{ab} 2\, \mathrm{I}_c^{\ t}\, \delta_{td}\delta^{da}\delta^{bc} = -2 \end{aligned}$$

Hence, the standard metric on $S\mathbb{H}$ can be seen as a supergeometric extension of the hyperbolic metric on $\mathbb{H}$ which has constant scalar curvature -2.

Proposition 10.2.8 *Let m be a metric compatible with the super Riemann surface structure on M and ∇ the corresponding standard connection. Let $\hat{m}$ be the metric that arises from m by a rescaling by λ^2 and a change of splitting given by $l \in \Gamma(\mathcal{D})$. The standard connection $\hat{\nabla}$ corresponding to $\hat{m}$ differs from ∇ by a differential form A with values in $\mathfrak{scl}$ given by:*

$$\left(\hat{\nabla} - \nabla\right) F_A = \begin{pmatrix} 2\frac{d\lambda}{\lambda}\,\mathrm{id}_a^{\ b} + 2\tau\, \mathrm{I}_a^{\ b} & \left(\nabla l - \frac{d\lambda}{\lambda} l\right)^\mu \gamma_{a\mu}^{\ \ \beta} \\ 0 & \frac{d\lambda}{\lambda}\,\mathrm{id}_\alpha^{\ \beta} + \tau\, \mathrm{I}_\alpha^{\ \beta} \end{pmatrix} \begin{pmatrix} F_b \\ F_\beta \end{pmatrix}$$

Here F_A is an m-orthonormal frame of TM and τ is a differential form such that

$$\langle F_\alpha, \tau \rangle = \frac{1}{3}\left(\frac{\mathrm{I}_\alpha^{\ \beta} F_\beta \lambda}{\lambda} - 4 l^\mu \delta_{\mu\alpha}\right).$$

Proof Since A is an even one-form with values in $\mathfrak{scl}$ it can be written as

$$\begin{pmatrix} 2\sigma\,\mathrm{id}_a^{\ b} + 2\tau\, \mathrm{I}_a^{\ b} & a^\mu \gamma_{a\mu}^{\ \ \beta} \\ 0 & \sigma\,\mathrm{id}_\alpha^{\ \beta} + \tau\, \mathrm{I}_\alpha^{\ \beta} \end{pmatrix},$$

where σ and τ are even local one-forms and $a = a^\mu F_\mu \in \Gamma\left(T^\vee M \otimes \mathcal{D}\right)$. Here the indices refer to an m-orthonormal frame F_A.

The form σ is determined by the compatibility of with the metric. Indeed, for all sections $X, Y \in \Gamma(\mathcal{D})$,

$$\begin{aligned}\lambda^2 m\left((\nabla + A)X, Y\right) + \lambda^2 m\left(X, (\nabla + A)Y\right) &= \hat{m}\left(\hat{\nabla}X, Y\right) + \hat{m}\left(X, \hat{\nabla}Y\right)\\ &= d\left(\hat{m}(X, Y)\right) = d\left(\lambda^2 m(X, Y)\right)\\ &= 2\lambda\, d\lambda m(X, Y) + \lambda^2\left(m(\nabla X, Y) + m(X, \nabla Y)\right).\end{aligned}$$

Hence, $\sigma = \frac{d\lambda}{\lambda}$. Similarly the section a is determined by looking at $X \in \Gamma\left(\mathcal{D}^\perp\right)$, $Y \in \Gamma(\mathcal{D})$:

$$\begin{aligned}&-\left(d\lambda^2\right) m\left(\gamma(X)l, Y\right) - \lambda^2\left(m\left(\nabla\gamma(X)l, Y\right) + m\left(\gamma(X)l, \nabla Y\right)\right)\\ &= d\left(\hat{m}(X, Y)\right) = \hat{m}\left(\hat{\nabla}X, Y\right) + \hat{m}\left(X, \hat{\nabla}Y\right)\\ &= \hat{m}\left(\nabla X + 2\frac{d\lambda}{\lambda}X + \gamma(X)a, Y\right) + \hat{m}\left(X, \nabla Y + \frac{d\lambda}{\lambda}Y\right)\\ &= -\lambda^2 m\left(\gamma(\nabla X)l + 2\frac{d\lambda}{\lambda}\gamma(X)l + \gamma(X)a, Y\right)\\ &\quad - \lambda^2 m\left(\gamma(X)l, \nabla Y + \frac{d\lambda}{\lambda}Y\right)\end{aligned}$$

Consequently,

$$a = \nabla l - \frac{d\lambda}{\lambda}l$$

The form τ is determined by the torsion-conditions of the standard connection. Denote the torsion tensor of ∇ by T and the torsion tensor of $\hat{\nabla}$ by $\hat{T}$. Then, for odd sections $X, Y \in \Gamma(\mathcal{D})$,

$$\begin{aligned}\hat{T}(X, Y) &= T(X, Y) + \langle X, A\rangle Y + \langle Y, A\rangle X\\ &= T(X, Y) + \frac{X\lambda}{\lambda}Y + \langle X, \tau\rangle \mathrm{I}Y + \frac{Y\lambda}{\lambda}X + \langle Y, \tau\rangle \mathrm{I}X\end{aligned}$$

Recall that an $\hat{m}$-orthogonal frame is given by

$$\hat{F}_a = \frac{1}{\lambda^2}\left(F_a + l^\mu \gamma_{a\mu}{}^\beta F_\beta\right), \qquad \hat{F}_\alpha = \frac{1}{\lambda}F_\alpha,$$

and the dual frames transform as

$$\hat{F}^a = \lambda^2 F^a, \qquad \hat{F}^\alpha = \lambda\left(F^\alpha - F^b l^\mu \gamma_{b\mu}{}^\alpha\right).$$

The orthogonal projection onto $\mathcal{D}$ of $\hat{T}\left(\hat{F}_\alpha, \hat{F}_\beta\right)$ with respect to $\hat{m}$ yields

$$\begin{aligned}
\hat{T}_{\alpha\beta}{}^\gamma &= \left\langle \hat{T}\left(\hat{F}_\alpha, \hat{F}_\beta\right), \hat{F}^\gamma \right\rangle \\
&= \frac{1}{\lambda}\left\langle T\left(F_\alpha, F_\beta\right) + \frac{F_\alpha\lambda}{\lambda} F_\beta + \langle F_\alpha, \tau\rangle \mathrm{I} F_\beta \right. \\
&\quad \left. + \frac{F_\beta\lambda}{\lambda} F_\alpha + \langle F_\beta, \tau\rangle \mathrm{I} F_\alpha, F^\gamma - F^c l^\mu \gamma_{c\mu}{}^\gamma \right\rangle \qquad (10.2.9) \\
&= \frac{1}{\lambda}\left(T_{\alpha\beta}{}^\gamma + 2\Gamma_{\alpha\beta}^d l^\mu \gamma_{d\mu}{}^\gamma + \frac{F_\alpha\lambda}{\lambda}\delta_\beta^\gamma + \langle F_\alpha, \tau\rangle \mathrm{I}_\beta{}^\gamma \right. \\
&\quad \left. + \frac{F_\beta\lambda}{\lambda}\delta_\alpha^\gamma + \langle F_\beta, \tau\rangle \mathrm{I}_\alpha{}^\gamma \right),
\end{aligned}$$

where the indices in the last line refer to the frame F_A. The contraction with $\mathrm{I}_\gamma{}^\beta$ yields

$$\langle F_\alpha, \tau\rangle = \frac{1}{3}\left(\frac{\mathrm{I}_\alpha{}^\beta F_\beta\lambda}{\lambda} - 4 l^\mu \delta_{\mu\alpha} \right).$$

Analogously, the term $\langle F_a, \tau\rangle$ can be determined as a function of λ, l and T from the condition that $\hat{T}_{12}{}^a$ vanishes. The precise form of τ will not be needed in the sequel. □

Corollary 10.2.10 *Let m be a superconformal metric on M. For any superconformal rescaling by λ^2 there exists a unique change of splitting l such that the torsion $\hat{T}$ of the standard connection to the resulting superconformal metric $\hat{m}$ fulfils $\hat{T}_{\alpha\beta}{}^\gamma = 0$.*

Proof Taking the trace of Eq. (10.2.9), we obtain

$$\begin{aligned}
\hat{T}_{\alpha\beta}{}^\gamma \delta_\gamma^\beta &= \frac{1}{\lambda}\left(T_{\alpha\beta}{}^\gamma \delta_\gamma^\beta + 2\Gamma_{\alpha\beta}^d l^\mu \gamma_{d\mu}{}^\gamma \delta_\gamma^\beta + 3\frac{F_\alpha\lambda}{\lambda} + \mathrm{I}_\alpha{}^\nu \langle F_\nu, \tau\rangle \right) \\
&= \frac{1}{\lambda}\left(T_{\alpha\beta}{}^\gamma \delta_\gamma^\beta - 4 l^\mu \varepsilon_{\mu\alpha} + 3\frac{F_\alpha\lambda}{\lambda} + \frac{1}{3}\mathrm{I}_\alpha{}^\nu \left(\mathrm{I}_\nu{}^\sigma \frac{F_\sigma\lambda}{\lambda} - 4 l^\mu \delta_{\mu\nu} \right) \right) \\
&= \frac{1}{\lambda}\left(T_{\alpha\beta}{}^\gamma \delta_\gamma^\beta - \frac{8}{3} l^\mu \varepsilon_{\mu\alpha} + \frac{8}{3}\frac{F_\alpha\lambda}{\lambda} \right).
\end{aligned}$$

Hence the vanishing of $\hat{T}_{\alpha\beta}{}^{\gamma}\delta_{\gamma}^{\beta}$ requires

$$l^{\mu} = -\left(\frac{3}{8}T_{\alpha\beta}{}^{\gamma}\delta_{\gamma}^{\beta} + \frac{F_{\alpha}\lambda}{\lambda}\right)\varepsilon^{\alpha\mu}. \tag{10.2.11}$$

By anti-symmetry of the torsion tensor, Proposition 10.1.5 and the torsion conditions of the standard connection, we have:

$$\hat{T}_{33}{}^{3} = \frac{1}{3}\hat{T}_{34}{}^{4} = \frac{1}{3}\hat{T}_{43}{}^{4} = \frac{1}{3}\hat{T}_{44}{}^{3}, \qquad \hat{T}_{44}{}^{4} = \frac{1}{3}\hat{T}_{43}{}^{3} = \frac{1}{3}\hat{T}_{34}{}^{3} = \frac{1}{3}\hat{T}_{33}{}^{4}.$$

Consequently, $\hat{T}_{\alpha\beta}{}^{\gamma}\delta_{\gamma}^{\beta} = 0$ implies $\hat{T}_{\alpha\beta}{}^{\gamma} = 0$. □

Corollary 10.2.12 *Let m be a superconformal metric such that for the torsion T of the standard connection holds $T_{\alpha\beta}{}^{\gamma} = 0$. Furthermore let $\hat{m}$ be the superconformal metric arising from rescaling by λ and such that for the torsion $\hat{T}$ of the standard connection holds $\hat{T}_{\alpha\beta}{}^{\gamma} = 0$. The scalar curvature of the superconformal metric $\hat{m}$ is given by*

$$\hat{R}\Big|_{\mathcal{D}} = \frac{1}{\lambda^2}\left(R|_{\mathcal{D}} - 2\frac{\Delta^{\mathcal{D}}\lambda}{\lambda}\right),$$

where $\Delta^{\mathcal{D}}$ is the second order differential operator, given with respect to the orthonormal frames F_A by $\Delta^{\mathcal{D}} = \varepsilon^{\alpha\beta}\left(F_{\alpha}F_{\beta} + (\operatorname{div} F_{\alpha})\, F_{\beta}\right)$.

Proof We use the notation from Proposition 10.2.8, in particular $\hat{\nabla} - \nabla = A$ where the differential form A with values in $\mathfrak{scl}$ operates on sections of $\mathcal{D}$ by $A = \sigma\,\mathrm{id} + \tau\,\mathrm{I}$. It follows that $\nabla\langle F_{\alpha}, A\rangle = d\langle F_{\alpha}, \sigma\rangle\,\mathrm{id} + d\langle F_{\alpha}, \tau\rangle\,\mathrm{I}$. Consequently, the scalar curvature transforms as follows:

$$\begin{aligned}
\hat{R}\Big|_{\mathcal{D}} &= \hat{m}\left(\hat{R}\left(\hat{F}_{\alpha}, \hat{F}_{\beta}\right)\hat{F}_{\gamma}, \hat{F}_{\delta}\right)\hat{m}^{\delta\alpha}\hat{m}^{\beta\gamma} \\
&= \frac{1}{\lambda^2}m\left(R\left(F_{\alpha}, F_{\beta}\right)F_{\gamma} + \left(\nabla_{F_{\alpha}}\left\langle F_{\beta}, A\right\rangle\right)F_{\gamma} + \left(\nabla_{F_{\beta}}\left\langle F_{\alpha}, A\right\rangle\right)F_{\gamma}\right. \\
&\qquad + \left\langle F_{\alpha}, A\right\rangle\left\langle F_{\beta}, A\right\rangle F_{\gamma} + \left\langle F_{\beta}, A\right\rangle\left\langle F_{\alpha}, A\right\rangle F_{\gamma} \\
&\qquad \left. - \left\langle [F_{\alpha}, F_{\beta}], A\right\rangle F_{\gamma}, F_{\delta}\right)m^{\delta\alpha}m^{\beta\gamma} \\
&= \frac{1}{\lambda^2}\left(R|_{\mathcal{D}} + \delta^{\alpha\beta}\left(2F_{\alpha}\left\langle F_{\beta}, \tau\right\rangle - \left\langle [F_{\alpha}, F_{\beta}], \tau\right\rangle\right)\right)
\end{aligned}$$

Now, since the torsion tensor of the standard connection for $\hat{m}$ fulfills $\hat{T}_{\alpha\beta}{}^{\gamma} = 0$, by Eq. (10.2.11),

$$l^{\mu} = -\frac{F_{\alpha}\lambda}{\lambda}\varepsilon^{\alpha\mu},$$

and consequently,

$$\langle F_\alpha, \tau\rangle = \frac{1}{3}\left(\frac{\mathrm{I}\, F_\alpha \lambda}{\lambda} - 4 l^\mu \delta_{\mu\alpha}\right) = -\frac{\mathrm{I}_\alpha{}^\beta\, F_\beta \lambda}{\lambda}.$$

Hence,

$$\begin{aligned}\lambda^2\, \hat{R}\Big|_{\mathcal{D}} &= R|_{\mathcal{D}} + \delta^{\alpha\beta}\left(2F_\alpha \langle F_\beta, \tau\rangle - \langle [F_\alpha, F_\beta], \tau\rangle\right)\\ &= R|_{\mathcal{D}} - \left(\frac{2\varepsilon^{\alpha\beta} F_\alpha F_\beta \lambda}{\lambda} - \delta^{\alpha\beta} d^\gamma_{\alpha\beta} \frac{\mathrm{I}_\gamma{}^\tau\, F_\tau \lambda}{\lambda}\right).\end{aligned}$$

By Lemma 10.2.13 below we have $-\delta^{\alpha\beta} d^\gamma_{\alpha\beta}\, \mathrm{I}_\gamma{}^\tau\, F_\tau = 2\varepsilon^{\alpha\beta}\left(\operatorname{div}_{[F^\bullet]} F_\alpha\right) F_\beta$ and hence

$$\lambda^2\, \hat{R}\Big|_{\mathcal{D}} = R|_{\mathcal{D}} - 2\frac{\Delta^{\mathcal{D}}\lambda}{\lambda}.$$ □

Lemma 10.2.13 *Let F_A be a* U(1)*-frame. Then the divergence of F_α with respect to the volume form $[F^\bullet]$ is given by $\operatorname{div}_{[F^\bullet]} F_\alpha = -\frac{1}{2}\delta^{\mu\nu} d^\tau_{\mu\nu}\delta_{\tau\alpha}$.*

Proof In the proof of this lemma we work in superconformal coordinates and assume $F_\alpha = \frac{1}{\lambda} D_\alpha$ and $F_a = \frac{1}{\lambda^2}\left(\partial_{x^a} + l^\mu \gamma_{a\mu}{}^\sigma D_\sigma\right)$. Then $\delta^{\mu\nu} d^\tau_{\mu\nu}\delta_{\tau\alpha} = 2D_\alpha \frac{1}{\lambda}$, and

$$\begin{pmatrix} F_a \\ F_\alpha \end{pmatrix} = \begin{pmatrix} \frac{1}{\lambda^2}\delta_a^b & 0 \\ 0 & \frac{1}{\lambda}\delta_\alpha^\beta \end{pmatrix} \begin{pmatrix} \delta_b^c & l^\mu \gamma_{b\mu}{}^\gamma \\ 0 & \delta_\beta^\gamma \end{pmatrix} \begin{pmatrix} \delta_c^d & 0 \\ \eta^\mu \Gamma_{\mu\gamma}^d & \delta_\gamma^\delta \end{pmatrix} \begin{pmatrix} \partial_{x^d} \\ \partial_{x^\delta} \end{pmatrix}.$$

Consequently, $\operatorname{Ber}[F^\bullet] = \frac{1}{\lambda^2}$. Now, we use the rules given in Proposition 8.1.2:

$$\begin{aligned}\left(\operatorname{div}_{[F^\bullet]} F_\alpha\right)[F^\bullet] &= L_{F_\alpha}[F^\bullet] = L_{F_\alpha^B \partial_{X^B}} (\operatorname{Ber} F)^{-1}[dX^\bullet]\\ &= \sum_B \partial_{X^B}\left(F_\alpha{}^B (\operatorname{Ber} F)^{-1}\right)[dX^\bullet]\\ &= \left(\sum_B \partial_{X^B} F_\alpha{}^B - \frac{F_\alpha \operatorname{Ber} F}{\operatorname{Ber} F}\right)[F^\bullet]\\ &= \left(\partial_{x^b}\left(\frac{1}{\lambda}\eta^\mu \Gamma^b_{\mu\alpha}\right) + \partial_{\eta^\beta}\left(\frac{1}{\lambda}\delta_\alpha^\beta\right) - \lambda^2\left(F_\alpha \frac{1}{\lambda^2}\right)\right)[F^\bullet]\\ &= -\left(D_\alpha \frac{1}{\lambda}\right)[F^\bullet] = -\frac{1}{2}\delta^{\mu\nu} d^\tau_{\mu\nu}\delta_{\tau\alpha}[F^\bullet]\end{aligned}$$ □

Example 10.2.14 Let (x^a, η^α) be real superconformal coordinates on $S\mathbb{H}$ and m the standard metric, that is ∂_{x^a} and D_α are m-orthonormal. The connection ∇ is trivial and has no curvature and no torsion except $T_{\alpha\beta}{}^c$. Denote by $\hat{m}$ the hyperbolic metric on $S\mathbb{H}$. The metric $\hat{m}$ arises from m by a superconformal rescaling by $\lambda = \frac{1}{\sqrt{x^2+\eta^3\eta^4}}$ and a change of splitting by $l^\nu = \frac{D_\mu\lambda}{\lambda}\varepsilon^{\mu\nu}$. By Corollary 10.2.10 the torsion tensor $\hat{T}$ of the standard connection to $\hat{m}$ has $T_{\alpha\beta}{}^\gamma = 0$. By Corollary 10.2.12, the $\mathcal{D}$-component of the scalar curvature is given by

$$\begin{aligned}\hat{R}\Big|_{\mathcal{D}} &= -2\frac{\Delta^{\mathcal{D}}\lambda}{\lambda^3} = -2\left(\sqrt{x^2+\eta^3\eta^4}\right)^3 \varepsilon^{\alpha\beta} D_\alpha D_\beta \frac{1}{\sqrt{x^2+\eta^3\eta^4}} \\ &= \left(\sqrt{x^2+\eta^3\eta^4}\right)^3 \varepsilon^{\alpha\beta} D_\alpha \left(\frac{2\eta^4\delta_{\beta 3}}{\left(\sqrt{x^2+\eta^3\eta^4}\right)^3}\right) = -2\end{aligned}$$

This confirms the calculations of Example 10.2.5.

Recall that any Riemann surface of genus strictly larger than one possesses a unique metric of constant scalar curvature -2. Analogously, it would be desirable to use conditions on torsion and curvature to single out a unique superconformal metric on a given super Riemann surfaces. Corollaries 10.2.10 and 10.2.12 indicate that a good candidate is a superconformal metric such that the standard connection has $T_{\alpha\beta}{}^\gamma = 0$ and $R|_{\mathcal{D}} = -2$. Any quotient of $S\mathbb{H}$ by a subgroup of Sp(2|1)-action, that is by Hodgkin (1987a) any super Riemann surface, possesses such a metric. However, at the moment it seems that uniqueness cannot be obtained because the theory of partial differential equations for odd derivatives is not sufficiently developed to assume uniqueness of the scaling factor. In particular an analogue of the maximum principle for $\Delta^{\mathcal{D}}$ would be desirable.

Chapter 11
Metrics and Gravitinos

Let $i: |M| \to M$ be an embedding of an underlying even manifold into a super Riemann surface M. In this chapter we are concerned with the structure induced on $|M|$. We will show that a given U(1)-structure on M induces a metric g, a spinor bundle S and a differential form χ with values in S, called gravitino, on $|M|$. Different U(1)-structures on M induce metrics and gravitinos which differ only by conformal and super Weyl transformations. Furthermore, the triple (g, S, χ) on $|M|$ is sufficient to reconstruct the super Riemann surface M. Supersymmetry of metric and gravitino are interpreted as an infinitesimal change of the embedding i. From this point of view we are able to give a description of the infinitesimal deformations of a super Riemann surface in terms of metric and gravitino.

The description of super Riemann surfaces in terms of metrics and gravitinos is generally accepted in the more physics oriented literature and attributed to Howe (1979). The construction given there is dependent on conditions on the torsion of U(1)-connections on M and predates even the definition of super Riemann surfaces. As those torsion constraints are not completely invariant on M, see Chap. 10, we propose here a different, more geometric way to look at metric and gravitino. Exploiting the theory of families of super Riemann surfaces, the theorem of Giddings and Nelson and the concept of underlying even manifold, we will be able to give global and coordinate independent statements. It is only in the reconstruction of the super Riemann surface from metric and gravitino that we use a local method similar to Howe (1979).

E. Keßler, *Supergeometry, Super Riemann Surfaces and the Superconformal Action Functional*, Lecture Notes in Mathematics 2230,
https://doi.org/10.1007/978-3-030-13758-8_11

11.1 Metric, Spinor Bundle and Gravitino

Let M be a super Riemann surface, where we understand implicitly, as always, that M is a family of super Riemann surfaces over a base B. By the definition of super Riemann surfaces we have a short exact sequence:

$$\begin{array}{ccccccccc} 0 & \longrightarrow & \mathcal{D} & \xrightarrow{\iota_{\mathcal{D}}} & TM & \xrightarrow{p_{TM/\mathcal{D}}} & \left(TM/\mathcal{D}\right) & \longrightarrow & 0 \\ & & \nwarrow & & \parallel & & \swarrow & & \\ & & p_{\mathcal{D}} & & \mathcal{D}\oplus\mathcal{D}^{\perp} & \xleftarrow{\iota_{TM/\mathcal{D}}} & & & \end{array} \tag{11.1.1}$$

The choice of a U(1)-structure yields additionally the dotted structures: $TM = \mathcal{D}\oplus\mathcal{D}^{\perp}$ and the splitting maps $\iota_{TM/\mathcal{D}}$ and $p_{\mathcal{D}}$. In this section we are interested in the pullback of this split short exact sequence (11.1.1) along an embedding of the underlying even manifold $i\colon |M| \to M$.

Proposition 11.1.2 *The composition*

$$T|M| \xrightarrow{di} i^*TM \xrightarrow{i^*p_{TM/\mathcal{D}}} i^*\left(TM/\mathcal{D}\right)$$

is an isomorphism of vector bundles.

Proof Let (x^a, η^α) be the real and imaginary part of some superconformal coordinates (z, θ). The functions $y^a = i^\# x^a$ yield local coordinates for $|M|$. In those coordinates we have that locally $i^*TM = \left\langle i^*\partial_{x^a}, i^*\partial_{\eta^\alpha}\right\rangle_{\mathcal{O}_{|M|}}$ and $T|M| = \left\langle \partial_{y^a}\right\rangle_{\mathcal{O}_{|M|}}$. The subbundle $i^*\mathcal{D} \subset i^*TM$ is generated by $i^*\left(\partial_{\eta^\alpha} + \eta^\mu \Gamma_{\mu\alpha}^k \partial_{x^k}\right)$ and consequently the quotient $i^*\left(TM/\mathcal{D}\right)$ is generated by $i^*\partial_{x^a}$. The map di is given by

$$\partial_{y^a} \mapsto i^*\partial_{x^a} + \left(\partial_{y^a} i^\# \eta^\alpha\right) i^*\partial_{\eta^\alpha}$$

and consequently yields a local isomorphism $T|M| \simeq i^*\left(TM/\mathcal{D}\right)$. □

Definition 11.1.3 As a consequence of Proposition 11.1.2, the pullback of the sequence (11.1.1) is given by

$$\begin{array}{ccccccccc} 0 & \longrightarrow & S & \underset{p_S}{\overset{\iota_S}{\rightleftarrows}} & i^*TM & \underset{\iota_{T|M|}}{\overset{p_{T|M|}}{\rightleftarrows}} & T|M| & \longrightarrow & 0 \end{array} \tag{11.1.4}$$

with $di\colon T|M| \to i^*TM$.

The bundles $S = i^*\mathcal{D}, i^*TM$ and $T|M|$ are complex line bundles such that $S \otimes_{\mathbb{C}} S = T|M|$. The pullback of the maps of the short exact sequence (11.1.1) are denoted $\iota_S = i^*\iota_{\mathcal{D}}$ and $\iota_{T|M|} = i^*\iota_{TM/\mathcal{D}}$. The map di induces a splitting of the short exact sequence (11.1.4).

If we have chosen a U(1)-structure on TM the vector bundle $T|M|$ can also be identified with $i^*\mathcal{D}^\perp$. The pullback of the maps given by the metric splittings in (11.1.1) are denoted $p_S = i^*p_{\mathcal{D}}$ and $\iota_{T|M|} = i^*\iota_{TM/\mathcal{D}}$. They give a second splitting of (11.1.4).

Definition 11.1.5 (Metric on $T|M|$ and S) Let m be a metric compatible with the super Riemann surface structure on M. We denote by g the metric on $T|M|$ induced from the metric $i^*m|_{\mathcal{D}^\perp}$ on $i^*\mathcal{D}^\perp$ via the identification of $T|M| \simeq i^*\mathcal{D}^\perp$ from Proposition 11.1.2. Similarly we denote by $g_S = i^*m|_{\mathcal{D}}$ the metric on $S = i^*\mathcal{D}$.

More explicitly, for vector fields X and Y on $|M|$ we have

$$g(X, Y) = i^*m\left(\left(i^* p_{\mathcal{D}^\perp} \circ di\right) X, \left(i^* p_{\mathcal{D}^\perp} \circ di\right) Y\right).$$

Hence, g is different from the pullback m_i of m along i. By definition, the vector bundle S is of real rank $0|2$, hence for any odd sections s and s' of S, we have $g_S(s, s') = -g_S(s', s)$. In addition, g and g_s are compatible with the almost complex structures induced from $\mathcal{D}$ and $\mathcal{D}^\perp$.

Notice that by the properties of the U(1)-structure $P_{\mathrm{U}(1)}$ on M, the bundle S is the associated bundle to $i^*P_{\mathrm{U}(1)}$ via the defining representation ρ. The bundle $T|M|$ is also an associated vector bundle to $i^*P_{\mathrm{U}(1)}$, however, via the representation ρ^2. Thus $i^*P_{\mathrm{U}(1)}$ is a spin structure on $|M|$ and S is a spinor bundle to the metric g. Furthermore, as explained in Appendix A, on S there is an almost complex structure I that is compatible with g_S and compatible Clifford maps $\gamma\colon T|M| \otimes S \to S$ and $\Gamma\colon S \otimes S \to T|M|$.

Let us denote a hermitian frame on S by

$$s = \frac{1}{2}(s_3 - \mathrm{i}s_4)$$

and its square by $f = s \otimes s$. Notice that the real spinor frames will be numbered by $\alpha = 3, 4$. The frame f is an hermitian orthonormal frame with respect to the hermitian form on M. Its real and imaginary part are given by

$$f = \frac{1}{2}(f_1 - \mathrm{i}f_2)$$

and orthonormal with respect to g. In the following, we always assume that the frames f_a and s_α are related by $f = s \otimes s$; that is they are induced by a cross-section of $i^*P_{\mathrm{U}(1)} \to |M|$. Notice that any SCL-frame F_A on M induces such a pair of frames by $f_a = i^*F_a$ and $s_\alpha = i^*F_\alpha$.

Definition 11.1.6 (Gravitino) The gravitino χ is defined as the difference of the splittings in the short exact sequence (11.1.4),

$$\chi = \iota_{T|M|} - di = -p_S \circ di.$$

Hence the gravitino in an even differential form with values in S that is defined in terms of the embedding i and a chosen U(1)-structure.

With respect to the frame (f_a, s_α), the gravitino χ can be expanded as

$$\chi(f_a) = \chi_a{}^\beta s_\beta.$$

As χ and f_a are even, and s_α are odd frames, the coefficients χ_a^β of χ are odd functions on $|M|$. This is another motivation to work with families of supermanifolds. Furthermore, the fact that χ_a^β is odd is a first indication that on trivial families of super Riemann surfaces the gravitino χ can be assumed to be zero (compare Proposition 11.1.11 below).

Example 11.1.7 (Metric and Gravitino on $\mathbb{C}^{1|1}$) Let (x^a, η^α) be real and imaginary part of superconformal coordinates (z, θ). Any U(1)-structure on $\mathbb{C}^{1|1}$ can be given by the frames

$$F_a = \frac{1}{\lambda^2}\left(\partial_{x^a} + l^\mu \gamma_{\mu a}{}^\tau \left(\partial_{\eta^\tau} + \Gamma^t_{\tau\sigma}\eta^\sigma \partial_{x^t}\right)\right), \quad F_\alpha = \frac{1}{\lambda}\left(\partial_{\eta^\alpha} + \Gamma^c_{\alpha\beta}\eta^\beta \partial_{x^c}\right),$$

for some even function λ and odd functions l^μ. Suppose that the embedding $i\colon \mathbb{R}^{2|0} \to \mathbb{C}^{1|1}$ is given by

$$i^\# x^a = y^a, \qquad i^\# \eta^\alpha = Q^\alpha(y),$$

where y^a are coordinates on $\mathbb{R}^{2|0}$ and Q^α are odd functions on $\mathbb{R}^{2|0}$. Recall that we always implicitly work over a base B. The differential di is given by

$$di\left(\partial_{y^a}\right) = i^*\partial_{x^a} + \left(\partial_{y^a} Q^\beta\right) i^*\partial_{\eta^\beta} = i^*\left(\lambda^2 F_a\right) + i^\#\lambda\left(\partial_{y^a} Q^\beta - i^\# l^\mu \gamma_{a\mu}{}^\beta\right) i^* F_\beta.$$

Consequently, the metrics g and g_S are given by

$$g(\partial_{y^a}, \partial_{y^b}) = i^\#\lambda^4 \delta_{ab}, \qquad g_S(i^*\partial_{\eta^\alpha}, i^*\partial_{\eta^\beta}) = i^\#\lambda^2 \varepsilon_{\alpha\beta}$$

and the gravitino is given by

$$\chi(\partial_{y^a}) = i^\#\lambda\left(i^\# l^\mu \gamma_{a\mu}{}^\beta - \partial_{y^a} Q^\beta\right) s_\beta.$$

Hence, for the standard metric on $\mathbb{C}^{1|1}$, that is $\lambda = 1$ and $l^\mu = 0$, we obtain the Euclidean metric g and $\chi = -\left(\partial_{y^a} Q^\beta\right) s_\beta$. In the case of a trivial family over the point $\mathbb{R}^{0|0}$ the gravitino has to vanish. For the standard metric on $S\mathbb{H}$, given by

$$\lambda = \frac{1}{\sqrt{x^2 + \eta^3\eta^4}}, \qquad s^\mu = -\frac{\eta^4\delta^{\mu 3}}{x^2 + \eta^3\eta^4},$$

we obtain

$$g(\partial_{y^a}, \partial_{y^b}) = \frac{1}{\left(y^2 + Q^3Q^4\right)^2}\delta_{ab},$$

$$\chi(\partial_{y^a}) = -\frac{1}{\sqrt{x^2 + Q^3Q^4}}\left(\frac{Q^4\gamma_{a3}{}^\beta}{y^2 + Q^3Q^4} + \partial_{y^a} Q^\beta\right) s_\beta.$$

In particular for $Q^\alpha = 0$, we obtain an isometric embedding $\mathbb{H} \to S\mathbb{H}$ with $\chi = 0$.

The definition of metric and gravitino given here is a purely geometric one. Example 11.1.7 shows explicitly how metric and gravitino depend on the U(1)-structure and the embedding i. However, in order to justify the name gravitino, it should have the expected physical properties mentioned in the Introduction: In Chap. 12 we will see that the gravitino appears in a geometrically defined action functional. Supersymmetry of metric and gravitino will be established in Sect. 11.4 as an infinitesimal change of the embedding i. In the remainder of this section we will interpret the conformal and super Weyl transformations of the metric and gravitino in terms of the super Riemann surface M. Furthermore, we explain under which conditions the gravitino can be assumed to be zero.

Proposition 11.1.8 *Different* U(1)*-structures on M do not necessarily induce different* U(1)*-structures on i^*TM. If the induced* U(1)*-structures on i^*TM differ, the corresponding metrics and gravitinos on $|M|$ differ by conformal and super Weyl transformations.*

Proof Assume that a U(1)-structure on M is given by a metric m. Any other metric $\tilde{m}$ may differ from m in in the metric on $\mathcal{D}$ and in the splitting $TM = \mathcal{D} \oplus \mathcal{D}^\perp$, see Proposition 9.5.3. That is, there exists a $\lambda \in \mathcal{O}_M$ and a $l \in \mathcal{D}$ such that

$$\tilde{m}|_\mathcal{D} = \lambda^2\, m|_\mathcal{D}, \qquad \tilde{p}_\mathcal{D}(X) - p_\mathcal{D}(X) = \gamma(p_\mathcal{D}X)l.$$

Here $\tilde{p}_\mathcal{D}$ and $p_\mathcal{D}$ denote the orthogonal projections on $\mathcal{D}$ for $\tilde{m}$ and m respectively. The scaling of the induced metrics on ${}^{TM}\!/_{\mathcal{D}}$ is given by λ^4. Hence the metrics $\tilde{g}$ and $\tilde{g}_S$ induced by $\tilde{m}$ are given by

$$\tilde{g}_S = \left(i^\#\lambda\right)^2 g_S, \qquad \tilde{g} = \left(i^\#\lambda\right)^4 g.$$

The gravitino is not affected by the rescaling but rather by the change of the splitting. By definition, the gravitino is given by $\chi(X) = -i^* p_{\mathcal{D}} \, di X$. Consequently, for all X we obtain

$$\tilde{\chi}(X) - \chi(X) = -\left(i^* \tilde{p}_{\mathcal{D}} - i^* p_{\mathcal{D}}\right) di X = \gamma(X) i^* l$$

which is precisely a super Weyl transformation. □

Remark 11.1.9 Notice that, in general, one cannot keep sections of the spinor bundle fixed while varying the metric. Rather, the spinor bundle depends on the metric. In the proof above, we have identified the spinor bundle S_g with $S_{i^\# \lambda^2 g}$ with the help of $i^* \operatorname{id}_{\mathcal{D}}$ to express the effect of conformal rescaling on the gravitino. More generally, one has to use the isometries $b\colon (T|M|, g) \to (T|M|, i^\# \lambda^4 g)$ and $\beta\colon S_g \to S_{i^\# \lambda^4 g}$ introduced in Appendix A.4. Then, the effect of conformal rescaling of the gravitino is given by

$$\tilde{\chi} = i^\# \lambda^{-1} \left(\left(b^\vee\right)^{-1} \otimes \beta \right) \chi \in \Gamma\left(T^\vee |M| \otimes S_{i^\# \lambda^4 g}\right),$$

which coincides with χ when identifying $T^\vee|M| \otimes S_g$ and $T^\vee|M| \otimes S_{i^\# \lambda^4 g}$ with the help of $\operatorname{id}_{T^\vee|M|} \otimes i^* \operatorname{id}_{\mathcal{D}}$, see Example A.4.1.

It is explained in Appendix A.3 that the decomposition of $T^\vee|M| \otimes S^\vee$ in vector bundles associated to irreducible representations of the spin group is given by $T^\vee M \otimes S^\vee = S^\vee \oplus S^{\vee \otimes 3}$. After identifying the gravitino with a section of $T^\vee|M| \otimes S^\vee$ with the help of g_S, the Wess–Zumino transformations changes the part of type $\frac{1}{2}$.

In contrast to $\iota_{T|M|}$, the map di does not respect the almost complex structures on $T|M|$ and i^*TM. Hence in general, the gravitino may also have a part of type $\frac{3}{2}$. However, if the gravitino is zero, di needs to coincide with $\iota_{T|M|}$. In that case $di = \iota_{T|M|}$ is compatible with the almost complex structures on M and $|M|$ respectively, see Definition 11.1.6, and hence i is holomorphic. In the construction of embeddings of the underlying even manifold, in Theorem 3.3.7, we have used a partition of unity. Since partitions of unity are in general not holomorphic, we cannot expect i to be holomorphic. The existence of a holomorphic $i\colon |M| \to M$ has strong geometric consequences, since holomorphic maps are more rigid:

Proposition 11.1.10 *Let M be a super Riemann surface with a holomorphic embedding $i\colon |M| \to M$. Then M is a complex relative split supermanifold.*

Proof We can cover $\|M\|$ by holomorphic coordinate charts U_α with holomorphic coordinates y_α on $U_\alpha \subset |M|$ and $(z_\alpha, \theta_\alpha)$ on $U_\alpha \subset M$ such that the embedding i is given by

$$i^\# z_\alpha = y_\alpha, \qquad\qquad i^\# \theta_\alpha = 0.$$

Up to a holomorphic coordinate change, we can assume that $(z_\alpha, \theta_\alpha)$ are superconformal coordinates, see Lemma 9.1.3. In particular the coordinate change in the proof of Lemma 9.1.3 preserves $i^\#\theta_\alpha = 0$. As explained in Sect. 9.1, all coordinate changes between superconformal coordinates are of the form

$$z_\beta = f_{\alpha\beta}(z_\alpha) + \theta_\alpha \zeta_{\alpha\beta}(z_\alpha) \qquad \theta_\beta = \xi_{\alpha\beta}(z_\alpha) + \theta_\alpha g_{\alpha\beta}(z_\alpha)$$

with

$$\zeta_{\alpha\beta} = g_{\alpha\beta}\xi_{\alpha\beta}, \qquad g_{\alpha\beta}^2 = f' + \xi_{\alpha\beta}\xi'_{\alpha\beta}.$$

However, as $i^\#\theta_\alpha = i^\#\theta_\beta = 0$, we have that $\xi_{\alpha\beta} = 0$ for all α, β and hence also $\zeta_{\alpha\beta} = 0$. Consequently M is in the form of a relative complex split supermanifold for the line bundle $S = i^*\mathcal{D}$ over $|M|$. □

Proposition 11.1.11 *Let M be a super Riemann surface. There exists a* U(1)*-structure on M and an embedding $i\colon |M| \to M$ such that $\chi = 0$ if and only if M is a complex relative split supermanifold.*

Proof We have already seen that the vanishing of the gravitino implies that i is holomorphic, hence M is split by Proposition 11.1.10.

If M is complex relative split, there exists a holomorphic $i\colon |M| \to M$ and the splitting p defined by $0 = \chi = p - di$ can be extended to a splitting on M, as there exists an inclusion $\mathcal{O}_{|M|} \hookrightarrow \mathcal{O}_M = \bigwedge \Gamma(S)$. By Proposition 9.5.3, an arbitrary choice of hermitian form on $\mathcal{D}$ is sufficient to determine a U(1)-structure on M. □

Even in the case where M is not split, it is at least locally split. Hence, the gravitino can be assumed to be zero locally:

Corollary 11.1.12 *Let $p \in \|M\|$. There exists a neighbourhood $U \subset \|M\|$, a* U(1)*-structure on the super Riemann surface M, and an embedding $i\colon |M| \to M$ such that $\chi|_U = 0$.*

11.2 Wess–Zumino Coordinates

Wess–Zumino coordinates are coordinates on the super Riemann surface that are particularly well adapted to an embedding $i\colon |M| \to M$ and a particular frame F_A. The choice of Wess–Zumino coordinates breaks the superdiffeomorphism invariance and is thus the first of two steps to describe a super Riemann surface M via its underlying even manifold $|M|$.

Definition 11.2.1 Let F_A be a SCL-frame over $U \subset M$. Let (x^a, η^α) be coordinates on $U \subset M$ and y^a coordinates on $U \subset |M|$. The coordinates y^a and $X^A = (x^a, \eta^\alpha)$

are called Wess–Zumino coordinates for i and F_A if the following conditions hold:

- The embedding $i\colon |M| \to M$ is given in the coordinates y^a and (x^a, η^α) by

$$i^\# x^a = y^a, \qquad i^\# \eta^\alpha = 0. \tag{11.2.2}$$

- The coefficients of $F_A = F_A{}^B \partial_{X^B}$ fulfil

$$\begin{aligned} F_\alpha{}^b &= \eta^\mu{}_\mu F_\alpha{}^b + \eta^3\eta^4{}_{34}F_\alpha{}^b, \\ F_\alpha{}^\beta &= \delta_\alpha^\beta + \eta^\mu{}_\mu F_\alpha{}^\beta + \eta^3\eta^4{}_{34}F_\alpha{}^\beta, \end{aligned} \tag{11.2.3}$$

  where the degree one coefficients are symmetric in the lower indices

$$\varepsilon^{\mu\alpha}{}_\mu F_\alpha{}^B = 0 \tag{11.2.4}$$

  and the degree two coefficients ${}_{34}F_\alpha{}^B$ are arbitrary.

Proposition 11.2.5 *Let $U \subseteq M$ be an open coordinate neighbourhood and F_A a* SCL*-frame over U. For any coordinates y^a on $U \subset |M|$ there are unique coordinates $X^A = (x^a, \eta^\alpha)$ on $U \subset M$, such that y^a and X^A are Wess–Zumino coordinates for i and F_A.*

Proof Let $(\tilde{x}^a, \tilde{\eta}^\alpha)$ be arbitrary coordinates on $U \subset M$ such that i is given in those coordinates by

$$i^\# \tilde{x}^a = y^a, \qquad i^\# \tilde{\eta}^\alpha = 0.$$

Any other coordinates (x^a, η^α) fulfilling the conditions (11.2.2) are related to $(\tilde{x}^a, \tilde{\eta}^\alpha)$ by a coordinate change of the type

$$\begin{aligned} x^a &= \tilde{x}^a + \tilde{\eta}^\mu{}_\mu f(\tilde{x})^a + \tilde{\eta}^3\tilde{\eta}^4{}_{34}f(\tilde{x})^a, \\ \eta^\alpha &= \tilde{\eta}^\mu{}_\mu f(\tilde{x})^\alpha + \tilde{\eta}^3\tilde{\eta}^4{}_{34}f(\tilde{x})^\alpha. \end{aligned}$$

All the coefficient functions are completely determined by their pullback along i. Notice that for the coordinate change to be invertible, the even matrix ${}_\mu f(\tilde{x})^\alpha$ needs to be invertible. The inverse of the coordinate change starts

$$\begin{aligned} \tilde{x}^a &= x^a + \eta^\mu{}_\mu g(x)^a + \eta^3\eta^4 \ldots \\ \tilde{\eta}^\alpha &= \eta^\mu{}_\mu g(x)^\alpha + \eta^3\eta^4 \ldots \end{aligned}$$

where ${}_\mu f(\tilde{x})^\beta{}_\beta g(\tilde{x})^\nu = \delta_\mu^\nu$ and ${}_\mu g(x)^c$ is completely determined by ${}_\mu f(\tilde{x})^C$.

Let us denote the frame coefficients of F_A with respect to ∂_{X^A} by ${F_A}^B$, those with respect to $\partial_{\tilde{X}^A}$ by ${\tilde{F}_A}^B$. Then it holds that

$$F_A = {\tilde{F}_A}^B \partial_{\tilde{X}^B} = {\tilde{F}_A}^B \left(\frac{\partial X^C}{\partial \tilde{X}^B}\right) \partial_{X^C} = {F_A}^C \partial_{X^C}.$$

The conditions (11.2.3) on the frame coefficients are then given by

$$0 = i^\# \left({\tilde{F}_\alpha}^B \left(\frac{\partial x^c}{\partial \tilde{X}^B}\right)\right) = \left(i^\# {\tilde{F}_\alpha}^c\right) + \left(i^\# {\tilde{F}_\alpha}^\beta\right)\left(i^\# {}_\beta f^c\right),$$

$$\delta_\alpha^\gamma = i^\# \left({\tilde{F}_\alpha}^B \left(\frac{\partial \eta^\gamma}{\partial \tilde{X}^B}\right)\right) = \left(i^\# {\tilde{F}_\alpha}^\beta\right)\left(i^\# {}_\beta f^\alpha\right).$$

As ${\tilde{F}_\alpha}^\beta$ is invertible those equations are solvable for $i^\# {}_\mu f^C$ and hence determine ${}_\mu f^C$.

The conditions (11.2.4) are given by

$$\begin{aligned}
0 = \varepsilon^{\mu\alpha} i^\# \left(\partial_{\eta^\mu} {F_\alpha}^C\right) &= \varepsilon^{\mu\alpha} i^\# \left(\left({}_\mu g^d \partial_{\tilde{x}^d} + {}_\mu g^\delta \partial_{\tilde{\eta}^\delta}\right)\left({\tilde{F}_\alpha}^B \left(\frac{\partial X^C}{\partial \tilde{X}^B}\right)\right)\right) \\
&= \varepsilon^{\mu\alpha} \left(i^\# \left(\left({}_\mu g^d \partial_{\tilde{x}^d} + {}_\mu g^\delta \partial_{\tilde{\eta}^\delta}\right) {\tilde{F}_\alpha}^b\right) \delta_b^C \right. \\
&\quad + i^\# \left(\left({}_\mu g^d \partial_{\tilde{x}^d} + {}_\mu g^\delta \partial_{\tilde{\eta}^\delta}\right) {\tilde{F}_\alpha}^\beta\right)\left(i^\# {}_\mu f^C\right) - i^\# \left({\tilde{F}_\alpha}^b {}_\mu g^\delta \left(\partial_{\tilde{x}^b} {}_\delta f^c\right)\right) \\
&\quad \left. - i^\# \left({\tilde{F}_\alpha}^\beta {}_\mu g^\delta \varepsilon_{\beta\delta}\right)\left(i^\# {}_{34} f^C\right)\right)
\end{aligned}$$

Those equations are solvable for $i^\# {}_{34} f^C$ and hence determine ${}_{34} f^C$. Consequently there are unique coordinates X^A such that y^a and X^A are Wess–Zumino coordinates for i and F_A. □

Remark 11.2.6 The Wess–Zumino coordinates should be seen as a theoretical tool. We will see later that they are part of the reconstruction of the super Riemann surface from metric and gravitino on $|M|$. However, they are rather bad coordinates to describe the super Riemann surface. If i is not holomorphic, they are not holomorphic coordinates and not very well suited to describe the superconformal structure. Already for the standard frames $(\partial_z, \partial_\theta + \theta\partial_z)$ and an arbitrary non-holomorphic embedding i the coordinate change between superconformal coordinates and Wess–Zumino coordinates possesses a full expansion in the odd directions.

Remark 11.2.7 The conditions on the frame expansion, Eqs. (11.2.3) and (11.2.4), together with the fact that they can be achieved, are to be found in Howe (1979, Equation 3.3) and D'Hoker and Phong (1988, Chapter III.C). In the latter reference those conditions are called Wess–Zumino gauge, which motivates our choice to call those particular coordinates Wess–Zumino coordinates.

The conditions (11.2.2) concerning the embedding i, are not mentioned in the given sources. They correspond, however, to the intuitive notion of setting $\eta = 0$.

Remark 11.2.8 It has been noticed already in D'Hoker and Phong (1988, Chapter III.C.) that the choice of Wess–Zumino coordinates relates the odd coordinates to spinors. Indeed, for the odd part η^α of Wess–Zumino coordinates, the frame $i^*\partial_{\eta^\alpha}$ is a spinor frame.

Corollary 11.2.9 *$U \subset M$ be a coordinate neighbourhood of the super Riemann surface M, $i\colon |M| \to M$ an underlying even manifold and F_A a* SCL*-frame over U. Any function $f \in \mathcal{O}_M(U)$ is completely determined by the functions*

$$i^\# f, \qquad i^\# F_\alpha f, \text{ and } \qquad i^\# F_4 F_3 f = -\frac{1}{2}\varepsilon^{\alpha\beta} i^\# F_\alpha F_\beta f.$$

Similarly, any function $g \in \mathcal{O}_M(U) \otimes \mathbb{C}$ is completely determined by the functions

$$i^\# g, \qquad i^\# F_+ g, \qquad i^\# F_- g, \text{ and } \qquad i^\# F_- F_+ g.$$

Proof Assume that the function f is given in the Wess–Zumino coordinates y^a on $U \subset |M|$ and $X^A = (x^a, \eta^\alpha)$ on $U \subset M$ by

$$f = {}_0f + \eta^\alpha {}_\alpha f + \eta^3\eta^4 {}_{34}f.$$

From the Wess–Zumino conditions it is clear that $i^\# f = i^\# {}_0 f$ and $i^\# F_\alpha f = i^\# {}_\alpha f$. For the highest order terms one gets from the Wess–Zumino conditions:

$$\begin{aligned}
i^\# F_4 F_3 f &= -\frac{1}{2}\varepsilon^{\alpha\beta} F_\alpha F_\beta f \\
&= -\frac{1}{2}\varepsilon^{\alpha\beta} i^\# \left(\partial_{\eta^\alpha} + \eta \ldots\right)\left(\partial_{\eta^\beta} + \eta^\mu {}_\mu F_\beta{}^C \partial_{X^C} + \eta^3\eta^4 \ldots\right) f \\
&= -\frac{1}{2}\varepsilon^{\alpha\beta}\left(\partial_{\eta^\alpha}\partial_{\eta^\beta} + {}_\alpha F_\beta{}^C \partial_{X^C}\right)\left({}_0 f + \eta^\alpha {}_\alpha f + \eta^3\eta^4 {}_{34} f\right) = i^\# {}_{34} f
\end{aligned}$$

As we are in a coordinate neighbourhood, the functions ${}_0 f$, ${}_\mu f$ and ${}_{34} f$ and hence f are completely determined by $i^\# f$, $i^\# {}_\mu f$ and $i^\# {}_{34} f$. □

Corollary 11.2.10 *Let y^a and $X^A = (x^a, \eta^\alpha)$ be Wess–Zumino coordinates for the* U(1)*-frame F_A and the embedding $i\colon |M| \to M$. Let us write*

$$F_A = \left({}_0F_A{}^B + \eta^\mu {}_\mu F_A{}^B + \eta^3\eta^4 {}_{34}F_A{}^B\right)\partial_{X^B}.$$

Recall that $i^* F_A$ induces an orthonormal frame f_a on $|M|$ and a spinor frame $s_\alpha = i^* F_\alpha$. Then it holds that

$$f_a = {}_0F_a{}^b \partial_{y^b}, \qquad \chi(f_a) = {}_0F_a{}^\beta s_\beta.$$

Proof Recall the short exact sequence (11.1.4)

$$0 \longrightarrow S \longrightarrow i^*TM \longrightarrow T|M| \longrightarrow 0.$$

In the proof of Proposition 11.1.2, we have seen a local description of the three bundles in terms of coordinate vector fields. Due to the Wess–Zumino conditions we know that S is generated locally by $s_\alpha = i^*\partial_{\eta^\alpha}$. Also, i^*TM is generated locally by $i^*\partial_{X^A}$ and $T|M|$ is generated by ∂_{y^a}. The inclusion of S into i^*TM is given by $s_\alpha \mapsto i^*\partial_{\eta^\alpha}$ and the projection on $T|M|$ is given by $i^*\partial_{x^a} \mapsto \partial_{y^a}$. The metric splitting $\iota_{T|M|}$ identifies f_a with i^*F_a and hence

$$f_a = {}_0F_a{}^b\partial_{y^b}.$$

The gravitino is locally described by

$$\chi(f_a) = \left(\iota_{T|M|} - di\right)(f_a) = i^*F_a - {}_0F_a{}^b i^*\partial_{x^b} = {}_0F_a{}^\beta s_\beta. \qquad \square$$

11.3 Wess–Zumino Frames

Wess–Zumino frames are SCL-frames on a super Riemann surface M with particular conditions on their structure coefficients. Their advantage is that they are completely determined by data on $|M|$, and thus allow for a reconstruction of the super Riemann surface structure from data on $|M|$. Furthermore, we will use Wess–Zumino frames together with Wess–Zumino coordinates in Chap. 12 for the calculation of the component action.

Definition 11.3.1 A SCL-frame $F_z, F_{\overline{z}}, F_+, F_-$ is called a Wess–Zumino frame provided that

$$i^\# d_{+-}^{+} = 0, \qquad i^\# F_+ d_{+-}^{+} = 0, \qquad d_{++}^{+} = 0. \tag{11.3.2}$$

Recall that d_{AB}^C denote the structure functions, that is $[F_A, F_B] = d_{AB}^C F_C$.

Proposition 11.3.3 *Let F_A be a SCL-frame over the open set $U \subset M$. There exists a unique Wess–Zumino frame $\tilde{F}_A$ over U such that $i^*F_A = i^*\tilde{F}_A$.*

Proof The two SCL-frames F_A and $\tilde{F}_A$ are related by a transformation from SCL, that is

$$\begin{pmatrix} \tilde{F}_z \\ \tilde{F}_+ \end{pmatrix} = \begin{pmatrix} T^2 & V \\ 0 & T \end{pmatrix} \begin{pmatrix} F_z \\ F_+ \end{pmatrix}$$

such that $i^\# T = 1$ and $i^\# V = 0$. Let us denote the structure coefficients of the frame F_A by d_{AB}^C and the structure coefficients of $\tilde{F}_A$ by $\tilde{d}_{AB}^C$. The coefficients $\tilde{d}_{AB}^C$

can be calculated from d_{AB}^{C} and T, V as follows:

$$\begin{aligned}
[\tilde{F}_+, \tilde{F}_+] &= [TF_+, TF_+] = 2T\,(F_+T) + T^2\left(2F_z + d_{++}^{+}F_+\right)\\
&= 2\tilde{F}_z + \left(2\,(F_+T) - V + Td_{++}^{+}\right)\tilde{F}_+\\
[\tilde{F}_+, \tilde{F}_-] &= [TF_+, \overline{T}F_-]\\
&= A\left(F_+\overline{T}\right)F_- + \overline{T}\,(F_-T)\,F_+ + T\overline{T}\left(d_{+-}^{+}F_+ + d_{+-}^{-}F_-\right)\\
&= \left(\overline{T}d_{+-}^{+} + \frac{\overline{T}}{T}F_-T\right)\tilde{F}_+ + \left(\frac{T}{\overline{\overline{T}}}F_+\overline{T}+\right)F_-
\end{aligned}$$

The conditions for the Wess–Zumino frames, Eq. (11.3.2), yield

$$\begin{aligned}
0 &= i^{\#}\tilde{d}_{+-}^{+} = i^{\#}\left(\overline{T}d_{+-}^{+} + \frac{\overline{T}}{T}F_-T\right) = i^{\#}d_{+-}^{+} + i^{\#}F_-T\\
0 &= i^{\#}\tilde{d}_{++}^{+} = 2F_+T + d_{++}^{+}\\
0 &= i^{\#}\tilde{F}_+\tilde{d}_{+-}^{+} = i^{\#}F_+\left(\overline{T}d_{+-}^{+} + \frac{\overline{T}}{T}F_-T\right)\\
&= i^{\#}\left(\left(F_+\overline{T}\right)d_{+-}^{+} + F_+d_{+-}^{+} + \left(F_+\frac{\overline{T}}{T}\right)F_T + F_+F_-T\right)
\end{aligned}$$

Those equations determine T completely by Corollary 11.2.9. The function V is determined by the conditions that also the higher order terms of $\tilde{d}_{++}^{+}$ vanish. □

Notice that the procedure given in the proof of Proposition 11.3.3 does not respect U(1)-structures. That is for a given U(1)-structure on M and a U(1)-frame F_A the frame $\tilde{F}_A$ is in general not a U(1)-frame for the same U(1)-structure. However, as $i^*F_A = i^*\tilde{F}_A$, the induced U(1)-structures on i^*TM and hence also the induced metric frame and gravitino on $|M|$ are the same. It is the merit of the Wess–Zumino frame conditions that they single out a unique SCL-frame on M that is completely determined by data on $|M|$:

Lemma 11.3.4 *Let $i\colon |M| \to M$ be an underlying even manifold for a super Riemann surface M, F_A a local Wess–Zumino frame and $X^A = (x^a, \eta^\alpha)$ Wess–Zumino coordinates for F_A. We denote the frame coefficients by $F_A^{\ B}$, that is $F_A = F_A^{\ B}\partial_{X^B}$, and their expansion in orders of η as follows:*

$$F_A^{\ B} = {}_0F_A^{\ B} + \eta^\mu\,{}_\mu F_A^{\ B} + \eta^3\eta^4\,{}_{34}F_A^{\ B}$$

All frame coefficients of F_A are completely determined by its independent components ${}_0F_a{}^b$ and ${}_0F_a{}^\beta$. Recall from Corollary 11.2.10 that $f_a = f_a{}^b\partial_{x^b} = {}_0F_a{}^b\partial_{x^b}$, $s_\alpha = i^*F_\alpha$ and $\chi(f_a) = \chi_a{}^\beta s_\beta = {}_0F_a{}^\beta s_\beta$. Then

$$\begin{aligned}
F_a &= \left(\delta_a^s + \eta^\mu \Gamma_{\mu\nu}^s \chi_a{}^\nu + \eta^3\eta^4\,{}_{34}\overline{F}_a{}^s\right) f_s{}^b\partial_{x^b} \\
&\quad + \left(\chi_a{}^\beta + \eta^\mu\,{}_\mu F_a{}^\beta + \eta^3\eta^4\,{}_{34}F_a{}^\beta\right)\partial_{\eta^\beta}, \\
F_\alpha &= \left(\eta^\mu \Gamma_{\mu\alpha}^s + \eta^3\eta^4\gamma_\alpha^t{}^\lambda \Gamma_{\lambda\tau}^s \chi_t{}^\tau\right) f_s{}^b\partial_{x^b} \\
&\quad + \left(\delta_\alpha^\beta + \eta^\mu \Gamma_{\mu\alpha}^s \chi_s{}^\beta + \eta^3\eta^4\,{}_{34}F_\alpha{}^\beta\right)\partial_{\eta^\beta}.
\end{aligned}$$

The remaining coefficients are given by

$$\begin{aligned}
{}_{34}\overline{F}_a{}^s &= -\left(\|Q\chi\|^2\delta_a^s + 2g_S\left((P\chi)_a, (Q\chi)_t\right)\delta^{ts}\right), \\
{}_\mu F_a{}^\beta &= \Gamma_{\mu\tau}^s \chi_a{}^\tau \chi_s{}^\beta + \frac{1}{4}\gamma_{a\mu}{}^\lambda \gamma_\lambda^t{}^\beta \left(\mathrm{I}_t^a\,\omega_a^{LC} + 2g_S\left(\delta_\gamma\chi, \chi_t\right)\right), \\
{}_{34}F_a{}^\beta &= \|Q\chi\|^2 (Q\chi)_a{}^\beta - 4g_S\left((P\chi)_a, (Q\chi)_t\right)\delta^{tr}\chi_r{}^\beta \\
&\quad - \frac{1}{2}(P\chi)_a{}^\kappa \gamma_\kappa^t{}^\beta \mathrm{I}_t^r\,\omega_r^{LC}, \\
{}_{34}F_\alpha{}^\beta &= -\left(\|Q\chi\|^2\delta_\alpha^\beta + \frac{1}{2}\gamma_\alpha^t{}^\mu \mathrm{I}_\mu{}^\beta\,\omega_t^{LC}\right).
\end{aligned}$$

Here, we denote by $Q\chi \in \Gamma\left(T^\vee|M| \otimes S\right)$ the $\frac{3}{2}$-part of the gravitino, see Appendix A.3. That is,

$$Q\chi = (Q\chi)_t\, f^t = (Q\chi)_t{}^\tau\, f^t \otimes s_\tau = \frac{1}{2}\chi_s{}^\sigma \gamma_{t\sigma}{}^\mu \gamma^s_\mu{}^\tau f^t \otimes s_\tau,$$

and similarly for the $\frac{1}{2}$-part of the gravitino, $P\chi = \chi - Q\chi$. The term $\delta_\gamma\chi = \chi_k{}^\kappa\gamma^k_\kappa{}^\lambda s_\lambda \in \Gamma(S)$ is also called γ-trace of the gravitino. Furthermore, we denote by $\omega_a^{LC} = -\varepsilon^{bd} f_b{}^c\left(\partial_{x^c} f_d{}^m\right){}_0E_m{}^n\delta_{na}$, such that $\nabla_{f_a} f_b = \omega_a^{LC}\,\mathrm{I}\, f_b$ denotes the Levi-Civita covariant derivative with respect to the metric defined by the orthonormal frame f_a.

Hence locally, the data of an orthonormal frame f_a, a spin frame s_α and gravitino χ determine the super Riemann surface structure. The following Theorem 11.3.5 shows that also globally the data of a metric, a spinor bundle and a gravitino is sufficient to completely determine a super Riemann surface with underlying even manifold.

The proof of Lemma 11.3.4 is straightforward but cumbersome and postponed to Chap. 13. The proof proceeds by calculating explicitly the commutators for frames in Wess–Zumino coordinates and solving the equations imposed by integrability conditions and the Wess–Zumino conditions. Evidence is given by counting the degrees of freedom. The frame coefficients of F_A consist of 16 functions from $\mathcal{O}_M$, hence 64 functions from $\mathcal{O}_{|M|}$, whereas the frame f_a and gravitino χ consists of 8 functions from $\mathcal{O}_{|M|}$. The number of constraints sums up to $56 = 64 - 8$: The integrability conditions as in Lemma 9.2.8 amount to 32 independent conditions, the conditions for Wess–Zumino coordinates, Eqs. (11.2.3) and (11.2.4), are given by 12 equations and the conditions for Wess–Zumino frames, Eq. (11.3.2), give another 12 equations.

Theorem 11.3.5 *Let $|M|$ be a family of supermanifolds over B of dimension* $2|0$ *together with a metric g, a spinor bundle S and a gravitino field χ. There is a unique family of super Riemann surfaces M over B together with an inclusion $i\colon |M| \to M$, such that the metric $\tilde{g}$ and the gravitino $\tilde{\chi}$, induced on $|M|$ by the choice of a* $\mathrm{U}(1)$*-structure on M, differs from g and χ, respectively, only by a conformal and a super Weyl transformation.*

Proof Let $|M|$ be covered by open coordinate sets V_k with coordinates y_k^a. Choose local $\mathrm{U}(1)$-frames $s_{k\alpha}$ of S and f_{ka} of $T|M|$ such that for the corresponding complex frames it holds that $s_k \otimes_{\mathbb{C}} s_k \mapsto f_k$. We construct the super Riemann surface M via the patching of local constructions, exploiting the uniqueness of Wess–Zumino coordinates and Wess–Zumino frames.

Locally, over each V_k, the super Riemann surface M is constructed as a split supermanifold by setting $\mathcal{O}_M|_{V_k} = \bigwedge(\Gamma_{V_k}(S^\vee))$ with coordinates $x_k^a = y_k^a$ and $\eta_k^\alpha = s_k^\alpha$, where s_k^α is the canonical dual basis to $s_{k\alpha}$. Denote by F_{kA} the Wess–Zumino frame constructed from the coefficients of the frame f_{ka} and the gravitino χ, such that $X_k^A = (x_k^a, \eta_k^\alpha)$ are Wess–Zumino coordinates for F_{kA}. That is, the frame coefficients are given by Lemma 11.3.4. The SCL-orbit of F_{kA} in the frame bundle of TV_k gives a SCL-reduction of the structure group of TV_k. This SCL-reduction is automatically integrable, as the frame F_{kA} is constructed by help of the integrability conditions. The map i is locally constructed via $i^\# x_k^a = y_k^a$ and $i^\# \eta_k^\alpha = 0$.

It remains to glue together the constructions in order to obtain a well-defined super Riemann surface over the topological space $\|M\|$. That is, for $V_k \cap V_l \neq \emptyset$ we have to give a coordinate change between X_k^A and X_l^B and show that the SCL-structures on V_k and V_l coincide. Let us denote the coordinate changes on $|M|$ by $y_k^a = {}_0f_{kl}(y_l)^a$ and the frame changes by $s_k = g_{kl}s_l$. On $V_k \cap V_l$ there is a unique Wess–Zumino frame $\tilde{F}_{kA}$ in the SCL-orbit of F_{lA} such that $i^*\tilde{F}_{k+} = g_{kl}F_{l+}$ and $i^*\tilde{F}_{kz} = g_{kl}^2 i^* F_{lz}$ by Proposition 11.3.3. Furthermore there are unique Wess–Zumino coordinates $\tilde{X}_k^A$ for $\tilde{F}_{kA}$ such that that $i^\#\tilde{x}_k^a = y_k^a$, where an explicit formula for the coordinate change from X_l^A can be derived as in Proposition 11.2.5. The glueing proceeds by identifying X_k^A with $\tilde{X}_k^A$ and $\tilde{F}_{kA}$ with F_{kA}. □

Notice that it is not possible to reconstruct the U(1)-structure on M from the gravitino and metric on $|M|$. The reason is that the transformation between the Wess–Zumino frames F_{kA} and F_{lA} cannot be chosen to be a U(1)-transformation, as already mentioned after Proposition 11.3.3. This is also in line with the fact that different U(1)-structures on M can induce the same metric and gravitino on $|M|$, as explained in Sect. 11.1.

Corollary 11.3.6 *For any base B there is a bijection of sets*

$$\{i\colon |M| \to M \mid M \text{ super Riemann surface over } B\} \longleftrightarrow \{(|M|, g, S, \chi)\}\big/_{\text{Weyl, SWeyl}}.$$

On the right hand side we have the set of tuples consisting of a $2|0$-dimensional manifold $|M|$ over B, a metric g, a spinor bundle S and a gravitino χ on $|M|$ up to Weyl and super Weyl transformations.

Remark 11.3.7 It has been conjectured (see D'Hoker and Phong (1988, Equation 3.85) and Jost (2009, Section 2.4.7)) that the bijection of sets from Corollary 11.3.6 descends to a bijection

$$\{M \text{ super Riemann surface}\}\big/_{\mathrm{Diff}_B(M)} \longleftrightarrow \{(|M|, g, S, \chi)\}\big/_{\text{Weyl, SWeyl, }\mathrm{Diff}_B(|M|),\text{ SUSY}, \mathbb{Z}_2}. \tag{11.3.8}$$

The quotient by "SUSY" on the right hand side is to identify tuples that belong to different embeddings i. The quotient by $\mathbb{Z}_2$ identifies $(|M|, g, S, \chi)$ with $(|M|, g, S, -\chi)$.

This conjecture is underpinned by Corollary 3.3.14 and Proposition 11.4.2 that describe the action of $\mathrm{Diff}_B(|M|)$ and SUSY on g, S and χ. Furthermore, Theorem 11.5.4 proves an infinitesimal variant of this conjecture.

In Sachse (2009) it is proven that (a cover of) the left hand side of (11.3.8) can be identified with (a cover of) the moduli space of super Riemann surfaces. Hence establishing (11.3.8) would lead to a description of the supermoduli space in terms of metrics and gravitino similar to Teichmüller theory of Riemann surfaces. In the case of Riemann surfaces, the so called Teichmüller space is defined as

$$\{[g] \text{ conformal class of metrics on } |M|\}\big/_{\mathrm{Diff}_0(|M|)}.$$

Here the group of diffeomorphism homotopic to the identity $\mathrm{Diff}_0(|M|)$ acts on conformal classes via pullback. It is a Theorem due to Oswald Teichmüller that the Teichmüller space is diffeomorphic to $\mathbb{R}^{6g-6}$ and a cover of the moduli space of Riemann surfaces. For an overview on Teichmüller theory consult (Jost 2006; Tromba 1992).

Remark 11.3.9 The Wess–Zumino frame conditions serve two purposes: on one hand they allow to prove Lemma 11.3.4 and thus Theorem 11.3.5, on the other hand they will serve as a tool to perform certain calculations later on. It may well be possible that different Wess–Zumino frame conditions, that is different restriction on the commutators of the SCL-frames, can serve the same purpose. However, Lemma 11.3.4 does not hold for an arbitrary SCL-frame.

In Howe (1979, Equation 3.5) and D'Hoker and Phong (1988, Equation 3.32) formulas similar to Lemma 11.3.4 are presented without further constraints. Possibly, additional constraints were used implicitly by relying on the full torsion constraints as shown in Eq. (10.0.1), which are not completely SCL-invariant.

11.4 Superdiffeomorphisms and Supersymmetry

In the last sections we have seen how a U(1)-structure on a super Riemann surface structure M induces a metric and gravitino on an underlying even manifold $i\colon |M| \to M$. In this section we consider the action of a superdiffeomorphism $\Xi \in \operatorname{Diff}(M)$ on the super Riemann surface M and the its effect on the metric and gravitino fields. It turns out that the action on the metric and gravitino can be described as a combination of a diffeomorphism of $|M|$ and a change of the embedding i. An infinitesimal change of the embedding i yields the well-known supersymmetry transformations of metric and gravitino.

We denote the U(1)-principal bundle on M by P. The principal bundle P on the supermanifold M determines the super Riemann surface structure on M and the supermetric m. For any diffeomorphism $\Xi \in \operatorname{Diff}(M)$ the U(1)-structure P_Ξ defines an isomorphic super Riemann surface structure on M and a metric m_Ξ compatible with the super Riemann surface structure defined by P_Ξ such that (M, m) and (M, m_Ξ) are isometrically isomorphic. Let us denote frames associated to P by $F_A[P]$ and those associated to P_Ξ by $F_A[P_\Xi]$. They are related by

$$d\Xi F_A[P_\Xi] = \Xi^* F_A[P]. \tag{11.4.1}$$

The frames $F_A[P]$ and $F_A[P_\Xi]$ induce different metrics, spinor bundles and gravitinos on $|M|$. We will also indicate their dependence on the U(1)-structure and embedding $i\colon |M| \to M$ in square brackets, for example $S[i, P]$, $g[i, P]$ and $\chi[i, P]$.

Proposition 11.4.2 *Let* $\Xi \in \mathrm{Diff}(M)$ *and* $i\colon |M| \to M$ *be given and* $\xi \in \mathrm{Diff}(|M|)$ *and* $j\colon |M| \to M$ *such that* $\Xi \circ i = j \circ \xi$ *(see Corollary 3.3.14). Then* $d\Xi$ *induces an isomorphism of hermitian vector bundles*

$$s\colon S[i, P_\Xi] \to \xi^* S[j, P]$$

and it holds that

$$\begin{aligned} g[i, P_\Xi] &= g[j, P]_\xi, \\ s \circ \chi[i, P_\Xi] &= \xi^* \left(\chi[j, P]\right) \circ d\xi. \end{aligned}$$

Proof By the definition of P_Ξ the map $d\Xi$ maps $\mathcal{D}[P_\Xi]$ to $\Xi^*\mathcal{D}[P]$. Taking the pullback along i yields

$$\begin{aligned} i^* d\Xi\big|_{\mathcal{D}[P_\Xi]} : S[i, P_\Xi] = i^*\mathcal{D}[P_\Xi] &\to i^*\Xi^*\mathcal{D}[P] = \xi^* j^*\mathcal{D}[P] = \xi^* S[j, P] \\ i^* F_\alpha[P_\Xi] &\mapsto i^* d\Xi\, F_\alpha[P_\Xi] {=} i^*\Xi^* F_\alpha[P] {=} \xi^* j^* F_\alpha[P] \end{aligned}$$

which is invertible as Ξ is invertible. Setting $s = i^* d\Xi$ it is obvious that s is an isomorphism of hermitian vector bundles.

In order to calculate the effect of Ξ on g and χ notice that

$$\begin{aligned} d\Xi \circ p_{\mathcal{D}}[P_\Xi] &= \Xi^* p_{\mathcal{D}}[P] \circ d\Xi, \\ d\Xi \circ p_{\mathcal{D}^\perp}[P_\Xi] &= \Xi^* p_{\mathcal{D}^\perp}[P] \circ d\Xi. \end{aligned}$$

By the definition of $g[i, P_\Xi]$, we have for all $X, Y \in \Gamma(T|M|)$

$$\begin{aligned} g[i, P_\Xi](X, Y) &= i^* m[P_\Xi] \left(i^* p_{\mathcal{D}^\perp} \circ di X, i^* p_{\mathcal{D}^\perp} \circ di Y\right) \\ &= i^*\Xi^* m[P] \left(i^* d\Xi \circ i^* p_{\mathcal{D}^\perp} \circ di X, i^* d\Xi \circ i^* p_{\mathcal{D}^\perp} \circ di Y\right) \\ &= i^*\Xi^* m[P] \left(i^*\Xi^* p_{\mathcal{D}^\perp} \circ i^* d\Xi \circ di X, i^*\Xi^* p_{\mathcal{D}^\perp} \circ i^* d\Xi \circ di Y\right) \\ &= \xi^* j^* m[P] \left(\xi^* j^* p_{\mathcal{D}^\perp} \circ \xi^* dj \circ d\xi X, \xi^* j^* p_{\mathcal{D}^\perp} \circ \xi^* dj \circ d\xi Y\right) \\ &= \xi^* g[j, P] \left(d\xi X, d\xi Y\right) = g[j, P]_\xi \left(X, Y\right). \end{aligned}$$

Remark that this also proves $\xi^* f_a[j, P] = d\xi f_a[i, P_\Xi]$ for the corresponding orthonormal frames. The definition of the gravitino yields

$$\begin{aligned} \xi^*\chi[j, P]\, d\xi &= -\xi^* j^* p_{\mathcal{D}}[P] \circ \xi^* dj \circ d\xi = -i^*\Xi^* p_{\mathcal{D}}[P] \circ i^* d\Xi \circ di \\ &= -i^* d\Xi \circ i^* p_{\mathcal{D}}[P_\Xi] \circ di = s \circ \chi[i, P_\Xi]. \end{aligned}$$ □

Proposition 11.4.2 shows that the effect of a superdiffeomorphism Ξ on the metric g and gravitino χ defined with respect to an embedding i is the same as the combination of a diffeomorphism on $|M|$ with the change of the embedding from i to j. We will now proceed to consider the three special cases, where either $\xi = \mathrm{id}_{|M|}$ or $i = j$.

The case $\xi = \mathrm{id}_{|M|}$ and $i = j$ is mostly irrelevant since in this case s is an hermitian automorphism of S. We will not need to consider this case except for the following automorphism of the super Riemann surface M:

Example 11.4.3 Let M be a family of super Riemann surfaces over $B = pt$. The map that sends any odd coordinate to its negative is a diffeomorphism Ξ. As $B = pt$ the embedding $i: |M| \to M$ is unique and $\Xi \circ i = i$, hence $\xi = \mathrm{id}_{|M|}$. However, the induced map $s: S \to S$ is the map that multiplies every section with -1.

Let us now assume that in Proposition 11.4.2 we have $i = j$, that is, $\Xi \circ i = i \circ \xi$. In order to simplify notation, we will write $g = g[i, P]$, $\chi = \chi[i, P]$ and $S_g = S[i, P]$. Consequently, $g[i, P_\Xi] = g_\xi$ and the map $s: S_{g_\xi} \to \xi^* S_g$ coincides with the map ξ_S identifying the corresponding spinor bundles, see Appendix A.4. The gravitino

$$\chi_\xi = \chi[i, P_\Xi] = \xi_S^{-1} \circ \xi^* \chi \circ d\xi \in \Gamma\left(T^\vee M \otimes S_{g_\xi}\right)$$

will be called the pullback of χ along ξ. We are now interested in the infinitesimal change of metric and gravitino. That is we assume that there are time indexed families of diffeomorphisms $\Xi: M \to M$ and $\xi_\bullet: \mathbb{R} \times |M| \to |M|$ such that $\Xi_0 = \mathrm{id}_M$, $\xi_0 = \mathrm{id}_{|M|}$ and $\Xi_t \circ i = i \circ \xi_t$. It follows $i^*\left(\frac{d}{dt}\big|_{t=0} \Xi_t\right) = di\, \frac{d}{dt}\big|_{t=0} \xi_t$ and we write $X = \frac{d}{dt}\big|_{t=0} \xi_t \in \Gamma(T|M|)$. In order to compare compare χ_{ξ_t} with χ we identify $T^\vee|M| \otimes S_{g_{\xi_t}}$ with $T^\vee|M| \otimes S_g$ via $b_t^\vee \otimes \beta_t^{-1}$ and obtain

$$\begin{aligned} \delta_X g &= \left.\frac{d}{dt}\right|_{t=0} g_{\xi_t} = L_X g, \\ \delta_X \chi &= \left.\frac{d}{dt}\right|_{t=0} b_t^\vee \otimes (\beta_t)^{-1} \chi_{\xi_t} = \mathcal{L}_X \chi. \end{aligned} \tag{11.4.4}$$

Here, $\mathcal{L}$ denotes the Bourguignon–Gauduchon Lie-derivative.

In the case that the superdiffeomorphism Ξ induces only a change of embedding, that is $\Xi \circ i = j$, for some embedding of the underlying manifold $j: |M| \to M$, we have

$$g[i, P_\Xi] = g[j, P], \qquad s \circ \chi[i, P_\Xi] = \chi[j, P],$$

where $s: S[i, P_\Xi] \to S[j, P]$ is a hermitian isomorphism. Hence, in this case the superdiffeomorphism Ξ is equivalent to the change of embedding from i to j. In the remainder of this section, we will show that an infinitesimal change of the embedding induces the well-known supersymmetry transformations:

Definition 11.4.5 The following transformations are called supersymmetry transformations of the metric g and the gravitino χ induced by the spinor field q:

$$\left(\mathrm{susy}_q\, g\right)(X, Y) = 2g_S(q, \gamma(X)\chi(Y) + \gamma(Y)\chi(X)), \tag{11.4.6}$$

$$\left(\mathrm{susy}_q\, \chi\right)(X) = -\nabla^{LC}_X q - \frac{1}{2} g_S\left(\chi(X), \chi(f_l)\right)\gamma^l q, \tag{11.4.7}$$

where ∇^{LC} denotes the Levi-Civita covariant derivative on $|M|$ and its lift to S respectively.

Recall that the normal bundle $\mathcal{N}_{|M|/M}$ of i is defined as the quotient bundle in

$$0 \longrightarrow T|M| \overset{di}{\longrightarrow} i^*TM \longrightarrow \mathcal{N}_{|M|/M} \longrightarrow 0. \tag{11.4.8}$$

We know that $\mathcal{N}_{|M|/M}$ is isomorphic to S and that the short exact sequence (11.4.8) has two splittings, given by either by the orthogonal complement of $T|M|$ or the inclusion of S in TM given by the super Riemann surface structure. We will always use the latter one, that is we assume that any even section N of $\mathcal{N}_{|M|/M}$ is a section of i^*TM that can locally be written as

$$N = N^\mu i^* F_\mu$$

for some odd local functions N^μ on $|M|$.

Let now $\Xi_\bullet\colon \mathbb{R}\times M \to M$ be a time-indexed family of superdiffeomorphisms of M and $j_\bullet\colon \mathbb{R}\times |M| \to M$ a time-indexed family of embeddings of the underlying even manifold such that $\Xi_0 = \mathrm{id}_M$, $j_0 = i$ and $\Xi_t \circ i = j_t$ for all t. The time derivative $q = \frac{d}{dt}\big|_{t=0} j_t$ is a section of i^*TM. Notice that the vector field q is is not necessarily a section of $\mathcal{N}_{|M|/M}$, rather it may contain a nilpotent part lying in $T|M|$. However, we will assume without loss of generality that q is a section of $\mathcal{N}_{|M|/M}$, because we have dealt with the vector fields of the form $di X$ before.

Proposition 11.4.9 *Let the time-indexed family of metrics g_t, spinor bundles S_{g_t} and gravitinos χ_t be obtained from the superconformal metric m on the super Riemann surface M and the family of embeddings $j_t\colon |M| \to M$. Furthermore, let ∇ be a* U(1)*-connection on M and T its torsion tensor. Then,*

$$\begin{aligned} \frac{d}{dt}\bigg|_{t=0} g_t(X, Y) &= i^*m\left(i^*p_{\mathcal{D}^\perp} i^*T(q, di X), i^*p_{\mathcal{D}^\perp}\, di Y\right) \\ &\quad + i^*m\left(i^*p_{\mathcal{D}^\perp}\, di X, i^*p_{\mathcal{D}^\perp} i^*T(q, di Y)\right), \end{aligned} \tag{11.4.10}$$

$$\begin{aligned} \left(\frac{d}{dt}\bigg|_{t=0} b_t^\vee \otimes (\beta_t)^{-1}\chi_t\right)(X) &= \left\langle \chi(X), \left\langle q, i^*\nabla^{TM} F^\alpha\right\rangle\right\rangle i^* F_\alpha \\ &\quad + \chi\left(\frac{d}{dt} b_t X\bigg|_{t=0}\right) - \nabla^{i^*TM}_X q - i^* p_{\mathcal{D}} i^* T(q, di X), \end{aligned} \tag{11.4.11}$$

for all $X, Y \in \Gamma(T|M|)$ and m-orthonormal frames F_A. The term $\frac{d}{dt} b_t X\big|_{t=0}$ in Eq. (11.4.11) can be calculated from $\frac{d}{dt} g_t$.

Proof Note first that for any time-dependent vector field X_t on $T|M|$, we have

$$\nabla^{j_t^* TM}_{\partial_t} \, d j_t X_t\Big|_{t=0} = \left(j_t^* T(d j_t \partial_t, d j_t X_t) + \nabla^{j_t^* TM}_{X_t} \, d j_t \partial_t + d j_t[\partial_t, X_t]\right)\Big|_{t=0}$$
$$= i^* T(q, d i X) + \nabla^{i^* TM}_X q + d i \, \frac{d}{dt} X_t\Big|_{t=0}.$$

Consequently, for time-independent vector fields X and Y:

$$\frac{d}{dt}\Big|_{t=0} g_t(X, Y) = \frac{d}{dt}\Big|_{t=0} j_t^* m\left(j_t^* p_{\mathcal{D}^\perp} \, d j_t X, j_t^* p_{\mathcal{D}^\perp} \, d j_t Y\right)$$
$$= i^* m\left(i^* p_{\mathcal{D}^\perp}\left(\nabla^{j_t^* TM}_{\partial_t} \, d j_t X\right)\Big|_{t=0}, i^* p_{\mathcal{D}^\perp} \, d i Y\right)$$
$$+ i^* m\left(i^* p_{\mathcal{D}^\perp} \, d i X, i^* p_{\mathcal{D}^\perp}\left(\nabla^{j_t^* TM}_{\partial_t} \, d j_t Y\right)\Big|_{t=0}\right)$$
$$= i^* m\left(i^* p_{\mathcal{D}^\perp} i^* T(q, d i X), i^* p_{\mathcal{D}^\perp} \, d i Y\right)$$
$$+ i^* m\left(i^* p_{\mathcal{D}^\perp} \, d i X, i^* p_{\mathcal{D}^\perp} i^* T(q, d i Y)\right).$$

To calculate the variation of the gravitino we use a basis expression. Let us choose a U(1)-frame F_A of TM and notice that $\beta_t i^* F_\alpha = j_t^* F_\alpha$. Then,

$$\left(b_t^\vee \otimes (\beta_t)^{-1} \chi_t\right)(X) = \langle \chi_t(b_t X), j_t^* F^\alpha\rangle (\beta_t)^{-1} j_t^* F_\alpha = \langle \chi_t(b_t X), j_t^* F^\alpha\rangle i^* F_\alpha$$

and obtain

$$\left(\frac{d}{dt}\Big|_{t=0} b_t^\vee \otimes (\beta_t)^{-1} \chi_t\right)(X) = \frac{d}{dt}\Big|_{t=0} \langle \chi_t(b_t X), j_t^* F^\alpha\rangle i^* F_\alpha$$
$$= - \frac{d}{dt}\Big|_{t=0} \langle d j_t b_t X, j_t^* F^\alpha\rangle i^* F_\alpha$$
$$= \left(\left\langle \chi(X), \nabla^{j_t^* TM}_{\partial_t} j_t^* F^\alpha\Big|_{t=0}\right\rangle - \left\langle \nabla^{j_t^* TM}_{\partial_t} (d j_t b_t X)\Big|_{t=0}, i^* F^\alpha\right\rangle\right) i^* F_\alpha$$
$$= \Bigg(\left\langle \chi(X), \left\langle q, i^* \nabla^{TM} F^\alpha\right\rangle\right\rangle$$
$$- \left\langle i^* T(q, d i X) + \nabla^{i^* TM}_X q + d i \, \frac{d}{dt} b_t X\Big|_{t=0}, i^* F^\alpha\right\rangle\Bigg) i^* F_\alpha$$
$$= \left\langle \chi(X), \left\langle q, i^* \nabla^{TM} F^\alpha\right\rangle\right\rangle i^* F_\alpha + \chi\left(\frac{d}{dt} b_t X\Big|_{t=0}\right) - \nabla^{i^* TM}_X q$$
$$- i^* p_{\mathcal{D}} i^* T(q, d i X)$$

□

Note that the supersymmetry variation of metric and gravitino given in Eqs. (11.4.6) and (11.4.7) do not depend on the full U(1)-structure on TM but only on g and χ. In contrast, the expressions in Eqs. (11.4.10) and (11.4.11) might depend on the full U(1)-structure and be, in general, not expressible in terms of q, metric and gravitino only. To explore this aspect further, let $\tilde{m}$ represent another U(1)-structure on M such that $i^*m = i^*\tilde{m}$. In particular, $\tilde{m}$ is given by a rescaling by λ^2 such that $i^*\lambda = 1$ and a change of splitting l such that $i^*l = 0$. By Proposition 10.2.8, any U(1)-connection $\tilde{\nabla}$ for $\tilde{m}$ differs from ∇ by a differential form A with values in $\mathfrak{scl}$, such that

$$i^*Ai^*F_A = \begin{pmatrix} 2i^*\frac{d\lambda}{\lambda}\,\mathrm{id}_a{}^b + 2i^*\tau\,\mathrm{I}_a{}^b & i^*\left(\nabla l + \frac{d\lambda}{\lambda}l\right)^\mu \gamma_{a\mu}{}^\beta \\ 0 & i^*\frac{d\lambda}{\lambda}\,\mathrm{id}_\alpha{}^\beta + i^*\tau\,\mathrm{I}_\alpha{}^\beta \end{pmatrix} \begin{pmatrix} i^*F_b \\ i^*F_\beta \end{pmatrix}.$$

Here F_A is an m-orthonormal frame, $i^*\tau \in \Gamma\left(i^*T^\vee M\right)$.

Let now

$$\tilde{g}_t(X, Y) = j_t^*m\left(j_t^*p_{\mathcal{D}^\perp}\,dj_tX,\, j_t^*p_{\mathcal{D}^\perp}\,dj_tY\right), \qquad \tilde{\chi}_t = -j_t^*p_{\mathcal{D}}\,dj_t,$$

with $\tilde{g}_0 = g$ and $\tilde{\chi}_0 = \chi$. We introduce the following shorthand notation for the variations calculated with respect to the different U(1)-structures given by m and $\tilde{m}$:

$$\delta_q g[m] = \left.\frac{d}{dt}\right|_{t=0} g_t, \qquad \delta_q\chi[m] = \left.\frac{d}{dt}\right|_{t=0} b_t^\vee \otimes (\beta_t)^{-1}\chi_t,$$

$$\delta_q g[\tilde{m}] = \left.\frac{d}{dt}\right|_{t=0} \tilde{g}_t, \qquad \delta_q\chi[\tilde{m}] = \left.\frac{d}{dt}\right|_{t=0} \tilde{b}_t^\vee \otimes \left(\tilde{\beta}_t\right)^{-1}\tilde{\chi}_t,$$

where we denote by $\tilde{b}_t\colon (T|M|, g) \to (T|M|, \tilde{g}_t)$ and $\tilde{\beta}_t\colon S_g \to S_{\tilde{g}_t}$ the canonical isometries.

Lemma 11.4.12 *Let $\tilde{m}$ arise from a superconformal rescaling by λ^2 and the superconformal change of splitting l such that $i^*\lambda = 1$ and $i^*l = 0$ as before. Then*

$$\delta_q g[\tilde{m}] - \delta_q g[m] = 4\langle q, i^*\,d\lambda\rangle\, g,$$
$$\left(\delta_q\chi[\tilde{m}] - \delta_q\chi[m]\right)(X) = -\gamma(X)\langle q, i^*\nabla l\rangle - \langle q, i^*\,d\lambda\rangle\,\chi(X).$$

Proof Let $\tilde{T}$ denote the torsion tensor of $\tilde{\nabla}$. The difference $\tilde{T}-T$ has been expressed in terms of A in Sect. 10.1. We use the expansions $q = q^\mu i^* F_\mu$ and $X = X^a f_a$ and $dif_a = i^* F_a - \chi_a{}^\alpha i^* F_\alpha$. We express the torsion tensors with respect to the basis $i^* F_A$. Then,

$$\begin{aligned}
&i^* p_{\mathcal{D}\perp} i^* \left(\tilde{T} - T\right)(q, diX) \\
&= q^\mu X^a \left(i^* \tilde{T}_{\mu a}{}^c - i^* T_{\mu a}{}^c + \chi_a{}^\nu \left(i^* \tilde{T}_{\mu\nu}{}^c - i^* T_{\mu\nu}{}^c\right)\right) i^* F_c \\
&= q^\mu X^a 2 i^* \left(\frac{d\lambda}{\lambda} \mathrm{id}_a{}^c + \tau\, \mathrm{I}_a{}^c\right)_\mu i^* F_c \\
&= 2\left(\langle q, i^* d\lambda\rangle + \langle q, i^* \tau\rangle \mathrm{I}\right) i^* p_{\mathcal{D}\perp}\, diX.
\end{aligned}$$

We obtain

$$\begin{aligned}
\delta_q g[\tilde{m}] - \delta_q g[m] &= i^* m\left(i^* p_{\mathcal{D}\perp} i^* \left(\tilde{T} - T\right)(q, diX), i^* p_{\mathcal{D}\perp}\, diY\right) \\
&\quad + i^* m\left(i^* p_{\mathcal{D}\perp}\, diX, i^* p_{\mathcal{D}\perp} i^* \left(\tilde{T} - T\right)(q, diY)\right) \\
&= 2 i^* m\left(\left(\langle q, i^* d\lambda\rangle + \langle q, i^* \tau\rangle \mathrm{I}\right) i^* p_{\mathcal{D}\perp}\, diX, i^* p_{\mathcal{D}\perp}\, diY\right) \\
&\quad + 2 i^* m\left(i^* p_{\mathcal{D}\perp}\, diX, \left(\langle q, i^* d\lambda\rangle + \langle q, i^* \tau\rangle \mathrm{I}\right) i^* p_{\mathcal{D}\perp}\, diY\right) \\
&= 4\langle q, i^* d\lambda\rangle g.
\end{aligned}$$

Consequently, we have $\frac{d}{dt}\tilde{b}_t X\Big|_{t=0} - \frac{d}{dt} b_t X\Big|_{t=0} = -2\langle q, i^* d\lambda\rangle X$. In addition,

$$\begin{aligned}
&i^* p_{\mathcal{D}} i^* \left(\tilde{T} - T\right)(q, dif_a) \\
&= q^\mu \left(i^* \tilde{T}_{\mu a}{}^\gamma - i^* T_{\mu a}{}^\gamma + \chi_a{}^\alpha \left(i^* \tilde{T}_{\mu\alpha}{}^\gamma - i^* T_{\mu\alpha}{}^\gamma\right)\right) i^* F_\gamma \\
&= \langle q, (i^* \nabla l)^\mu\rangle \gamma_{a\mu}{}^\gamma i^* F_\gamma - \left(\langle i^* F_a, i^* d\lambda\rangle + \langle i^* F_a, i^* \tau\rangle \mathrm{I}\right) q \\
&\quad + q^\mu \chi_a{}^\alpha \left(\left(i^* d\lambda\right)_\mu \delta_\alpha^\gamma + i^* \tau_\mu \mathrm{I}_\alpha{}^\gamma + \left(i^* d\lambda\right)_\alpha \delta_\mu^\gamma + i^* \tau_\alpha \mathrm{I}_\mu{}^\gamma\right) i^* F_\gamma \\
&= \gamma(f_a)\langle q, i^* \nabla l\rangle - \left(\langle dif_a, i^* d\lambda\rangle + \langle dif_a, i^* \tau\rangle \mathrm{I}\right) q \\
&\quad - \left(\langle q, i^* d\lambda\rangle + \langle q, i^* \tau\rangle \mathrm{I}\right) \chi(f_a).
\end{aligned}$$

Furthermore,

$$\tilde{\nabla}_{f_a}^{i^* TM} q - \nabla_{f_a}^{i^* TM} q = \left(\langle dif_a, i^* d\lambda\rangle + \langle dif_a, i^* \tau\rangle \mathrm{I}\right) q.$$

Now assume define an $\tilde{m}$-orthonormal frame $\tilde{F}_A$ by $\tilde{F}_a = \lambda^{-2}\left(F_a + l^\mu \gamma_{a\mu}{}^\beta F_\beta\right)$ and $\tilde{F}_\alpha = \lambda^{-1} F_\alpha$. In particular, we have $i^* F_A = i^* \tilde{F}_A$. Then,

$$\begin{aligned}\left\langle q, i^* \tilde{\nabla} \tilde{F}^\alpha \right\rangle - \left\langle q, i^* \nabla F^\alpha \right\rangle &= \left\langle q, i^* \left(\tilde{\nabla} \lambda \left(F^\alpha - F^b l^\mu \gamma_{b\mu}{}^\alpha\right) - \nabla F^\alpha\right)\right\rangle \\ &= i^* F^\alpha \left\langle q, i^* d\lambda\right\rangle - i^* F^\beta \left(\langle q, d\lambda\rangle \delta^\alpha_\beta + \left\langle q, i^*\tau\right\rangle \mathrm{I}_\beta{}^\alpha\right) \\ &\quad - i^* F^b \left\langle q, \gamma_b i^* \nabla l\right\rangle - i^* F^b \left\langle q, \gamma_b i^* \nabla l\right\rangle \\ &= -\left\langle q, i^*\tau\right\rangle i^* F^\beta \,\mathrm{I}_\beta{}^\alpha - 2 i^* F^b \left\langle q, \gamma_b i^* \nabla l\right\rangle.\end{aligned}$$

Summing up, we obtain:

$$\begin{aligned}\left(\delta_q \chi[\tilde{m}] - \delta_q \chi[m]\right)(X) &= -\left\langle q, i^*\tau\right\rangle \mathrm{I}\,\chi(X) - 2\left\langle q, i^* d\lambda\right\rangle \chi(X) \\ &\quad - \left(\left\langle di X, i^* d\lambda\right\rangle + \left\langle di X, i^*\tau\right\rangle \mathrm{I}\right) q \\ &\quad - \gamma(X)\left\langle q, i^* \nabla l\right\rangle + \left(\left\langle di X, i^* d\lambda\right\rangle + \left\langle di X, i^*\tau\right\rangle \mathrm{I}\right) q \\ &\quad + \left(\left\langle q, i^* d\lambda\right\rangle + \left\langle q, i^*\tau\right\rangle \mathrm{I}\right)\chi(X) \\ &= -\gamma(X)\left\langle q, i^* \nabla l\right\rangle - \left\langle q, i^* d\lambda\right\rangle \chi(X).\end{aligned}$$

This shows the claim. □

The above Lemma 11.4.12 shows that $\delta_q g[m]$ and $\delta_q \chi[m]$ really depend on the U(1)-structure m and not only on the metric g and gravitino χ. In contrast, the supersymmetry transformations given in Definition 11.4.5 depend only on g and χ. Hence the following is the best we can expect:

Proposition 11.4.13 *For every metric m compatible with the super Riemann surface structure on M the infinitesimal variations induced on metric and gravitino coincide with the supersymmetry transformations up to an infinitesimal Weyl and super Weyl transformation. That is, there is an $\sigma \in \mathcal{N}_{M/|M|}{}^\vee$ and an $s \in \mathcal{N}_{M/|M|}{}^\vee \otimes S$ such that*

$$\begin{aligned}\delta_q g[m] &= \mathrm{susy}_q\, g + 4\langle q, \sigma\rangle g, \\ \left(\delta_q \chi[m]\right)(X) &= \left(\mathrm{susy}_q\, \chi\right)(X) - \gamma(X)\langle q, s\rangle - \langle q, \sigma\rangle \chi(X),\end{aligned}$$

for every $X \in T|M|$.

For the proof of Proposition 11.4.13, it remains to show the following lemma treating the local case for Wess–Zumino pairs.

Lemma 11.4.14 *Let F_A be the Wess–Zumino frame determined by f_a, s_α and χ and m the superconformal metric determined by F_A. There exists a $s \in \mathcal{N}_{M/|M|}{}^\vee \otimes S$*

such that

$$\delta_q g[m] = \operatorname{susy}_q g,$$
$$(\delta_q \chi[m])(X) = (\operatorname{susy}_q \chi)(X) - \gamma(X)\langle q, s\rangle .$$

We postpone the proof of Lemma 11.4.14 to Sect. 13.8.

Remark 11.4.15 The supersymmetry transformations of metric and gravitino can be found in the literature in the following form:

$$\operatorname{susy}_q f_a = -2gs\left(q, \gamma^c \chi(f_a)\right) f_c,$$
$$\operatorname{susy}_q \chi(Y) = -\nabla^{LC}_Y q + gs\left(\delta_\gamma \chi, \chi(\mathrm{I}Y)\right)\mathrm{I} q,$$

see, for example Brink et al. (1976, Equation (16)), Deser and Zumino (1976, Equation (5)) and Jost (2009, Equations 2.4.145 and 2.4.146). Here the metric g is encoded by an orthonormal frame f_a. It can be directly verified that the above supersymmetry variation of the frame f_a induce the supersymmetry variation $\operatorname{susy}_q g$ given in Eq. (11.4.6). Conversely, $\operatorname{susy}_q f_a$ coincides with $\frac{d}{dt}\big|_{t=0} b_t f_a$ up to an infinitesimal U(1)-transformation.

For the supersymmetry of the gravitino, it holds

$$(\operatorname{susy}_q \chi)(X) = \operatorname{susy}_q \chi(X) + \chi\left(\left.\frac{d}{dt}\right|_{t=0} b_t X\right),$$

compare the proof of Lemma 11.4.14 in Sect. 13.8. Thus, $\operatorname{susy}_q \chi(X)$ coincides with $\frac{d}{dt}\big|_{t=0} \beta_t(\chi_t(X))$ up to infinitesimal conformal and super Weyl transformations.

Notice that we have given a supergeometric explanation to the terms $\operatorname{susy}_q g$ and $\operatorname{susy}_q \chi$ in this section. In contrast, in the literature cited before, the supersymmetry transformations are introduced as particular symmetry transformations that leave the superconformal action functional (1.1.2) invariant.

11.5 Infinitesimal Deformations of Super Riemann Surfaces

By Theorem 11.3.5 a super Riemann surface is completely determined by a metric g, the spinor bundle S and the gravitino χ. Infinitesimal deformations of super Riemann surfaces are hence encoded in infinitesimal deformations of the metric g and the gravitino χ. Notice that the spinor bundle S cannot be deformed infinitesimally. However, certain infinitesimal deformations of g and χ are not infinitesimal deformations of the super Riemann surface structure because different metrics and gravitinos give rise to the same super Riemann surface.

For the rest of this section we will consider g as a section of $T^\vee|M| \odot T^\vee|M|$ and χ as a section of $T^\vee|M| \otimes S^\vee$ by identifying S and $S^\vee$ with the help of g_S. Consequently an infinitesimal deformations of the metric is given by $h \in \Gamma\left(T^\vee|M| \odot T^\vee|M|\right)$; an infinitesimal deformation of the gravitino is given by $\rho \in \Gamma\left(T^\vee|M| \otimes S^\vee\right)$. The infinitesimal deformations of metric and gravitino that do not deform the super Riemann surface structure are given by

- Infinitesimal conformal transformations:

$$(h, \rho) = (4\sigma g, -\sigma\chi)$$

  Here σ denotes a function on $|M|$.
- Infinitesimal super Weyl transformations:

$$(h, \rho) = (0, \gamma s)$$

  Here s denotes a section of $S^\vee$.
- Infinitesimal diffeomorphisms, see Eq. (11.4.4):

$$(h, \rho) = (L_Y g, \mathcal{L}_Y \chi)$$

  Here Y is a vector field on $|M|$.
- Supersymmetry, see Proposition 11.4.13:

$$(h, \rho) = (\operatorname{susy}_q g, \operatorname{susy}_q \chi)$$

  The supersymmetry parameter q is spinor, that is $q \in \Gamma(S)$.

Let us denote by

$$TD \subset \Gamma\left(T^\vee|M| \odot T^\vee|M| \oplus T^\vee|M| \otimes S^\vee\right)$$

the subspace trivial deformations, that is the subspace generated by infinitesimal conformal, infinitesimal super Weyl transformations, vector fields and supersymmetry. The quotient

$$\Gamma\left(T^\vee|M| \odot T^\vee|M| \oplus T^\vee|M| \otimes S^\vee\right)\big/_{TD} \tag{11.5.1}$$

contains all true infinitesimal deformations of the given super Riemann surface.

In order to study the space of true infinitesimal deformations of the super Riemann surface, we introduce the following generalization of the Weil–Petersson metric:

Definition 11.5.2 The super L^2-metric

$$\ell\left((h, \rho), (\tilde{h}, \tilde{\rho})\right) = \int_{|M|/B} g^\vee \otimes g^\vee\left(h, \tilde{h}\right) + g^\vee \otimes g_S^\vee(\rho, \tilde{\rho})\, dvol_g$$

is a symmetric, non-degenerate bilinear form on $\Gamma\left(T^\vee|M| \odot T^\vee|M| \oplus T^\vee|M| \otimes S^\vee\right)$. With respect to a local frame f_a of $T^\vee|M|$ and s_α of S it is given by

$$\ell\left((h,\rho),(\tilde{h},\tilde{\rho})\right) = \int_{|M|/B} g^{kl} g^{mn} h_{km} \tilde{h}_{ln} - g^{kl} g_S{}^{\mu\nu} \rho_{k\mu} \tilde{\rho}_{l\nu}\, dvol_g.$$

Note that we use the term L^2-metric for its similarity to the L^2-metric on the space of symmetric two-forms, see for example in Tromba (1992), Nevertheless, we will only apply it to smooth sections. The orthogonal complement $TD^\perp$ of TD with respect to the L^2-metric is naturally identified with the quotient (11.5.1) and hence contains the true deformations of the super Riemann surface. The L^2-metric induces a metric on the space $TD^\perp$ of infinitesimal deformations of the super Riemann surface. This induced metric might be an analogue of the Weil–Petersson metric for super Riemann surfaces. We have to leave the study of the super Weil–Petersson metric for later and will instead describe the space of infinitesimal deformations more precisely.

Proposition 11.5.3 *A tuple* $(h,\rho) \in \Gamma\left(T^\vee|M| \odot T^\vee|M| \oplus T^\vee|M| \otimes S^\vee\right)$ *lies in* $TD^\perp$ *if and only if*

$$\begin{aligned}
0 &= 4\operatorname{Tr}_g h - g^\vee \otimes g_S^\vee(\chi,\rho),\\
0 &= \delta_\gamma \rho,\\
0 &= 2\operatorname{div}_g h - \operatorname{div}_\chi \rho,\\
0 &= 4g^\vee \otimes g^\vee(h,\gamma\chi) + \operatorname{div}_g \rho - \frac{1}{2} g^{kl} g_S(\chi(f_k),\chi(f_a))\gamma^a \rho(f_l).
\end{aligned}$$

Recall that $\operatorname{div}_g h = \operatorname{Tr}_g \nabla^{LC} h \in \Gamma\left(T^\vee|M|\right)$, and $\operatorname{div}_\chi \rho \in \Gamma\left(T^\vee|M|\right)$ given by

$$\left\langle X, \operatorname{div}_\chi \rho\right\rangle = g^\vee \otimes g_S^\vee\left(\nabla^{LC}_X \chi, \rho\right) + \mathrm{I}X\left(g^\vee \otimes g_S^\vee\left(\frac{1}{2}\chi\circ \mathrm{I} - \frac{1}{4}\mathrm{I}\circ\chi, \rho\right)\right),$$

are L^2-adjoint to Lie derivatives, see Proposition A.4.4. Similarly, we define $\operatorname{div}_g \rho \in \Gamma\left(S^\vee\right)$ by $\operatorname{div}_g \rho = \operatorname{Tr}_g \nabla^{LC} \rho$.

Proof Let (h,ρ) be an element of $TD^\perp$. The condition that (h,ρ) is orthogonal to all conformal transformations implies for all $\sigma \in \mathcal{O}_{|M|}$

$$0 = \ell\left((h,\rho),(4\sigma g, -\sigma\chi)\right) = \int_{|M|/B} 4\sigma g^{kl} h_{kl} + \sigma g^{kl} g_S{}^{\mu\nu} \chi_{k\mu} \rho_{l\nu}\, dvol_g$$

and hence that the g-trace of h is prescribed.

Furthermore, for all $Y \in \Gamma(T|M|)$ it holds

$$\begin{aligned} 0 &= \ell\left((h,\rho),(L_Y g,\mathcal{L}_Y \chi)\right) \\ &= \int_{|M|/B} g^\vee \otimes g^\vee (h, L_Y g) + g^\vee \otimes g_S^\vee (\rho, \mathcal{L}_Y \chi)\, dvol_g \\ &= -\int_{|M|/B} \langle Y, 2\operatorname{div}_g h - \operatorname{div}_\chi \rho\rangle\, dvol_g \end{aligned}$$

and consequently $2\operatorname{div}_g h - \operatorname{div}_\chi \rho = 0$.

Let now s be a section of $S^\vee$. Then

$$\begin{aligned} 0 = \ell\left((h,\rho),(0,\gamma s)\right) &= \int_{|M|/B} g^\vee \otimes g_S^\vee (\rho, \gamma s)\, dvol_g \\ &= \int_{|M|/B} g_S^\vee \left(\delta_\gamma \rho, s\right) dvol_g. \end{aligned}$$

That is, the orthogonality to super Weyl transformations implies $\delta_\gamma \rho = 0$.

Finally, since (h, ρ) is also orthogonal to supersymmetry transformations, we obtain for all $q \in \Gamma(S)$

$$\begin{aligned} 0 &= \ell\left((h,\rho),(\operatorname{susy}_q g, \operatorname{susy}_q \chi)\right) \\ &= \int_{|M|/B} 4 g^{kl} g^{mn} h_{km} \langle q, \gamma_l \chi(f_n)\rangle \\ &\quad + g^{kl}\left\langle -\nabla_{f_k} q - \frac{1}{2} g_S(\chi(f_k), \chi(f_a))\gamma^a q, \rho(f_l)\right\rangle dvol_g \\ &= \int_{|M|/B} \langle q, 4 g^\vee \otimes g^\vee (h, \gamma\chi) + \operatorname{div}_g \rho\rangle \\ &\quad - \left\langle q, \frac{1}{2} g^{kl} g_S(\chi(f_k), \chi(f_a))\gamma^a \rho(f_l)\right\rangle dvol_g. \end{aligned}$$

□

Theorem 11.5.4 *Let M be a super Riemann surfaces that is given by g, S and $\chi = 0$ with respect to the underlying even manifold $i\colon |M| \to M$. The space of infinitesimal deformations of the super Riemann surface M can be identified with*

$$H^0(T^\vee|M| \otimes_{\mathbb{C}} T^\vee|M|) \oplus H^0(S^\vee \otimes_{\mathbb{C}} S^\vee \otimes_{\mathbb{C}} S^\vee)$$

Here H^0 denotes even holomorphic sections.

Proof By Proposition 11.5.3 infinitesimal deformations of the given super Riemann surface are given by $(h, \rho) \in \Gamma\left(T^\vee|M| \odot T^\vee|M| \oplus T^\vee|M| \otimes S^\vee\right)$ such that $\operatorname{Tr}_g h = 0$, $\operatorname{div}_g h = 0$, $\delta_\gamma \rho = 0$ and $\operatorname{div}_g \rho = 0$. Symmetric, trace-free and divergence-free bilinear forms on $T|M|$ can be identified with holomorphic quadratic differentials on $|M|$, see, for example, Jost (2001, Chapter 2.6). The identification is given in conformal coordinates $z = x^1 + \mathrm{i}x^2$ by

$$h_{ij}\, dx^i \otimes dx^j \mapsto \left(h_{11} - \mathrm{i}h_{12}\right) dz \otimes dz.$$

This identification yields a holomorphic section of $T^\vee|M| \otimes_{\mathbb{C}} T^\vee|M|$ because h is divergence-free. Indeed, for the orthonormal frame $f_a = \frac{1}{\sigma^2}\partial_{x^a}$, we obtain

$$\begin{aligned}
0 = (\mathrm{div}_g\, h)(f_e) &= \left(\delta^{ad}\nabla^{LC}_{f_a}\left(h(f_b, f_c) f^b \odot f^c\right)\right)(f_d, f_e) \\
&= \delta^{ad}\left(\left(f_a h(f_b, f_c)\right) f^b \odot f^c + 2h(f_b, f_c)\left(\nabla^{LC}_{f_a} f^b\right) \odot f^c\right)(f_d, f_e) \\
&= \delta^{ad}\left(\left(f_a h(f_d, f_e)\right) - 2h(f_b, f_e)\left\langle f_a, \omega^{LC}\right\rangle \mathrm{I}_d{}^b\right) \\
&= \delta^{ad}\frac{1}{\sigma^2}\left(\partial_{x^a}\left(\frac{1}{\sigma^4}h(\partial_{x^d}, \partial_{x^e})\right) - 2\frac{1}{\sigma^4}h(\partial_{x^b}, \partial_{x^e})\left(\sigma^2\, \mathrm{I}_a{}^n\, \partial_{x^n}\frac{1}{\sigma^2}\right)\mathrm{I}_d{}^b\right) \\
&= \delta^{ad}\frac{1}{\sigma^6}\partial_{x^a}\left(h(\partial_{x^d}, \partial_{x^e})\right).
\end{aligned}$$

Here $\left\langle \partial_{x^a}, \omega^{LC}\right\rangle = \sigma^2\, \mathrm{I}_a{}^b\, \partial_{x^b}\frac{1}{\sigma^2}$ is the one form such that $\nabla^{LC}X = \omega^{LC}\,\mathrm{I}\,X$. For $e = 1, 2$ respectively,

$$\begin{aligned}
0 &= \partial_{x^1}h_{11} + \partial_{x^2}h_{21} = \partial_{x^1}h_{11} + \partial_{x^2}h_{12}, \\
0 &= \partial_{x^1}h_{12} + \partial_{x^2}h_{22} = \partial_{x^1}h_{12} - \partial_{x^2}h_{11},
\end{aligned}$$

implies the holomorphicity of $(h_{11} - \mathrm{i}h_{12})\, dz \otimes_{\mathbb{C}} dz$.

To show that ρ can be identified with a holomorphic section of $S^\vee \otimes_{\mathbb{C}} S^\vee \otimes_{\mathbb{C}} S^\vee$ choose a local holomorphic frame t_+ such that $t_+ \otimes t_+ = \partial_z$ and $t_+ = \frac{1}{2}(t_3 - \mathrm{i}t_4)$. The equation $\delta_\gamma \rho = 0$ is equivalent to

$$\rho_{13} + \rho_{24} = 0, \qquad\qquad \rho_{14} - \rho_{23} = 0.$$

Consequently the map

$$\rho_{a\beta}\, dx^a \otimes t^\beta \mapsto \left(\rho_{13} - \mathrm{i}\rho_{14}\right) dz \otimes t^+$$

is well-defined independently of the holomorphic frame.

The section ρ is holomorphic because $\mathrm{div}_g\, \rho = 0$, analogous to h. For the proof notice that the metric g is given by $g\left(\partial_{x^i}, \partial_{x^j}\right) = \sigma^4\delta_{ij}$ and $g_S\left(t_\alpha, t_\beta\right) = \sigma^2\varepsilon_{\alpha\beta}$.

Consequently, $f_a = \frac{1}{\sigma^2}\partial_{x^a}$ and $s_\alpha = \frac{1}{\sigma}t_\alpha$ are orthonormal frames. We obtain

$$
\begin{aligned}
0 = \operatorname{div}_g \rho &= s^\mu \delta^{kl} \left(\nabla^{LC}_{f_l} \rho\right)_{k\mu} \\
&= \delta^{kl} s^\mu \left(f_l \left(\langle s_\mu, \rho(f_k)\rangle\right) + \left\langle f_l, \omega^{LC}\right\rangle \left(- \mathrm{I}_k{}^m \delta^\nu_\mu - \frac{1}{2} \mathrm{I}_\mu{}^\nu \delta^m_k \right) \langle s_\nu, \rho(f_m)\rangle \right) \\
&= \delta^{kl} \sigma t^\mu \left(\frac{1}{\sigma^2} \partial_{x^l} \left(\frac{1}{\sigma^3} \left\langle t_\mu, \rho(\partial_{x^k})\right\rangle \right) - \frac{3}{2} \mathrm{I}_l{}^b \left(\partial_{x^b} \frac{1}{\sigma^2} \right) \mathrm{I}_k{}^m \frac{1}{\sigma^3} \left\langle t_\mu, \rho(\partial_{x^m})\right\rangle \right) \\
&= \frac{1}{\sigma^4} \delta^{kl} t^\mu \partial_{x^l} \left(\left\langle t_\mu, \rho(\partial_{x^k})\right\rangle\right) \\
&= \frac{1}{\sigma^4} \left(t^3 \left(\partial_{x^1} \rho_{13} + \partial_{x^2} \rho_{14} \right) + t^4 \left(\partial_{x^1} \rho_{14} - \partial_{x^2} \rho_{13} \right) \right)
\end{aligned}
$$

Here we have used several times that ρ is a section of $T^\vee|M| \otimes_{\mathbb{C}} S^\vee$. □

Similar statements for trivial families can be found in LeBrun and Rothstein (1988), Crane and Rabin (1988) or Sachse (2009). The proof given here shows directly which deformations of metric and gravitino correspond to infinitesimal deformations of the given super Riemann surface. Notice that Theorem 11.5.4 is an infinitesimal variant of the conjecture presented in Remark 11.3.7.

The complex dimension of the infinitesimal deformation space can be calculated by the theorem of Riemann–Roch in the case of $B = \mathbb{R}^{0|0}$. The dimension is found to be $3p - 3|2p - 2$ for genus $p \geq 2$. Here we denote the holomorphic quadratic differentials as even deformations and the sections of $T^\vee|M| \otimes_{\mathbb{C}} S^\vee$ as odd deformations.

Chapter 12
The Superconformal Action Functional

This chapter treats the superconformal action functional. For any map $\Phi : M \to N$ from a super Riemann surface to an arbitrary Riemannian supermanifold N and any U(1)-metric m on M, the action functional $A(\Phi, m)$ is a function in $\mathcal{O}_B$ that depends on the super Riemann surface structure on M and the map Φ. We will see that the action functional $A(\Phi, m)$ is a natural generalization of the action functional of harmonic maps on Riemann surfaces in several aspects.

The first analogy between the harmonic action functional and the superconformal action functional are the conformal properties. The harmonic action functional on Riemann surfaces is conformally invariant, that is depends only on the conformal class or complex structure on the surface. The superconformal action functional $A(\Phi, m)$ is superconformally invariant. That is, $A(\Phi, m)$ does not depend on the given U(1)-structure but rather only on the SCL-structure or super Riemann surface structure. Furthermore, the harmonic action functional is diffeomorphism invariant, whereas the superconformal action functional is superdiffeomorphism invariant.

The conformal and diffeomorphism invariance of the harmonic action functional allow to use it as a tool to study the Teichmüller theory of Riemann surfaces, see Wolf (1989) and Tromba (1992). It is expected that the superconformal action functional can be used to study Teichüller theory of super Riemann surfaces, see D'Hoker and Phong (1988) and Jost (2009). In this text we can only give a first indication in this direction, namely that the superconformal action functional induces a linear functional on the space of infinitesimal deformations of super Riemann surfaces. This allows to identify the conserved quantities with elements of the infinitesimal deformation space of super Riemann surfaces. More global applications of the superconformal action functional must be left for later work.

However, the appearance of the superconformal action functional in physics predates the definition of super Riemann surfaces, see Deser and Zumino (1976) and Brink et al. (1976). There, the motivation was to study a supersymmetric extension of the harmonic action functional. In order to show that this supersymmetric

E. Keßler, *Supergeometry, Super Riemann Surfaces and the Superconformal Action Functional*, Lecture Notes in Mathematics 2230,
https://doi.org/10.1007/978-3-030-13758-8_12

extension of the harmonic action functional coincides with the superconformal action functional, we choose an underlying even manifold $i\colon |M| \to M$. The map $\Phi\colon M \to N$ decomposes into a map $\varphi\colon |M| \to N$, a twisted spinor ψ and a section $F \in \varphi^*TN$. The superconformal action $A(\Phi, m)$ then reduces to an action $A(\varphi, g, \psi, \chi, F)$ defined on $|M|$. In the case of a flat target and $F = 0$, the action functional reduces to the supersymmetric extension of the harmonic action functional discussed in string theory and supergravity (see Sect. 1.1). The supergeometric definition of the superconformal action functional allows for a consistent geometric description of the different symmetries of the action $A(\varphi, g, \psi, \chi, F)$. In particular, supersymmetry extends to superdiffeomorphism invariance.

In Sect. 12.1 we define $A(\Phi, m)$ and give basic properties. After the decomposition of Φ in component fields, given in Sect. 12.2, we can give the reduction of $A(\Phi, m)$ to an action functional $A(\varphi, g, \psi, \chi, F)$ on an underlying even manifold $i\colon |M| \to M$. This will be discussed in Sect. 12.3. In Sect. 12.4, we give an application of the superconformal action functional to infinitesimal deformations of super Riemann surfaces.

The action functional $A(\Phi, m)$ and several of its properties can be found in the literature, see, for example, D'Hoker and Phong (1988, Chapter III.D) and Giddings and Nelson (1988, Chapter 5). However, appropriate proofs are missing. To my knowledge, the exposition here is the first one to provide a fully geometric interpretation to all symmetries of the superconformal action functional via super Riemann surfaces.

12.1 SCL-Invariant Action

From now on, we assume that M is a fiberwise compact family of super Riemann surfaces over B. We denote the SCL-reduction of the frame bundle by P_{SCL}. A further reduction to a U(1)-principal bundle P is given by the choice of a supermetric m compatible with P_{SCL}, see Sect. 9.5. Let N be an arbitrary (super) manifold with Riemannian metric n and Levi-Civita covariant derivative ∇^{TN}. The manifold N and n are fixed throughout this chapter.

We are interested in an action functional $A(\Phi, m)$ that associates to any map $\Phi\colon M \to N$ and supermetric m a function in $\mathcal{O}_B$. The action is given by the following integral

$$A(\Phi, m) = \frac{1}{2} \int_{M/B} \| d\Phi|_{\mathcal{D}} \|^2_{m^\vee|_{\mathcal{D}^\vee} \otimes \Phi^* n} [dvol_m]. \tag{12.1.1}$$

This action can be viewed as a generalization of the ordinary harmonic action functional to super Riemann surfaces. Note that in contrast to the harmonic action functional the tangent map $d\Phi$ is restricted to the subbundle $\mathcal{D}$ in TM. The volume form $[dvol_m]$ is defined with respect to the Shander orientation given by the holomorphic structure on M, see Definition 8.2.3 and Example 8.2.4.

Given U(1)-frames F_A the action can be written as

$$A(\Phi, m) = \frac{1}{2}\int_{M/B} \varepsilon^{\alpha\beta}\,\Phi^*n\left(F_\alpha\Phi, F_\beta\Phi\right)[F^\bullet]. \qquad (12.1.2)$$

In this form the action functional $A(\Phi, m)$ can be found in the literature, see in particular Giddings and Nelson (1988) and D'Hoker and Phong (1988, Equation 3.40). This representation of $A(\Phi, m)$ is particularly suited to prove that $A(\Phi, m)$ is superconformally invariant. Indeed, as explained in Sect. 9.5, a U(1)-frame $\tilde{F}_A$ for $\tilde{m}$ differs from the frame F_A by the composition of a U(1)-transformation, a rescaling by λ and a change of splitting by l:

$$\begin{pmatrix}\tilde{F}_a\\ \tilde{F}_\alpha\end{pmatrix} = \begin{pmatrix}U^2 & 0\\ 0 & U\end{pmatrix}\begin{pmatrix}\mathrm{id}_{2\times 2} & \gamma l\\ 0 & \mathrm{id}_{2\times 2}\end{pmatrix}\begin{pmatrix}\lambda^{-2}\,\mathrm{id}_{2\times 2} & 0\\ 0 & \lambda^{-1}\,\mathrm{id}_{2\times 2}\end{pmatrix}\begin{pmatrix}F_a\\ F_\alpha\end{pmatrix} \qquad (12.1.3)$$

Hence the term $\| d\Phi|_{\mathcal{D}} \|^2$ rescales by the factor λ^{-2}. On the other hand, the volume form rescales by the inverse of the Berezinian of the transformation (12.1.3), that is λ^2. Hence, the integrand in Eq. (12.1.2) is not only U(1)- but SCL-invariant. A direct proof of this fact can also be found in Giddings and Nelson (1988).

In addition to the superconformal invariance, the action functional $A(\Phi, m)$ is superdiffeomorphism invariant. That is for any diffeomorphism $\Xi\colon M \to M$ we have that $A(\Phi \circ \Xi, m_\Xi) = A(\Phi, m)$. Here m_Ξ is the supermetric on M given by

$$m_\Xi\,(X, Y) = \Xi^* m(d\Xi X, d\Xi Y).$$

The supermetric m_Ξ is compatible with the super Riemann surface structure given by P_Ξ. The diffeomorphism invariance follows directly from the diffeomorphism invariance of the integral:

$$\begin{aligned} A(\Phi \circ \Xi, m_\Xi) &= \int_{M/B} \|d(\Phi \circ \Xi)|_{\mathcal{D}}\|^2_{m_\Xi^\vee \otimes (\Phi\circ\Xi)^* n}[dvol_{m_\Xi}]\\ &= \int_{M/B} \Xi^\# \| d\Phi|_{\mathcal{D}} \|^2_{m^\vee \otimes \Phi^* n}\,\mathrm{Ber}\,(d\Xi)\,\Xi^\#[dvol_m]\\ &= \int_{M/B} \| d\Phi|_{\mathcal{D}} \|^2_{m^\vee \otimes \Phi^* n}[dvol_m]\\ &= A(\Phi, m) \end{aligned}$$

As a consequence of the superconformal invariance, the action functional $A(\Phi, m)$ depends only on the super Riemann surface structure on M. If we want to emphasize the dependence on the super Riemann surface structure we can write $A(\Phi, P_{\mathrm{SCL}})$ where P_{SCL} is the SCL-reduction of the frame bundle. By the diffeomorphism invariance, the action functional descends to the quotient

$$\{M \text{ super Riemann surface}\} \big/ \mathrm{Diff}\, M.$$

It is thus defined on the moduli space of super Riemann surfaces, compare Remark 11.3.7. The analogy to the Teichmüller theory of Riemann surfaces via harmonic maps, as explained in Jost (2006), leads to the hope that $A(\Phi, P_{\mathrm{SCL}})$ is a useful tool to study the moduli space of super Riemann surfaces. We will come back to this issue in Sect. 12.4.

Now we turn to the Φ-dependence of $A(\Phi, m)$ and calculate the Euler–Lagrange equations of Φ.

Proposition 12.1.4 *The Euler–Lagrange equation of* (12.1.1) *for Φ is*

$$0 = \Delta^{\mathcal{D}}\Phi = \varepsilon^{\alpha\beta}\nabla^{\Phi^*TN}_{F_\alpha} F_\beta \Phi + \varepsilon^{\alpha\beta}\,(\operatorname{div} F_\alpha)\, F_\beta \Phi \tag{12.1.5}$$

We call the differential operator $\Delta^{\mathcal{D}}$, defined here, the $\mathcal{D}$-Laplace operator.

Proof Let $\Phi_\bullet \colon M \times \mathbb{R} \to N$ be a perturbation of $\Phi_0 = \Phi$. Let us denote $\partial_t \Phi_t|_{t=0} = X \in \Gamma(\Phi^*TN)$ and expand $A(\Phi, m)$ in t around 0:

$$\begin{aligned}
\left.\frac{d}{dt}\right|_{t=0} A(\Phi_t, m) &= \frac{1}{2}\left.\frac{d}{dt}\right|_{t=0} \int_{M/B} \varepsilon^{\alpha\beta}\Phi_t^* n\left(F_\alpha \Phi_t, F_\beta \Phi_t\right)[F^\bullet] \\
&= \frac{1}{2}\int_{M/B} \partial_t \varepsilon^{\alpha\beta}\Phi_t^* n\left(F_\alpha \Phi_t, F_\beta \Phi_t\right)[F^\bullet]\Big|_{t=0} \\
&= \int_{M/B} \varepsilon^{\alpha\beta}\Phi_t^* n\left(\nabla^{\Phi_t^*TN}_{\partial_t} F_\alpha \Phi_t, F_\beta \Phi_t\right)[F^\bullet]\Big|_{t=0} \\
&= \int_{M/B} \varepsilon^{\alpha\beta}\Phi_t^* n\left(\nabla^{\Phi_t^*TN}_{F_\alpha} \partial_t \Phi_t, F_\beta \Phi_t\right)[F^\bullet]\Big|_{t=0} \\
&= \int_{M/B} \varepsilon^{\alpha\beta}\Phi^* n\left(\nabla^{\Phi^*TN}_{F_\alpha} X, F_\beta \Phi\right)[F^\bullet] \\
&= -\int_{M/B} \varepsilon^{\alpha\beta}\left(\Phi^* n\left(X, \nabla^{\Phi^*TN}_{F_\alpha} F_\beta \Phi\right)[F^\bullet] - \Phi^* n\left(X, F_\beta \Phi\right) L_{F_\alpha}[F^\bullet]\right)
\end{aligned}$$

The result follows from the definition of divergence (Definition 8.1.4). □

Notice that under a SCL-transformation as in Eq. (12.1.3), the $\Delta^{\mathcal{D}}$-operator rescales by a factor λ^{-2}. Hence the $\Delta^{\mathcal{D}}$-operator is only U(1)-invariant. However, the Euler–Lagrange Eq. (12.1.5) is of course G-invariant like the action (12.1.1).

12.2 Component Fields of Φ

In this section we decompose the map $\Phi : M \to N$ into the component fields φ, ψ and F on $|M|$ and show that they determine Φ completely. Furthermore, we deduce the transformations induced on the component fields under a superdiffeomorphism. The supersymmetry transformations of the component fields are described as the effect of an infinitesimal variation of the embedding $i : |M| \to M$.

Definition 12.2.1 (Component Fields) The component fields of Φ with respect to $i : |M| \to M$ are $\varphi : |M| \to N$, $\psi \in \Gamma\left(S^\vee \otimes \varphi^* TN\right)$, and $F \in \Gamma\left(\varphi^* TN\right)$, defined by

$$\varphi = \Phi \circ i, \qquad \psi = s^\alpha \otimes i^* F_\alpha \Phi, \qquad F = -\frac{1}{2} i^* \Delta^{\mathcal{D}} \Phi.$$

Here $s_\alpha = i^* F_\alpha$ is the basis of S induced by the SCL-frame F_A and s^α its dual basis.

In contrast to ψ and F, the field φ does not depend on the metric m. A superconformal rescaling of m by λ^2 yields a rescaling of g by $i^\#\lambda^4$, that is, $\tilde{g} = i^\# \lambda^4 g$. As in Remark 11.1.9, we have two different ways to identify the spinor bundle S_g with $S_{\tilde{g}}$. If we identify S_g with $S_{\tilde{g}}$ and $S_g^\vee$ with $S_{\tilde{g}}^\vee$ via $i^* \operatorname{id}_{\mathcal{D}}$ and its dual, we have $\tilde{\psi} = \psi$. In contrast, if we use $\beta : S_g \to S_{\tilde{g}}$ or its dual, we obtain, $\tilde{\psi} = i^\#\lambda^{-1}\left(\beta^\vee\right)^{-1} \otimes \operatorname{id}_{\varphi^* TN} \psi$. The field F scales by $i^\#\lambda^{-2}$.

Let $X^A = (x^a, \eta^\alpha)$ be Wess–Zumino coordinates for the SCL-frame F_A on $U \subset M$ and $y^a = i^\# x^a$ be coordinates on $U \subset |M|$. For any coordinates Z^B on N the pullback along Φ can be expressed in those coordinates as follows:

$$\Phi^\# Z^B = \Phi^B = {}_0\Phi(x)^B + \eta^\mu{}_\mu\Phi(x)^B + \eta^3\eta^4{}_{34}\Phi(x)^B$$

We show that the coefficients ${}_0\Phi(x)^B$, ${}_\mu\Phi(x)^B$, and ${}_{34}\Phi(x)^B$ are coordinate expressions for the component fields φ, ψ and F respectively. Indeed, with the help of the condition (11.2.2), that is $i^\# x^a = y^a$ and $i^\#\eta^\alpha = 0$, we obtain

$$\varphi^\# Z^C = i^\#\left(\Phi^\# Z^C\right) = i^\#{}_0\Phi(x)^C = {}_0\Phi(y)^C.$$

The coefficient ${}_\mu\psi$ in $\psi = s^\mu \otimes {}_\mu\psi$ is a section of $\varphi^* TN$, that is, a derivation of functions on N with values in $\mathcal{O}_{|M|}$. Hence by the conditions on the frame coefficients for Wess–Zumino coordinates, Eq. (11.2.3),

$${}_\mu\psi(Z^B) = \left(i^* F_\mu \Phi\right) Z^B = i^\#\left(F_\mu\left(\Phi^\# Z^B\right)\right) = {}_\mu\Phi(y)^B.$$

Hence, ψ is locally given by

$$\psi = s^{\mu} \otimes {}_{\mu}\Phi(y)^{B} i^{*}\partial_{Z^B}.$$

In order to compare F and ${}_{34}\Phi^{B}$, recall the definition of the $\mathcal{D}$-Laplace operator given in Eq. (12.1.5):

$$F(Z^B) = -\frac{1}{2}\left(i^{\#}\Delta^{\mathcal{D}}\Phi\right)Z^B = \frac{1}{2}i^{\#}\varepsilon^{\mu\nu}\left(\nabla^{\Phi^{*}TN}_{F_\mu}F_\nu\Phi^{\#} + \left(\operatorname{div}F_\mu\right)F_\nu\Phi^{\#}\right)Z^B$$

As F rescales by $i^{\#}\lambda^{-2}$, we are free to assume that F_A is Wess–Zumino frame in the calculation of the divergence. By Lemma 13.7.4, the term $i^{\#}\left(\operatorname{div}F_\mu\right)$ vanishes. Consequently, the expression for F reduces to

$$\begin{aligned}
F(Z^B) &= -\frac{1}{2}i^{\#}\varepsilon^{\mu\nu}\left(\nabla^{\Phi^{*}TN}_{F_\mu}F_\nu\Phi^{\#}\right)Z^B \\
&= -\frac{1}{2}i^{\#}\varepsilon^{\mu\nu}\left(\nabla^{\Phi^{*}TN}_{F_\mu}\left(F_\nu\Phi^{\#}Z^C\right)\Phi^{*}\partial_{Z^C}\right)Z^B \\
&= -\frac{1}{2}i^{\#}\varepsilon^{\mu\nu}\left(\left(F_\mu F_\nu\Phi^{\#}Z^C\right)\Phi^{*}\partial_{Z^C}\right. \\
&\qquad \left. - (-1)^{p(C)}\left(F_\nu\Phi^{\#}Z^C\right)\nabla^{\Phi^{*}TN}_{F_\mu}\Phi^{*}\partial_{Z^C}\right)Z^B \\
&= {}_{34}\Phi^{B} + \frac{1}{2}\varepsilon^{\mu\nu}\,{}_{\mu}\psi(Z^A)\,{}_{\nu}\psi(Z^C)\,{}_{CA}\!\left(\varphi^{*}\omega^{N}\right)^{B}.
\end{aligned}$$

Here ω^N is the connection form of the Levi-Civita connection on TN with respect to the frame ∂_{Z^A}.

In summary, in the case of a target supermanifold N with trivial connection, we may write

$$\Phi = \varphi + \eta^{\mu}\,{}_{\mu}\psi + \eta^{3}\eta^{4}F.$$

Furthermore, we also proved the following:

Proposition 12.2.2 *Let $i\colon |M| \to M$ be an embedding of an underlying even manifold of a super Riemann surface. Furthermore, let $\varphi\colon |M| \to N$ be a map to an arbitrary Riemannian supermanifold N, and $\psi \in \Gamma\left(S^{\vee}\otimes\varphi^{*}TN\right)$ and $F \in \Gamma\left(\varphi^{*}TN\right)$ two sections. Then there exists a unique map $\Phi\colon M \to N$ such that the component fields of Φ coincide with φ, ψ and F.*

We now turn to the effect of superdiffeomorphisms on the component fields. As in Sect. 11.4, we note in square brackets the dependence on the embedding i, the field Φ and the U(1)-structure P, that is, $\varphi[i, \Phi]$, $\psi[i, \Phi, P]$ and $F[i, \Phi, P]$.

Proposition 12.2.3 *Let $\Xi : M \to M$ be a diffeomorphism and $i : |M| \to M$. Let furthermore ξ be a diffeomorphism of $|M|$ and $j : |M| \to M$ such that $\Xi \circ i = j \circ \xi$ (see Corollary 3.3.14). For the component fields of $\Phi \circ \Xi$ it holds that*

$$\begin{aligned} \varphi[i, \Phi \circ \Xi] &= \varphi[j, \Phi] \circ \xi, \\ \psi[i, \Phi \circ \Xi, P_\Xi] &= \xi^* \psi[j, \Phi, P] \circ \left(s \otimes \mathrm{id}_{\varphi[j,\Phi]^* TN}\right), \\ F[i, \Phi \circ \Xi, P_\Xi] &= \xi^* F[j, \Phi, P], \end{aligned}$$

where $s : S[i, P_\Xi] \to \xi^* S[j, P]$ is defined in Proposition 11.4.2.

Proof For φ one verifies easily that

$$\varphi[i, \Phi \circ \Xi] = \Phi \circ \Xi \circ i = \Phi \circ j \circ \xi = \varphi[j, \Phi] \circ \xi.$$

For ψ conclude from Eq. (11.4.1) that $F^A[P_\Xi] = \Xi^* F^A[P] \circ d\Xi$ and recall that $s = i^* d\Xi|_{\mathcal{D}[P_\Xi]}$. Then,

$$\begin{aligned} \psi[i, \Phi \circ \Xi, P_\Xi] &= i^* F^\alpha[P_\Xi] \otimes i^* \left(d(\Phi \circ \Xi) F_\alpha[P_\Xi]\right) \\ &= i^* \left(\Xi^* F^\alpha[P] \circ d\Xi\right) \otimes i^* \Xi^* d\Phi \, \Xi^* F_\alpha[P] \\ &= \xi^* j^* F^\alpha[P] \circ s \otimes \xi^* j^* d\Phi F_\alpha[P] \\ &= \xi^* \psi[j, \Phi, P] \circ \left(s \otimes \mathrm{id}_{\varphi[j,\Phi]^* TN}\right). \end{aligned}$$

Notice that $\Delta^{\mathcal{D}[P_\Xi]} \Phi \circ \Xi = \Xi^* \Delta^{\mathcal{D}[P]} \Phi$ because the action functional is diffeomorphism invariant. Hence,

$$\begin{aligned} F[i, \Phi \circ \Xi, P_\Xi] &= i^* \Delta^{\mathcal{D}[P_\Xi]} \Phi \circ \Xi = i^* \Xi^* \Delta^{\mathcal{D}[P]} \Phi \\ &= \xi^* j^* \Delta^{\mathcal{D}[P]} \Phi = \xi^* F[j, \Phi, P]. \end{aligned}$$

□

Hence, the action of a diffeomorphism $\Xi : M \to M$ on the component fields of Φ can be described as a combination of a change of the embedding i and a diffeomorphism ξ on $|M|$. We are now turning to the infinitesimal change of the embedding i. The result is similar to those for metric and gravitino. That is, the infinitesimal change of embedding in a normal direction $q \in \Gamma(S)$ induces the known supersymmetry transformations up to infinitesimal Weyl transformations.

Definition 12.2.4 We call

$$\begin{aligned} &\mathrm{susy}_q \varphi = \langle q, \psi \rangle \\ &\mathrm{susy}_q \psi = - \vee \left(\gamma^s q\right) \otimes \left(f_s \varphi + \langle \chi(f_s), \psi \rangle\right) - \vee q \otimes F \\ &\mathrm{susy}_q F = \left\langle q, \not{D}\psi + \|Q\chi\|^2 \psi - \frac{1}{3} S R^N(\psi) \right\rangle - 2\, d\varphi\left(\langle q, \vee Q\chi \rangle\right) + g_S\left(q, \delta_\gamma \chi\right) F \end{aligned}$$

the supersymmetry transformations of the fields φ, ψ and F in the direction $q \in \Gamma(S)$.

In the special case $F = 0$ the transformations $\operatorname{susy}_q \varphi$ and $\operatorname{susy}_q \psi$ coincide with the supersymmetry transformations given in Brink et al. (1976, Equation (16)) and Deser and Zumino (1976, Equation (15)). In the special case $\chi = 0$ and flat target supermanifold N the transformations $\operatorname{susy}_q \varphi$, $\operatorname{susy}_q \psi$ and $\operatorname{susy}_q F$ coincide with Jost (2009, Equations (2.4.53)–(2.4.55)).

Suppose now, as in Sect. 11.4, that $j_t \colon |M| \to M$ is a time indexed family of embeddings such that $j_0 = i$ and $\frac{d}{dt}\big|_{t=0} j_t = q \in \Gamma(S)$. This time indexed family induced time-indexed families of fields φ_t, $\psi_t \in \Gamma\left(S^\vee \otimes \varphi_t^* TN\right)$ and $F_t \in \Gamma\left(\varphi_t^* TN\right)$ such that $\varphi_0 = \varphi$, $\psi_0 = \psi$ and $F_0 = F$. For their time derivatives we introduce the shorthand notation

$$\begin{aligned}
\delta_q \varphi &= \left.\frac{d}{dt}\right|_{t=0} \varphi_t \in \Gamma\left(\varphi^* TN\right),\\
\delta_q \psi[m] &= \nabla^{S^\vee \otimes \varphi_t^* TN}_{\partial_t} \left.\left(\beta_t^\vee \otimes \mathrm{id}_{\varphi_t^* TN}\, \psi\right)\right|_{t=0} \in \Gamma\left(S^\vee \otimes \varphi^* TN\right),\\
\delta_q F[m] &= \left.\nabla^{\varphi_t^* TN}_{\partial_t} F_t\right|_{t=0} \in \Gamma\left(\varphi^* TN\right).
\end{aligned}$$

As indicated, the variations $\delta_q \psi[m]$ and $\delta_q F[m]$ depend on the chosen superconformal metric m.

Proposition 12.2.5 *For every superconformal metric m and every variation of the embedding i in the direction $q \in \Gamma(S)$, the induced variations of φ, ψ and F coincide with the supersymmetry transformations up to infinitesimal Weyl transformations. That is, for the $\sigma \in \Gamma\left(S^\vee\right)$ of Proposition 11.4.13 we have*

$$\begin{aligned}
\delta_q \varphi &= \operatorname{susy}_q \varphi,\\
\delta_q \psi[m] &= \operatorname{susy}_q \psi - \langle q, \sigma\rangle \psi,\\
\delta_q F[m] &= \operatorname{susy}_q F - 2\langle q, \sigma\rangle F.
\end{aligned}$$

Proof We obtain for the variation of φ:

$$\delta_q \varphi = \left.\frac{d}{dt}\right|_{t=0} \Phi \circ j_t = d(\Phi \circ j_t)\partial_t|_{t=0} = \left(i^* d\Phi\right) q = \langle q, \psi\rangle .$$

Let now F_A be an m-orthonormal frame and $s_\alpha = i^* F_\alpha$. For the variation of $\psi = s^\alpha \otimes i^* F_\alpha \Phi$ recall that $\beta_t s_\alpha = j_t^* F_\alpha$ and hence

$$\beta_t^\vee \otimes \mathrm{id}_{\varphi_t^* TN}\, \psi_t = s^\alpha \otimes j_t^* F_\alpha \Phi.$$

Consequently,

$$\delta_q \psi[m] = \nabla^{S \otimes \varphi_t^* TN}_{\partial_t} s^\alpha \otimes j_t^* F_\alpha \Phi = s^\alpha \otimes q^\mu i^* \nabla_{F_\mu} F_\alpha \Phi.$$

For the variation of F we obtain

$$\delta_q F[m] = -\frac{1}{2} q^\mu i^* \nabla_{F_\mu} \Delta^{\mathcal{D}} \Phi.$$

Let now $\tilde{m}$ be a superconformal metric arising from m by a superconformal rescaling by λ and superconformal change of splitting l such that $i^*\lambda = 1$ and $\tilde{F}_\alpha = \lambda^{-1} F_\alpha$. Then

$$\delta_q \psi[\tilde{m}] = s^\alpha \otimes q^\mu i^* \nabla_{\tilde{F}_\mu} \tilde{F}_\alpha \Phi = \delta_q \psi[m] - \langle q, i^* d\lambda \rangle \psi,$$

and since $\Delta^{\mathcal{D}}[\tilde{m}] = \lambda^{-2} \Delta^{\mathcal{D}}[m]$

$$\delta_q F[\tilde{m}] = -\frac{1}{2} q^\mu i^* \nabla_{F_\mu} \lambda^{-2} \Delta^{\mathcal{D}} \Phi = \delta_q F[m] - 2 \langle q, i^* d\lambda \rangle F.$$

To complete the proof, it remains to calculate $\delta_q \psi[m]$ and $\delta_q F[m]$ for m the superconformal metric defined by a Wess–Zumino frame F_A. We conclude with the help of Lemma 13.6.2:

$$\begin{aligned}
\delta_q \psi[m] &= s^\alpha \otimes \langle q, i^* \nabla F_\alpha \Phi \rangle = s^\alpha \otimes \left(q^\mu i^* \nabla_{F_\mu} F_\alpha \Phi \right) \\
&= s^\alpha \otimes \left(q^\mu \left(\Gamma^s_{\mu\alpha} \left(f_s \varphi + \langle \chi(f_s), \psi \rangle \right) + \varepsilon_{\mu\alpha} F \right) \right) \\
&= - \vee \left(\gamma^s q \right) \otimes \left(f_s \varphi + \langle \chi(f_s), \psi \rangle \right) - \vee q \otimes F
\end{aligned}$$

For $\delta_q F[m]$ we conclude with the help of Proposition 13.10.1:

$$\begin{aligned}
\delta_q F[m] = & -\frac{1}{2} \left\langle q, i^* \nabla^{\Phi^* TN} \Delta^{\mathcal{D}} \Phi \right\rangle \\
= & \left\langle q, \not{D} \psi + \|Q\chi\|^2 \psi - \frac{1}{3} SR^N(\psi) \right\rangle \\
& - 2\, d\varphi\left(\langle q, \vee Q\chi \rangle \right) + g_S \left(q, \delta_\gamma \chi \right) F
\end{aligned}$$

□

12.3 Component Action

In this section we explain how the action functional (1.1.2), that is, the supersymmetric extension of the harmonic action functional, is related to super Riemann surfaces. Roughly, it is the SCL-invariant action functional from Sect. 12.1 where the integration over the odd coordinates has been carried out.

Theorem 12.3.1 *Let M be a fiberwise compact family of super Riemann surfaces and $i: |M| \to M$ an underlying even manifold. We denote by g, χ, and g_S respectively the metric, gravitino and spinor metric on $|M|$. Let $\Phi: M \to N$ be* *a morphism to a Riemannian supermanifold (N, n) and φ, ψ, and F its component fields. The action functional $A(\varphi, g, \psi, \chi, F)$ defined by*

$$\begin{aligned} A(\varphi, g, \psi, \chi, F) = \int_{|M|/B} \Big(& \|d\varphi\|^2_{g^\vee \otimes \varphi^* n} + g_S^\vee \otimes \varphi^* n \left(\not{D}\psi, \psi \right) - \|F\|^2_{\varphi^* n} \\ & + 4 g^\vee \otimes \varphi^* n \left(d\varphi, \langle Q\chi, \psi \rangle \right) + \|Q\chi\|^2_{g^\vee \otimes g_S} \|\psi\|^2_{g_S^\vee \otimes \varphi^* n} \\ & - \frac{1}{6} g_S^\vee \otimes \varphi^* n \left(SR^N(\psi), \psi \right) \Big) \, dvol_g \end{aligned} \tag{12.3.2}$$

equals $A(\Phi, m)$.

Here, by slight abuse of notation we denote by $\langle Q\chi, \psi \rangle \in \Gamma\left(T^\vee |M| \otimes \varphi^* TN\right)$ the contraction of $Q\chi$ with ψ along the spinor factor. Furthermore,

$$SR^N(\psi) = s^\alpha \otimes \left(g_S{}^{\mu\nu} R^N \left({}_\alpha\psi, {}_\mu\psi \right) {}_\nu\psi \right)$$

is the contraction of the pullback of the curvature tensor R^N of the Levi-Civita connection on N along φ with $\psi = s^\alpha{}_\alpha\psi$.

Variants of Theorem 12.3.1 are claimed in the literature, see, for example, Howe (1979, Section 5) and D'Hoker and Phong (1988, Chapter III.D). Our proof of Theorem 12.3.1 is based on Proposition 8.4.4 and inspired by similar calculations in Deligne and Freed (1999b). One uses crucially that integration in the odd directions is locally a derivation. A local expression for the action in Wess–Zumino coordinates $X^A = (x^a, \eta^\alpha)$ for the Wess–Zumino frame F_A is given by

$$\begin{aligned} A(\Phi, m) &= \frac{1}{2} \int_{M/B} \varepsilon^{\alpha\beta} \Phi^* n \left(F_\alpha \Phi, F_\beta \Phi \right) (\operatorname{Ber} F)^{-1} [dX^\bullet] \\ &= \frac{1}{2} \int_{|M|/B} i^* \partial_{\eta^3} \partial_{\eta^4} \left(\varepsilon^{\alpha\beta} \Phi^* n \left(F_\alpha \Phi, F_\beta \Phi \right) (\operatorname{Ber} F)^{-1} \right) dx^1 \, dx^2 \\ &= \frac{1}{4} \int_{|M|/B} i^* \varepsilon^{\mu\nu} F_\mu F_\nu \left(\varepsilon^{\alpha\beta} \Phi^* n \left(F_\alpha \Phi, F_\beta \Phi \right) (\operatorname{Ber} F)^{-1} \right) dx^1 \, dx^2. \end{aligned}$$

The expansion of the last expression is given in terms of component fields of Φ (compare Definition 12.2.1), and commutators of F_α and derivatives of Ber F. By Lemma 11.3.4 the coordinate expansion of F_α, its commutators and the Berezinian are determined by g and χ. A detailed calculation is given in Sect. 13.9.

The action functional $A(\varphi, g, \psi, \chi, F)$ is a non-linear generalization of the action functional introduced by Brink et al. (1976) and Deser and Zumino (1976). Indeed, in the case of the flat target $N = \mathbb{R}$ and $F = 0$ the action $A(\Phi, m)$ reduces to the action functional (1.1.2). It is important to remember that $|M|$ is a family of supermanifolds of dimension $2|0$ over an arbitrary base B. If B is assumed to be trivial, the fields ψ and χ are zero. The functional $A(\varphi, g, \psi, \chi, F)$ then reduces to a trivial extension of the functional of harmonic maps on $|M|$. In the case $\chi = 0$ and $F = 0$, the action functional $A(\varphi, g, \psi, \chi, F)$ reduces to a functional that resembles the functional of Dirac-harmonic maps introduced in Chen et al. (2006). However, in Chen et al. (2006) different conventions are used, in particular for the Clifford relation and g_S, to be able to define the functional without use of anti-commuting variables.

The symmetries of $A(\Phi, m)$ induce symmetries of $A(\varphi, g, \psi, \chi, F)$ which are generalizations of the symmetries listed in the introduction. From the explicit formula in Eq. (12.3.2) one can read that $A(\varphi, g, \psi, \chi, F)$ is diffeomorphism invariant; that is for every diffeomorphism $\xi\colon |M| \to |M|$ it holds that

$$A(\varphi \circ \xi, g_\xi, \psi_\xi, \xi^*\chi \circ d\xi, \xi^* F) = A(\varphi, g, \psi, \chi, F).$$

Furthermore, as the reduction of $A(\Phi, m)$ to $A(\varphi, g, \psi, \chi, F)$ uses an explicit embedding $i\colon |M| \to M$ that is not part of the data, we have

$$A(\varphi[i], g[i], \psi[i], \chi[i], F[i]) = A(\Phi, m) = A(\varphi[j], g[j], \psi[j], \chi[j], F[j])$$

for any other embedding $j\colon |M| \to M$. Here we used again square brackets to note the embedding used for the definition of the component fields.

It was explained in Propositions 11.4.13 and 12.2.5 that an infinitesimal change of the embedding $i\colon |M| \to M$ induces the supersymmetry transformations of the component fields. Consequently, the action functional $A(\varphi, g, \psi, \chi, F)$ is invariant under the following infinitesimal transformations:

$$\begin{aligned}
\left(\operatorname{susy}_q g\right)(X, Y) &= 2 g_S(q, \gamma(X)\chi(Y) + \gamma(Y)\chi(X)),\\
\left(\operatorname{susy}_q \chi\right)(X) &= -\nabla^{LC}_X q - \frac{1}{2} g_S\left(\chi(X), \chi(f_l)\right)\gamma^l q,\\
\operatorname{susy}_q \varphi &= \langle q, \psi\rangle,\\
\operatorname{susy}_q \psi &= -\vee\left(\gamma^s q\right) \otimes \left(f_s\varphi + \langle\chi(f_s), \psi\rangle\right) - \vee q \otimes F,\\
\operatorname{susy}_q F &= \left\langle q, D\!\!\!/\,\psi + \|Q\chi\|^2\psi - \frac{1}{3} SR^N(\psi)\right\rangle\\
&\quad - 2\, d\varphi\left(\langle q, \vee Q\chi\rangle\right) + g_S\left(q, \delta_\gamma \chi\right) F.
\end{aligned}$$

Supersymmetry was the main reason to introduce $A(\varphi, g, \psi, \chi)$ in Brink et al. (1976); Deser and Zumino (1976). There, the supersymmetry transformations were introduced for physical reasons and lacking a geometric interpretation. Here, the supersymmetry transformations are interpreted either as the action of a diffeomorphism of M or a variation of the embedding i. Hence, for a geometric interpretation of the supersymmetry transformations, supergeometry is indispensable. We will, however, give a direct verification of the supersymmetry of $A(\varphi, g, \psi, \chi)$ in Appendix B.

In Propositions 11.4.2 and 12.2.3 we have seen that a diffeomorphism on M induces changes on the metric, gravitino and component fields of Φ that correspond to a combination of a diffeomorphism on $|M|$, a change of embedding of the underlying even manifold and the map s. In Example 11.4.3 the map s is given by multiplication of the spinor bundle with -1. The corresponding invariance of the action is

$$A(\varphi, g, \psi, \chi, F) = A(\varphi, g, -\psi, -\chi, F),$$

an instance of gauge invariance.

In addition to the diffeomorphism invariance, the action $A(\Phi, m)$ is also superconformally invariant, hence invariant under rescaling of the metric and superconformal change of splitting. The invariance of $A(\Phi, m)$ under superconformal rescaling yields the conformal invariance of the component action:

$$A(\varphi, \lambda^4 g, \lambda^{-1}(\beta^\vee)^{-1} \otimes \mathrm{id}_{\varphi^*TN} \psi, \lambda^{-1}(b^\vee)^{-1} \otimes \beta\chi, \lambda^{-2}F) = A(\varphi, g, \psi, \chi, F),$$

or if we identify S_g with $S_{\lambda^4 g}$ with the help of $i^\# \mathrm{id}_\mathcal{D}$, as explained in Remark 11.1.9,

$$A(\varphi, g, \psi, \chi, F) = A(\varphi, \lambda^2 g, \psi, \chi, \lambda F).$$

The invariance under superconformal change of splitting yields the super Weyl invariance:

$$A(\varphi, g, \psi, \chi, F) = A(\varphi, g, \psi, \chi + \gamma t, F).$$

In summary, the different symmetries of the action functional $A(\varphi, g, \psi, \chi, F)$ that may seem at first glance somewhat arbitrary are explained by the rather simple geometric properties of $A(\Phi, m)$.

Proposition 12.3.3 *The Euler–Lagrange equation* $\Delta^\mathcal{D}\Phi = 0$ *of the action functional* (12.1.1) *is equivalent to the following coupled system of equations:*

$$F = 0, \tag{12.3.4a}$$

$$\not{D}^{LC}\psi = 2\, d\varphi\,(\vee Q\chi) - \|Q\chi\|^2\psi + \frac{1}{3}SR^N(\psi), \tag{12.3.4b}$$

$$\begin{aligned}\tau(\varphi) = & -2\,\mathrm{Tr}_g \nabla \langle Q\chi, \psi\rangle - \frac{1}{6} g_S{}^{\mu\nu} g_S{}^{\alpha\beta} \left(\nabla_{\mu\psi} R^N\right)({}_\nu\psi, {}_\alpha\psi)\,{}_\beta\psi \\ & - \frac{1}{2} g_S{}^{\alpha\beta} \gamma^k{}_\beta{}^\nu R^N\left({}_\nu\psi, {}_\alpha\psi\right) f_k\varphi.\end{aligned} \tag{12.3.4c}$$

Here $\tau(\varphi) = \mathrm{Tr}_g \nabla\, d\varphi \in \Gamma(\varphi^*TN)$ *is the tension field of* φ*, that is, if* $\tau(\varphi) = 0$ *then* φ *is harmonic.*

Proof This result follows directly from the calculation of the component fields of $\Delta^{\mathcal{D}} \Phi$ in Proposition 13.10.1. Indeed, for $0 = i^* \Delta^{\mathcal{D}} \Phi = -2F$ yields $F = 0$. Inserting $F = 0$ in the first order component of $\Delta^{\mathcal{D}} \Phi$ yields the Dirac-equation. Inserting $F = 0$ and the Dirac-equation into the second order component of $\Delta^{\mathcal{D}} \Phi$ yields the Laplace equation for φ. However, it remains to check that the following curvature terms cancel:

$$\varepsilon^{\alpha\beta} \gamma^{k}_{\beta}{}^{\nu} R^N \left({}_{\nu}\psi, {}_{\alpha}\psi\right) \chi_k{}^{\kappa}{}_{\kappa}\psi - \frac{2}{3} \chi_k{}^{\kappa} \gamma^{k}_{\kappa}{}^{\mu} \varepsilon^{\alpha\beta} R^N \left({}_{\mu}\psi, {}_{\alpha}\psi\right){}_{\beta}\psi = 0$$

To this end, note that by the anti-symmetry of the curvature tensor in the first two entries and the Bianchi identity we obtain:

$$\begin{aligned}
R^N \left({}_3\psi, {}_3\psi\right){}_3\psi &= R^N \left({}_4\psi, {}_4\psi\right){}_4\psi = 0, \\
R^N \left({}_4\psi, {}_3\psi\right){}_3\psi &= R^N \left({}_3\psi, {}_4\psi\right){}_3\psi = -\frac{1}{2} R^N \left({}_3\psi, {}_3\psi\right){}_4\psi, \\
R^N \left({}_4\psi, {}_3\psi\right){}_4\psi &= R^N \left({}_3\psi, {}_4\psi\right){}_4\psi = -\frac{1}{2} R^N \left({}_4\psi, {}_4\psi\right){}_3\psi.
\end{aligned}$$

It follows

$$\begin{aligned}
&\varepsilon^{\alpha\beta} \gamma^{k}_{\beta}{}^{\nu} R^N \left({}_{\nu}\psi, {}_{\alpha}\psi\right) \chi_k{}^{\kappa}{}_{\kappa}\psi - \frac{2}{3} \chi_k{}^{\kappa} \gamma^{k}_{\kappa}{}^{\mu} \varepsilon^{\alpha\beta} R^N \left({}_{\mu}\psi, {}_{\alpha}\psi\right){}_{\beta}\psi \\
&= \chi_1{}^3 \left(-R^N \left({}_4\psi, {}_4\psi\right){}_3\psi - \frac{2}{3} \left(R^N \left({}_4\psi, {}_3\psi\right){}_4\psi - R^N \left({}_4\psi, {}_4\psi\right){}_3\psi \right) \right) \\
&\quad + \chi_2{}^3 \left(2R^N \left({}_4\psi, {}_3\psi\right){}_4\psi + \frac{2}{3} \left(R^N \left({}_3\psi, {}_3\psi\right){}_4\psi - R^N \left({}_3\psi, {}_4\psi\right){}_3\psi \right) \right) \\
&\quad + \chi_1{}^4 \left(R^N \left({}_3\psi, {}_3\psi\right){}_4\psi - \frac{2}{3} \left(R^N \left({}_3\psi, {}_3\psi\right){}_4\psi - R^N \left({}_3\psi, {}_4\psi\right){}_3\psi \right) \right) \\
&\quad + \chi_2{}^4 \left(2R^N \left({}_4\psi, {}_3\psi\right){}_4\psi - \frac{2}{3} \left(R^N \left({}_4\psi, {}_3\psi\right){}_4\psi - R^N \left({}_4\psi, {}_4\psi\right){}_3\psi \right) \right) \\
&= 0
\end{aligned}$$

□

Another, equivalent way to obtain the Euler–Lagrange equations for the component fields is to vary them directly in (12.3.2). This approach is taken in Jost et al. (2018a) for the model with only commuting variables.

12.4 Energy-Momentum Tensor and Supercurrent

We have already noticed that due to its SCL-invariance and diffeomorphism invariance the action functional $A(\Phi, m)$ for fixed $\Phi\colon M \to N$ can be seen as a functional on

$$\left\{M \text{ super Riemann surface}\right\} \big/ \operatorname{Diff} M$$

and hence on the moduli space of super Riemann surfaces. In principle the action functional $A(\varphi, g, \psi, \chi, F)$ is also a functional on the moduli space of super Riemann surfaces because $A(\Phi, m) = A(\varphi, g, \psi, \chi, F)$ by Theorem 12.3.1. A proof of the conjecture in Remark 11.3.7 would then allow for an application of methods from geometric analysis to study the moduli space of super Riemann surfaces.

However, due to Theorem 11.5.4 we can already give some properties of the action functional $A(\varphi, g, \psi, \chi, F)$ that are related to infinitesimal deformations of a given super Riemann surface M. Let $i\colon |M| \to M$ and $\Phi\colon M \to N$ be given. The geometry of M is then determined by the metric g and the gravitino χ on $|M|$. In the remainder of this section, we will study the variation of the action under infinitesimal deformations of metric and gravitino. We will assume, as in Sect. 11.5, that χ is a section of $T^\vee|M| \otimes S^\vee$ and use metric dualization with respect to g_S when necessary.

To calculate the variation of the action functional under a variation of the metric, we have to keep in mind that we cannot keep the spinor bundle fixed under a variation of the metric. Let us denote the spinor bundle to the metric g_t by S_{g_t} and recall that there are isometries $b_t\colon (T|M|, g) \to (T|M|, g_t)$ and $\beta_t\colon (S_g, g_S) \to (S_{g_t}, g_{S,t})$, as explained in the Appendix A.4.

Proposition 12.4.1 (Energy-Momentum Tensor) *For $t \in \mathbb{R}$ let g_t be a time indexed family of metrics such that $g_0 = g$ and $\left.\frac{d}{dt}\right|_{t=0} g_t = h$.*

$$\begin{aligned}\left.\frac{d}{dt}\right|_{t=0} & A\left(\varphi, g_t, \left(\left(\beta_t^\vee\right)^{-1} \otimes \mathrm{id}_{\varphi^*TN}\right)\psi, \left(b_t^\vee \otimes \left(\beta_t^\vee\right)^{-1}\right)\chi, F\right) \\ &= \int_{|M|/B} g^\vee \otimes g^\vee (h, T)\, dvol_g.\end{aligned}$$

Here, $T \in T^\vee|M| \odot T^\vee|M|$ is given by

$$\begin{aligned}T(X, Y) = {} & \varphi^*n\,(d\varphi X, d\varphi Y) \\ & - \frac{1}{4} g_S^\vee \otimes \varphi^*n \left(\psi, \gamma(X)\nabla_Y^{S^\vee\otimes\varphi^*TN}\psi + \gamma(Y)\nabla_X^{S^\vee\otimes\varphi^*TN}\psi\right) \\ & + g_S^\vee \otimes \varphi^*n\,((Q\chi)(X) \otimes d\varphi(Y) + (Q\chi)(Y) \otimes d\varphi(X), \psi) \\ & - \frac{1}{2} g(X, Y) \Bigg(\|d\varphi\|^2_{g^\vee\otimes\varphi^*n} + g_S^\vee \otimes \varphi^*n\left(\not{D}\psi, \psi\right) - \|F\|^2_{\varphi^*n} \\ & + 4g^\vee \otimes \varphi^*n\,(d\varphi, \langle Q\chi, \psi\rangle) + \|Q\chi\|^2_{g^\vee\otimes g_S} \|\psi\|^2_{g_S^\vee\otimes\varphi^*n} \\ & - \frac{1}{6} g_S^\vee \otimes \varphi^*n\left(SR^N(\psi), \psi\right)\Bigg)\end{aligned}$$

and called energy-momentum tensor of $A(\varphi, g, \psi, \chi, F)$.

Proof For a g-orthonormal frame f_a we denote $h(f_a, f_b) = h_{ab}$ and its inverse by $h^{ab} = -g^{ac}h_{cd}g^{db}$ and $h_a{}^b = h_{ac}g^{cb}$. It follows that $\frac{d}{dt}\big|_{t=0} b_t f_a = -\frac{1}{2}h_a{}^b f_b$. We calculate the variation of the action term by term

i) The energy of the map can be analyzed as follows:

$$\frac{d}{dt}\bigg|_{t=0} \|d\varphi\|^2_{g_t^\vee\otimes\varphi^*n} = \frac{d}{dt}\bigg|_{t=0} g_t{}^{ab}\varphi^*n\,(d\varphi f_a, d\varphi f_b)$$
$$= h^{ab}\varphi^*n\,(d\varphi f_a, d\varphi f_b)\,.$$

ii) For $\psi = s^\alpha \otimes {}_\alpha\psi$ we have

$$\left(\left(\beta_t^\vee\right)^{-1} \otimes \mathrm{id}_{\varphi^*TN}\right)\psi = \left(\beta_t^\vee\right)^{-1} s^\alpha \otimes {}_\alpha\psi \in \Gamma\left(S_{g_t} \otimes \varphi^*TN\right).$$

Consequently, for the twisted Dirac-operator $\not{D}_{g_t}$ at time t:

$$\begin{aligned}
&g^\vee_{S,t} \otimes \varphi^*n\left(\left(\left(\beta_t^\vee\right)^{-1} \otimes \mathrm{id}_{\varphi^*TN}\right)\psi, \not{D}_{g_t}\left(\left(\beta_t^\vee\right)^{-1} \otimes \mathrm{id}_{\varphi^*TN}\right)\psi\right)\\
&= -\,g^\vee_{S,t}\left(\left(\beta_t^\vee\right)^{-1}s^\alpha, \not{\partial}_{g_t}\left(\beta_t^\vee\right)^{-1}s^\beta\right)\varphi^*n\left({}_\alpha\psi, {}_\beta\psi\right)\\
&\quad -\,g^\vee_{S,t}\left(\left(\beta_t^\vee\right)^{-1}s^\alpha, \gamma_t\left(\left(b_t^\vee\right)^{-1}f^k\right)\left(\beta_t^\vee\right)^{-1}s^\beta\right)\varphi^*n\left({}_\alpha\psi, \nabla^{\varphi^*TN}_{b_t f_k}\,{}_\beta\psi\right)\\
&= -\,g^\vee_S\left(s^\alpha, \beta_t^\vee\not{\partial}_{g_t}\left(\beta_t^\vee\right)^{-1}s^\beta\right)\varphi^*n\left({}_\alpha\psi, {}_\beta\psi\right)\\
&\quad -\,g^\vee_S\left(s^\alpha, \gamma(f^k)s^\beta\right)\varphi^*n\left({}_\alpha\psi, \nabla^{\varphi^*TN}_{b_t f_k}\,{}_\beta\psi\right)
\end{aligned}$$

Taking derivative with respect to t:

$$\begin{aligned}
&\frac{d}{dt}\bigg|_{t=0} g^\vee_{S,t} \otimes \varphi^*n\left(\left(\left(\beta_t^\vee\right)^{-1} \otimes \mathrm{id}_{\varphi^*TN}\right)\psi, \not{D}_{g_t}\left(\left(\beta_t^\vee\right)^{-1} \otimes \mathrm{id}_{\varphi^*TN}\right)\psi\right)\\
&= -\,g^\vee_S\left(s^\alpha, \frac{d}{dt}\bigg|_{t=0}\beta_t^\vee\not{\partial}_{g_t}\left(\beta_t^\vee\right)^{-1}s^\beta\right)\varphi^*n\left({}_\alpha\psi, {}_\beta\psi\right)\\
&\quad -\,g^\vee_S\left(s^\alpha, \gamma(f^k)s^\beta\right)\varphi^*n\left({}_\alpha\psi, \frac{d}{dt}\bigg|_{t=0}\nabla^{\varphi^*TN}_{b_t f_k}\,{}_\beta\psi\right)\\
&= \frac{1}{2}g^\vee_S\left(s^\alpha, \gamma(f^k)\nabla^{S^\vee}_{h_k{}^l f_l}s^\beta\right)\varphi^*n\left({}_\alpha\psi, {}_\beta\psi\right)\\
&\quad +\frac{1}{2}g^\vee_S\left(s^\alpha, \gamma(f^k)s^\beta\right)\varphi^*n\left({}_\alpha\psi, \nabla^{\varphi^*TN}_{h_k{}^l f_l}\,{}_\beta\psi\right)\\
&= -\,\frac{1}{4}h^{ab}g^\vee_S \otimes \varphi^*n\left(\psi, \gamma(f_a)\nabla^{S^\vee\otimes\varphi^*TN}_{f_b}\psi + \gamma(f_b)\nabla^{S^\vee\otimes\varphi^*TN}_{f_a}\psi\right).
\end{aligned}$$

where the last equality holds since h^{ab} is symmetric.

iii) The term $\|F\|^2_{\varphi^*n}$ does not depend on g.

iv) Next we consider the summand of the action functional pairing the gravitino with ψ. Actually, since

$$
\begin{aligned}
&4g_t^\vee \otimes \varphi^*n\left(d\varphi, g_S^\vee\left(Q_t\left(\left(b_t^\vee\right)^{-1} \otimes \left(\beta_t^\vee\right)^{-1}\right)\chi, \left(\left(\beta_t^\vee\right)^{-1} \otimes \mathrm{id}_{\varphi^*TN}\right)\psi\right)\right)\\
&\quad = 2\delta^{cd} g_{S,t}^\vee\left(\gamma_t(b_t f_c)\gamma_t(b_t f_b)\left(\beta_t^\vee\right)^{-1}\chi((b_t)^{-1}b_t f_d), \left(\beta_t^\vee\right)^{-1}s^\kappa\right)\\
&\qquad \cdot \delta^{ab}\varphi^*n\left(d\varphi(b_t f_a), {}_\kappa\psi\right)\\
&\quad = 2\delta^{cd} g_S^\vee\left(\gamma(f_c)\gamma(f_b)\chi(f_d), s^\kappa\right)\delta^{ab}\varphi^*n\left(d\varphi(b_t f_a), {}_\kappa\psi\right),
\end{aligned}
$$

we obtain

$$
\begin{aligned}
&\left.\frac{d}{dt}\right|_{t=0} 4g_t^\vee \otimes \varphi^*n\left(d\varphi, g_S^\vee\left(Q_t\left(\left(b_t^\vee\right)^{-1} \otimes \left(\beta_t^\vee\right)^{-1}\right)\chi, \left(\left(\beta_t^\vee\right)^{-1} \otimes \mathrm{id}\right)\psi\right)\right)\\
&\quad = -\delta^{cd} g_S^\vee\left(\gamma(f_c)\gamma(f_b)\chi(f_d), s^\kappa\right)\delta^{ab}\varphi^*n\left(d\varphi(h_a^{\ e} f_e), {}_\kappa\psi\right)\\
&\quad = h^{ab}g^{cd}\left(g_S^\vee\left(\gamma(f_c)\gamma(f_a)\chi(f_d), s^\kappa\right)\varphi^*n\left(d\varphi(f_b), {}_\kappa\psi\right)\right)\\
&\quad = h^{ab}g_S^\vee \otimes \varphi^*n\left((Q\chi)(f_a) \otimes d\varphi(f_b) + (Q\chi)(f_b) \otimes d\varphi(f_a), \psi\right).
\end{aligned}
$$

v) For the term quadratic in gravitino and ψ we conclude

$$
\begin{aligned}
&\left\|Q_t\left(\left(b_t^\vee\right)^{-1} \otimes \left(\beta_t^\vee\right)^{-1}\right)\chi\right\|^2_{g^\vee \otimes g_{S,t}^\vee}\left\|\left(\beta_t^\vee\right)^{-1} \otimes \mathrm{id}_{TN}\psi\right\|^2_{g_{S,t}\otimes\varphi^*n}\\
&\quad = \|Q\chi\|^2_{g^\vee\otimes g_S^\vee}\|\psi\|^2_{g_S^\vee\otimes\varphi^*n}
\end{aligned}
$$

and hence its time derivative vanishes.

vi) The curvature term write $\psi = s^\alpha \otimes {}_\alpha\psi$. Then

$$
\begin{aligned}
&-\frac{1}{6}g_{S,t}^\vee\otimes\varphi^*n\left(SR^N\left(\left(\beta_t^\vee\right)^{-1} \otimes \mathrm{id}_{TN}\psi\right), \left(\beta_t^\vee\right)^{-1} \otimes \mathrm{id}_{TN}\psi\right)\\
&\quad = \frac{1}{6}g_{S,t}^\vee\left(\left(\beta_t^\vee\right)^{-1}s^\alpha, \left(\beta_t^\vee\right)^{-1}s^\beta\right)g_{S,t}^\vee\left(\left(\beta_t^\vee\right)^{-1}s^\alpha, \left(\beta_t^\vee\right)^{-1}s^\beta\right)\\
&\qquad \cdot \varphi^*n\left({}_\alpha\psi, R^N\left({}_\beta\psi, {}_\mu\psi\right){}_\nu\psi\right)\\
&\quad = \frac{1}{6}g_S^\vee(s^\alpha, s^\beta)g_S^\vee(s^\alpha, s^\beta)\varphi^*n\left({}_\alpha\psi, R^N\left({}_\beta\psi, {}_\mu\psi\right){}_\nu\psi\right)\\
&\quad = -\frac{1}{6}g_S^\vee \otimes \varphi^*n\left(SR^N(\psi), \psi\right)
\end{aligned}
$$

Consequently, its time derivative vanishes.

vii) We still need to consider the change in the volume form. We have that

$$\left.\frac{d}{dt}\right|_{t=0} dvol_{g_t} = \frac{1}{2}\operatorname{Tr}_g(h)\, dvol_g = -\frac{1}{2}h^{ab}g_{ab}\, dvol_g.$$

Summing up the contributions from i) to vii) yields the result. □

Proposition 12.4.2 (Supercurrent) *For $t \in \mathbb{R}$ let $X_t \in \Gamma\left(T^\vee|M| \otimes S^\vee\right)$ be a time indexed family of gravitinos such that $X_0 = \chi$ and $\left.\frac{d}{dt}\right|_{t=0} X_t = \rho$. Then*

$$\left.\frac{d}{dt}\right|_{t=0} A(\varphi, g, \psi, X_t, F) = \int_{|M|/B} g^\vee \otimes g_S^\vee\, (\rho, J)\, dvol_g.$$

Here, $J \in T^\vee|M| \otimes S^\vee$ is given by

$$J = 4Q\varphi^*n\, (d\varphi, \psi) + 2Q\chi \|\psi\|^2_{g_S^\vee \otimes \varphi^* n}$$

and called supercurrent of $A(\varphi, g, \psi, \chi, F)$.

Proof

$$\begin{aligned}
\left.\frac{d}{dt}\right|_{t=0} A(\varphi, g, \psi, X_t, F) &= \left.\frac{d}{dt}\right|_{t=0} \int_{|M|/B} \Big(4g^\vee \otimes \varphi^* n\left(d\varphi, g_S^\vee\left(QX_t, \psi\right)\right) \\
&\quad + \|QX_t\|^2_{g^\vee \otimes g_S} \|\psi\|^2_{g_S^\vee \otimes \varphi^* n}\Big)\, dvol_g \\
&= \int_{|M|/B} g^\vee \otimes g_S^\vee \left(Q\rho, 4\varphi^*n\, (d\varphi, \psi) + 2Q\chi \|\psi\|^2_{g_S^\vee \otimes \varphi^* n}\right) dvol_g \\
&= \int_{|M|/B} g^\vee \otimes g_S^\vee \left(\rho, 4Q\varphi^*n\, (d\varphi, \psi) + 2Q\chi \|\psi\|^2_{g_S^\vee \otimes \varphi^* n}\right) dvol_g
\end{aligned}$$

□

Remark 12.4.3 In principle one could try to study the change of $A(\Phi, m)$ under an infinitesimal variation $H_a{}^\beta$ of the SCL-structure, see Sect. 9.3. The super energy-momentum tensor $T^{super} \in \Gamma\left({}^{TM}\!/_{\mathcal{D}}{}^\vee \otimes \mathcal{D}\right)$ is then defined by

$$\delta A(\Phi, m) = \int_{M/B} H_\beta{}^a{}_a{}^\beta T^{super}[dvol_m].$$

Since the variation of the super Riemann surface given by the (h, ρ) can also be given in terms of $H_a{}^\beta$, the tensor T^{super} contains T and J. Hence, the tensor T^{super} combines T and J in a supergeometric object. However, the additional degrees of freedom fixed by the choice of Wess–Zumino coordinates and Wess–Zumino frames turn T^{super} into a more complicated object.

Integration against T and J yields a linear functional L on the space of infinitesimal deformations (h, ρ) of g and χ:

$$L((h,\rho)) = \int_{|M|/B} g^\vee \otimes g^\vee\,(h,T) + g^\vee \otimes g_S^\vee\,(\rho, J)\,dvol_g$$

The functional L can be rewritten with the help of the super L^2-metric

$$L((h,\rho)) = \ell\,((h,\rho),(T,J))\,.$$

We will now see that the Noether theorem implies that the pair (T, J) has the same holomorphicity properties as the true infinitesimal deformations of the super Riemann surface in Sect. 9.3.

Proposition 12.4.4 *For all fields φ, ψ and F the following holds:*

$$\begin{aligned}4\,\mathrm{Tr}_g T &= g_S^\vee \otimes \varphi^* n\,(\psi, EL(\psi)) + g^\vee \otimes g_S^\vee\,(\chi, J) + 2\|F\|^2,\\ \delta_\gamma J &= 0.\end{aligned}$$

Here $EL(\psi)$ is a shorthand for the Euler–Lagrange equation of ψ given in Eq. (12.3.4b).

Proof Let $\lambda_t = e^{t\sigma}$ for $\sigma \in \mathcal{O}_M$ and $t \in \mathbb{R}$ and $b_t\colon (T|M|, g) \to (T|M|, g_t)$ and $\beta_t\colon S_g \to S_{g_t}$ the standard isometries. Taking the time derivative of

$$\begin{aligned}&A(\varphi, g, \psi, \chi, F)\\ &= A\left(\varphi, \lambda_t^4 g, \frac{1}{\lambda_t}\left(\left(\beta_t^\vee\right)^{-1} \otimes \mathrm{id}_{\varphi^*TN}\right)\psi, \frac{1}{\lambda_t}\left(\left(b_t^\vee\right)^{-1} \otimes \left(\beta_t^\vee\right)^{-1}\chi\right), \frac{1}{\lambda_t^2}F\right)\end{aligned}$$

yields

$$\begin{aligned}0 = \int_{|M|/B} \sigma\,\Big(4\,\mathrm{Tr}_g T &- g_S^\vee \otimes \varphi^* n\,(\psi, EL(\psi))\\ &- g^\vee \otimes g_S^\vee\,(\chi, J) - 2\|F\|^2\Big)\,dvol_g.\end{aligned}$$

The equation $\delta_\gamma J = 0$ is trivial because J is in the image of Q. But the following argument shows explicitly how the super Weyl symmetry implies $\delta_\gamma J = 0$. Taking time derivative of

$$A(\varphi, g, \psi, \chi, F) = A(\varphi, g, \psi, \chi + \gamma t s, F),$$

where $s \in \Gamma\left(S^\vee\right)$ yields

$$0 = \int_{|M|/B} g^\vee \otimes g_S^\vee\,(\gamma s, J)\,dvol_g = \int_{|M|/B} g_S^\vee\left(s, \delta_\gamma J\right)dvol_g.$$

□

Proposition 12.4.5 *If the fields* φ, ψ, *and* F *fulfil the Euler–Lagrange Equations* (12.3.4), *then*

$$0 = 2\,\mathrm{div}_g\, T - \mathrm{div}_\chi\, J,$$

$$0 = 4g^\vee \otimes g^\vee\,(T, \gamma\chi) + \mathrm{div}_g\, J - \frac{1}{2}g^{kl} g_S\,(\chi(f_k), \chi(f_a))\,\gamma^a J(f_l).$$

For the Definition of $\mathrm{div}_g\, T$, $\mathrm{div}_\chi\, J$ and $\mathrm{div}_g\, J$ see Proposition 11.5.3 and Appendix A.4.

Proof Suppose that $\xi_t \colon |M| \to |M|$ is a time-indexed family of diffeomorphisms such that $\xi_0 = \mathrm{id}_{|M|}$ and $\frac{d}{dt}\big|_{t=0}\xi_t = X \in \Gamma\,(T|M|)$. The time-derivative of the diffeomorphism invariance

$$A(\varphi, g, \psi, \chi, F) = A(\varphi \circ \xi_t, g_{\xi_t}, \psi_{\xi_t}, \xi_t^*\chi \circ d\xi_t, \xi_t^* F)$$

yields

$$\begin{aligned} 0 &= \int_{|M|/B} g^\vee \otimes g^\vee\,(L_X g, T) + g^\vee \otimes g_S^\vee\,(\mathcal{L}_X \chi, J)\, dvol_g \\ &= \int_{|M|/B} -2\langle X, \mathrm{div}_g\, T\rangle + \langle X, \mathrm{div}_\chi\, J\rangle\, dvol_g. \end{aligned}$$

Note that no terms involving time derivatives of $\varphi \circ \xi_t$, ψ_{ξ_t} or $\xi_t^* F$ appear, because we have assumed that φ, ψ, and F satisfy the Euler–Lagrange Equations (12.3.4).

Similarly, for a time-indexed family of embeddings $j_t \colon |M| \to M$ such that $j_0 = i$ and $\frac{d}{dt}\big|_{t=0} j_t = q \in i^*\mathcal{D}$, the time derivative of

$$A(\varphi[i], g[i], \psi[i], \chi[i], F[i]) = A(\varphi[j_t], g[j_t], \psi[j_t], \chi[j_t], F[j_t])$$

yields, up to the infinitesimal conformal and super Weyl transformations, as explained in Sect. 11.4

$$\begin{aligned} 0 &= \int_{|M|/B} g^\vee \otimes g^\vee\,\big(\mathrm{susy}_q\, g, T\big) + g^\vee \otimes g_S^\vee\,\big(\mathrm{susy}_q\, \chi, J\big)\, dvol_g \\ &= \int_{|M|/B} \Big\langle q, 4g^\vee \otimes g^\vee\,(T, \gamma\chi) + \mathrm{div}_g\, J \\ &\quad - \frac{1}{2} g^{kl} g_S\,(\chi(f_k), \chi(f_a))\,\gamma^a J(f_l)\Big\rangle\, dvol_g. \end{aligned}$$

The last equality is analogous to the proof of Proposition 11.5.3. □

The following Theorem is now a direct consequence of Theorem 11.5.4:

Theorem 12.4.6 *The pair* (T, J) *is orthogonal to the trivial deformations of* (g, χ) *if the fields* φ, ψ, *and* F *fulfil the Euler–Lagrange Equations* (12.3.4). *That is,* (T, J) *is orthogonal in the super* L^2*-metric to the infinitesimal conformal transformations* $(4\sigma g, -\sigma\chi)$, *the infinitesimal super Weyl transformations* $(0, \gamma s)$, *infinitesimal diffeomorphisms* $(L_Y g, \mathcal{L}_Y \chi)$ *and supersymmetries* $(\mathrm{susy}_q\, g, \mathrm{susy}_q\, \chi)$. *Consequently, in this case* (T, J) *is a true infinitesimal deformation of the given super Riemann surface.*

In particular, if $\chi = 0$ *the energy-momentum tensor is a holomorphic quadratic differential and* J *is a holomorphic section of* $S^\vee \otimes_{\mathbb{C}} S^\vee \otimes_{\mathbb{C}} S^\vee$.

Chapter 13
Computations in Wess–Zumino Gauge

This chapter regroups different calculations in the Wess–Zumino gauge. The proof of Lemma 11.3.4 spans Sects. 13.2–13.4. The calculation of the Berezinian in terms of metric and gravitino given in Sect. 13.7 and is a crucial ingredient to the proof of Theorem 12.3.1 in Sect. 13.9.

All calculations in this chapter are local in nature. That is, we work in a coordinate neighbourhood $U \subset M$ of a super Riemann surface M with local coordinates $X^A = (x^a, \eta^\alpha)$. From Sect. 13.2 on, we assume that X^A together with coordinates y^a of $U \subset |M|$ form Wess–Zumino coordinates for the SCL-frame F_A and the embedding $i\colon |M| \to M$. Recall from Definition 11.2.1 that the coordinates X^A and y^a are called Wess–Zumino coordinates for i and F_A if

$$i^\# x^a = y^a, \qquad\qquad i^\# \eta^\alpha = 0,$$

and the η-expansion

$$F_A{}^B = {}_0F_A{}^B + \eta^\mu {}_\mu F_A{}^B + \eta^3\eta^4 {}_{34}F_A{}^B,$$

of the frame coefficients of $F_A = F_A{}^B \partial_{X^B}$ fulfills

$${}_0F_\alpha{}^b = 0, \qquad\qquad {}_0F_\alpha{}^\beta = \delta_\alpha^\beta, \qquad\qquad \varepsilon^{\mu\alpha} {}_\mu F_\alpha{}^B = 0.$$

By Proposition 11.2.5 such coordinates always exist.

E. Keßler, *Supergeometry, Super Riemann Surfaces and the Superconformal Action Functional*, Lecture Notes in Mathematics 2230,
https://doi.org/10.1007/978-3-030-13758-8_13

13.1 Formulas in 2-Dimensional Superalgebras

This section gathers algebraic formulas valid in all superalgebras with two odd generators η^3, η^4, such as the smooth functions on $\mathbb{R}^{m|2}$.

Lemma 13.1.1 *It holds*

$$\begin{aligned} \eta^\mu \eta^\nu &= \eta^3 \eta^4 \varepsilon^{\mu\nu} & \partial_{\eta^\mu} \partial_{\eta^\nu} &= \varepsilon_{\mu\nu} \partial_{\eta^3} \partial_{\eta^4} \\ 2\eta^3 \eta^4 &= \varepsilon_{\mu\nu} \eta^\mu \eta^\nu & 2\partial_{\eta^3} \partial_{\eta^4} &= \varepsilon^{\mu\nu} \partial_{\eta^\mu} \partial_{\eta^\nu} \\ \partial_{\eta^\mu} \eta^3 \eta^4 &= \varepsilon_{\mu\nu} \eta^\nu & \varepsilon^{\mu\nu} \partial_{\eta^\mu} \partial_{\eta^\nu} \eta^3 \eta^4 &= -2 \end{aligned}$$

Lemma 13.1.2 *Let*

$$t = {}_0t + \eta^\mu\, {}_\mu t + \eta^3 \eta^4\, {}_{34}t$$

be even. Then

$$t^2 = {}_0t^2 + 2\eta^\mu\, {}_\mu t\, {}_0t + \eta^3 \eta^4 \left(2\, {}_0t\, {}_{34}t - \varepsilon^{\mu\nu}\, {}_\mu t\, {}_\nu t\right),$$

and t is invertible if and only if ${}_0t$ is invertible and it holds:

$$\frac{1}{t} = \frac{1}{{}_0t} - \eta^\mu \frac{{}_\mu t}{{}_0t^2} - \eta^3 \eta^4 \left(\frac{\varepsilon^{\mu\nu}\, {}_\mu t\, {}_\nu t}{{}_0t^3} + \frac{{}_{34}t}{{}_0t^2} \right) \tag{13.1.3}$$

$$\frac{1}{t^2} = \frac{1}{{}_0t^2} - 2\eta^\mu \frac{{}_\mu t}{{}_0t^3} - \eta^3 \eta^4 \left(3 \frac{\varepsilon^{\mu\nu}\, {}_\mu t\, {}_\nu t}{{}_0t^4} + 2 \frac{{}_{34}t}{{}_0t^3} \right)$$

Proof The formulas are verified as follows:

$$\begin{aligned} t^2 &= \left({}_0t + \eta^\mu\, {}_\mu t + \eta^3 \eta^4\, {}_{34}t\right)^2 = {}_0t^2 + 2\eta^\mu\, {}_\mu t\, {}_0t + 2\eta^3 \eta^4\, {}_0t\, {}_{34}t + \left(\eta^\mu\, {}_\mu t\right)\left(\eta^\nu\, {}_\nu t\right) \\ &= {}_0t^2 + 2\eta^\mu\, {}_\mu t\, {}_0t + \eta^3 \eta^4 \left(2\, {}_0t\, {}_{34}t - \varepsilon^{\mu\nu}\, {}_\mu t\, {}_\nu t\right), \end{aligned}$$

$$\begin{aligned} &\left({}_0t + \eta^\mu\, {}_\mu t + \eta^3 \eta^4\, {}_{34}t\right) \left(\frac{1}{{}_0t} - \eta^\mu \frac{{}_\mu t}{{}_0t^2} - \eta^3 \eta^4 \left(\frac{\varepsilon^{\mu\nu}\, {}_\mu t\, {}_\nu t}{{}_0t^3} + \frac{{}_{34}t}{{}_0t^2} \right) \right) \\ &= 1 + \eta^\mu \left(-\frac{{}_\mu t}{{}_0t} + \frac{{}_\mu t}{{}_0t} \right) \\ &\quad + \eta^3 \eta^4 \left(-{}_0t \left(\frac{\varepsilon^{\mu\nu}\, {}_\mu t\, {}_\nu t}{{}_0t^3} + \frac{{}_{34}t}{{}_0t^2} \right) + \frac{\varepsilon^{\mu\nu}\, {}_\mu t\, {}_\nu t}{{}_0t^2} + \frac{{}_{34}t}{{}_0t} \right) \\ &= 1. \end{aligned}$$

□

13.2 Inverse

Let us denote the inverse of the matrix of frame coefficients $F_A{}^B$ as $E_B{}^C$. Then it holds that

$$\partial_{X^A} = E_A{}^B F_B$$

and

$$F^B = dX^C E_C{}^B.$$

Lemma 13.2.1 *The inverse $E_A{}^B$ of the matrix of frame coefficients $F_A{}^B$ with respect to Wess–Zumino coordinates $X^B = (x^b, \eta^\beta)$ of a SCL-frame $F_A = F_A{}^B \partial_{X^B}$ on a super Riemann surface fulfills*

$$\begin{pmatrix} E_c{}^b & E_c{}^\beta \\ E_\gamma{}^b & E_\gamma{}^\beta \end{pmatrix} = \begin{pmatrix} \frac{1}{\det F_m{}^n}\, {}_0F_m{}^n \varepsilon^{mb}\varepsilon_{nc} & -\frac{1}{\det {}_0F_m{}^n}\, {}_0F_m{}^n \varepsilon^{mp}\varepsilon_{nc}\, {}_0F_p{}^\beta \\ 0 & \delta_\gamma^\beta \end{pmatrix}$$
$$+ \eta^\mu \begin{pmatrix} {}_\mu E_c{}^b & {}_\mu E_c{}^\beta \\ {}_\mu E_\gamma{}^b & {}_\mu E_\gamma{}^\beta \end{pmatrix} + \eta^3\eta^4 \begin{pmatrix} {}_{34}E_c{}^b & {}_{34}E_c{}^\beta \\ {}_{34}E_\gamma{}^b & {}_{34}E_\gamma{}^\beta \end{pmatrix},$$

where the first order terms are given by

$$\begin{aligned}
{}_\mu E_c{}^b &= -\, {}_0E_c{}^p \left({}_\mu F_p{}^a + {}_0F_p{}^\sigma\, {}_\mu F_\sigma{}^a\right) {}_0E_a{}^b, \\
{}_\mu E_c{}^\beta &= {}_0E_c{}^p \left(\left({}_\mu F_p{}^a + {}_0F_p{}^\sigma\, {}_\mu F_\sigma{}^a\right) {}_0E_a{}^s\, {}_0F_s{}^\beta - {}_\mu F_p{}^\beta - {}_0F_p{}^\sigma\, {}_\mu F_\sigma{}^\beta\right), \\
{}_\mu E_\gamma{}^b &= -\, {}_\mu F_\gamma{}^a\, {}_0E_a{}^b, \\
{}_\mu E_\gamma{}^\beta &= {}_\mu F_\gamma{}^a\, {}_0E_a{}^s\, {}_0F_s{}^\beta - {}_\mu F_\gamma{}^\beta,
\end{aligned}$$

and the second order terms are given by

$$\begin{aligned}
{}_{34}E_c{}^b = &-\, {}_0E_c{}^s \left({}_{34}F_s{}^a + \varepsilon^{\mu\nu} \left({}_\mu F_s{}^a + {}_0F_s{}^\tau\, {}_\mu F_\tau{}^a\right) {}_0E_a{}^t \right. \\
&\cdot \left({}_\nu F_t{}^a + {}_0F_t{}^\sigma\, {}_\nu F_\sigma{}^a\right) - {}_0F_s{}^\sigma\, {}_{34}F_\sigma{}^a \\
&\left. - \varepsilon^{\mu\nu} \left({}_\mu F_s{}^\sigma + {}_0F_s{}^\tau\, {}_\mu F_\tau{}^\sigma\right) {}_\nu F_\sigma{}^a\right) {}_0E_a{}^b, \\
{}_{34}E_c{}^\alpha = &\; {}_0E_c{}^p \left(\varepsilon^{\mu\nu} \left({}_\mu F_p{}^m + {}_0F_p{}^\tau\, {}_\mu F_\tau{}^m\right) {}_0E_m{}^t \right. \\
&\cdot \left(\left({}_\nu F_t{}^a + {}_0F_t{}^\sigma\, {}_\nu F_\sigma{}^a\right) {}_0E_a{}^b\, {}_0F_b{}^\alpha - {}_\nu F_t{}^\alpha - {}_0F_t{}^\beta\, {}_\nu F_\beta{}^\alpha\right) - {}_{34}F_p{}^\alpha \\
&+ \left({}_{34}F_p{}^a - {}_0F_p{}^\sigma\, {}_{34}F_\sigma{}^a\right) {}_0E_a{}^b\, {}_0F_b{}^\alpha - \varepsilon^{\mu\nu} \left({}_\mu F_p{}^\sigma + {}_0F_p{}^\tau\, {}_\mu F_\tau{}^\sigma\right) \\
&\left. \cdot \left({}_\nu F_\sigma{}^a\, {}_0E_a{}^b\, {}_0F_b{}^\alpha - {}_\nu F_\sigma{}^\alpha\right) + {}_0F_p{}^\beta\, {}_{34}F_\beta{}^\alpha\right),
\end{aligned}$$

$$
\begin{aligned}
{}_{34}E_\gamma{}^b &= \Big(\varepsilon^{\mu\nu}\,{}_\mu F_\gamma{}^s\,{}_0E_s{}^t\,{}_\nu F_t{}^a - {}_{34}F_\gamma{}^a \\
&\quad + \varepsilon^{\mu\nu}\Big({}_\mu F_\gamma{}^a\,{}_0E_a{}^s\,{}_0F_s{}^\tau - {}_\mu F_\gamma{}^\tau\Big){}_\nu F_\tau{}^a\Big)\,{}_0E_a{}^b, \\
{}_{34}E_\gamma{}^\alpha &= \varepsilon^{\mu\nu}\,{}_\mu F_\gamma{}^a\,{}_0E_a{}^b\,{}_\nu F_b{}^\alpha - \Big(\varepsilon^{\mu\nu}\,{}_\mu F_\gamma{}^s\,{}_0E_s{}^t\,{}_\nu F_t{}^a - {}_{34}F_\gamma{}^a \\
&\quad + \varepsilon^{\mu\nu}\Big({}_\mu F_\gamma{}^a\,{}_0E_a{}^s\,{}_0F_s{}^\tau - {}_\mu F_\gamma{}^\tau\Big){}_\nu F_\tau{}^a\Big)\,{}_0E_a{}^b\,{}_0F_b{}^\alpha - {}_{34}F_\gamma{}^\alpha \\
&\quad + \varepsilon^{\mu\nu}\Big({}_\mu F_\gamma{}^a\,{}_0E_a{}^s\,{}_0F_s{}^\beta - {}_\mu F_\gamma{}^\beta\Big){}_\nu F_\beta{}^\alpha.
\end{aligned}
$$

Proof As $E_C{}^B$ is the inverse of $F_A{}^B$ their matrix product is given by

$$
\begin{aligned}
\delta_C^A = E_C{}^B F_B{}^A &= \begin{pmatrix} E_c{}^b & E_c{}^\beta \\ E_\gamma{}^b & E_\gamma{}^\beta \end{pmatrix}\begin{pmatrix} F_b{}^a & F_b{}^\alpha \\ F_\beta{}^a & F_\beta{}^\alpha \end{pmatrix} \\
&= \begin{pmatrix} E_c{}^b F_b{}^a + E_c{}^\beta F_\beta{}^a & E_c{}^b F_b{}^\alpha + E_c{}^\beta F_\beta{}^\alpha \\ E_\gamma{}^b F_b{}^a + E_\gamma{}^\beta F_\beta{}^a & E_\gamma{}^b F_b{}^\alpha + E_\gamma{}^\beta F_\beta{}^\alpha \end{pmatrix}.
\end{aligned}
$$

Expanding in orders of η we get the following equations

$$
\begin{aligned}
\delta_c^a &= E_c{}^b F_b{}^a + E_c{}^\beta F_\beta{}^a \\
&= {}_0E_c{}^b\,{}_0F_b{}^a + \eta^\mu\Big({}_\mu E_c{}^b\,{}_0F_b{}^a + {}_0E_c{}^b\,{}_\mu F_b{}^a - {}_0E_c{}^\beta\,{}_\mu F_\beta{}^a\Big) \\
&\quad + \eta^3\eta^4\Big({}_0E_c{}^b\,{}_{34}F_b{}^a - \varepsilon^{\mu\nu}\,{}_\mu E_c{}^b\,{}_\nu F_b{}^a + {}_{34}E_c{}^b\,{}_0F_b{}^a + {}_0E_c{}^\beta\,{}_{34}F_\beta{}^a \\
&\quad + \varepsilon^{\mu\nu}\,{}_\mu E_c{}^\beta\,{}_\nu F_\beta{}^a\Big) \\
0 &= E_c{}^b F_b{}^\alpha + E_c{}^\beta F_\beta{}^\alpha \\
&= {}_0E_c{}^b\,{}_0F_b{}^\alpha + {}_0E_c{}^\alpha + \eta^\mu\Big({}_\mu E_c{}^b\,{}_0F_b{}^\alpha + {}_0E_c{}^b\,{}_\mu F_b{}^\alpha + {}_\mu E_c{}^\alpha \\
&\quad - {}_0E_c{}^\beta\,{}_\mu F_\beta{}^\alpha\Big) + \eta^3\eta^4\Big({}_0E_c{}^b\,{}_{34}F_b{}^\alpha - \varepsilon^{\mu\nu}\,{}_\mu E_c{}^b\,{}_\nu F_b{}^\alpha + {}_{34}E_c{}^b\,{}_0F_b{}^\alpha \\
&\quad + {}_0E_c{}^\beta\,{}_{34}F_\beta{}^\alpha + \varepsilon^{\mu\nu}\,{}_\mu E_c{}^\beta\,{}_\nu F_\beta{}^\alpha + {}_{34}E_c{}^\alpha\Big) \\
0 &= E_\gamma{}^b F_b{}^a + E_\gamma{}^\beta F_\beta{}^a \\
&= {}_0E_\gamma{}^b\,{}_0F_b{}^a + \eta^\mu\Big({}_\mu E_\gamma{}^b\,{}_0F_b{}^a - {}_0E_\gamma{}^b\,{}_\mu F_b{}^a + {}_0E_\gamma{}^\beta\,{}_\mu F_\beta{}^a\Big) \\
&\quad + \eta^3\eta^4\Big({}_0E_\gamma{}^b\,{}_{34}F_b{}^a + \varepsilon^{\mu\nu}\,{}_\mu E_\gamma{}^b\,{}_\nu F_b{}^a + {}_{34}E_\gamma{}^b\,{}_0F_b{}^a + {}_0E_\gamma{}^\beta\,{}_{34}F_\beta{}^a \\
&\quad - \varepsilon^{\mu\nu}\,{}_\mu E_\gamma{}^\beta\,{}_\nu F_\beta{}^a\Big)
\end{aligned}
$$

$$
\begin{aligned}
\delta^{\alpha}_{\gamma} &= E_{\gamma}{}^{b} F_{b}{}^{\alpha} + E_{\gamma}{}^{\beta} F_{\beta}{}^{\alpha} \\
&= {}_0E_{\gamma}{}^{b}\, {}_0F_{b}{}^{\alpha} + {}_0E_{\gamma}{}^{\alpha} + \eta^{\mu} \left({}_{\mu}E_{\gamma}{}^{b}\, {}_0F_{b}{}^{\alpha} - {}_0E_{\gamma}{}^{b}\, {}_{\mu}F_{b}{}^{\alpha} + {}_{\mu}E_{\gamma}{}^{\alpha} \right. \\
&\quad \left. + {}_0E_{\gamma}{}^{\beta}\, {}_{\mu}F_{\beta}{}^{\alpha} \right) + \eta^3\eta^4 \left({}_0E_{\gamma}{}^{b}\, {}_{34}F_{b}{}^{\alpha} + \varepsilon^{\mu\nu}\, {}_{\mu}E_{\gamma}{}^{b}\, {}_{\nu}F_{b}{}^{\alpha} + {}_{34}E_{\gamma}{}^{b}\, {}_0F_{b}{}^{\alpha} \right. \\
&\quad \left. + {}_0E_{\gamma}{}^{\beta}\, {}_{34}F_{\beta}{}^{\alpha} - \varepsilon^{\mu\nu}\, {}_{\mu}E_{\gamma}{}^{\beta}\, {}_{\nu}F_{\beta}{}^{\alpha} + {}_{34}E_{\gamma}{}^{\alpha} \right)
\end{aligned}
$$

Now we look at the constant terms of this expansion and conclude in that order:

$$
\begin{aligned}
{}_0E_{c}{}^{b} &= \frac{1}{\det {}_0F_{m}{}^{n}}\, {}_0F_{m}{}^{n} \varepsilon^{mb} \varepsilon_{nc} \\
{}_0E_{c}{}^{\beta} &= -\, {}_0E_{c}{}^{p}\, {}_0F_{p}{}^{\beta} \\
{}_0E_{\gamma}{}^{b} &= 0 \\
{}_0E_{\gamma}{}^{\beta} &= \delta^{\beta}_{\gamma}
\end{aligned}
$$

Looking at the first order terms and using the previous results one obtains

$$
\begin{aligned}
{}_{\mu}E_{c}{}^{b} &= \left(-\, {}_0E_{c}{}^{p}\, {}_{\mu}F_{p}{}^{a} + {}_0E_{c}{}^{\sigma}\, {}_{\mu}F_{\sigma}{}^{a} \right) {}_0E_{a}{}^{b} \\
&= -\left({}_0E_{c}{}^{p}\, {}_{\mu}F_{p}{}^{a} + {}_0E_{c}{}^{p}\, {}_0F_{p}{}^{\sigma}\, {}_{\mu}F_{\sigma}{}^{a} \right) {}_0E_{a}{}^{b} \\
&= -\, {}_0E_{c}{}^{p} \left({}_{\mu}F_{p}{}^{a} + {}_0F_{p}{}^{\sigma}\, {}_{\mu}F_{\sigma}{}^{a} \right) {}_0E_{a}{}^{b} \\
{}_{\mu}E_{c}{}^{\beta} &= -\, {}_{\mu}E_{c}{}^{s}\, {}_0F_{s}{}^{\beta} - {}_0E_{c}{}^{s}\, {}_{\mu}F_{s}{}^{\beta} + {}_0E_{c}{}^{\beta}\, {}_{\mu}F_{\beta}{}^{\alpha} \\
&= {}_0E_{c}{}^{p} \left({}_{\mu}F_{p}{}^{a} + {}_0F_{p}{}^{\sigma}\, {}_{\mu}F_{\sigma}{}^{a} \right) {}_0E_{a}{}^{s}\, {}_0F_{s}{}^{\beta} - {}_0E_{c}{}^{p}\, {}_{\mu}F_{p}{}^{\beta} \\
&\quad - {}_0E_{c}{}^{p}\, {}_0F_{p}{}^{\sigma}\, {}_{\mu}F_{\sigma}{}^{\beta} \\
&= {}_0E_{c}{}^{p} \left(\left({}_{\mu}F_{p}{}^{a} + {}_0F_{p}{}^{\sigma}\, {}_{\mu}F_{\sigma}{}^{a} \right) {}_0E_{a}{}^{s}\, {}_0F_{s}{}^{\beta} - {}_{\mu}F_{p}{}^{\beta} - {}_0F_{p}{}^{\sigma}\, {}_{\mu}F_{\sigma}{}^{\beta} \right) \\
{}_{\mu}E_{\gamma}{}^{b} &= -\, {}_{\mu}F_{\gamma}{}^{a}\, {}_0E_{a}{}^{b} \\
{}_{\mu}E_{\gamma}{}^{\beta} &= -\, {}_{\mu}E_{\gamma}{}^{s}\, {}_0F_{s}{}^{\beta} - {}_{\mu}F_{\gamma}{}^{\beta} = {}_{\mu}F_{\gamma}{}^{a}\, {}_0E_{a}{}^{s}\, {}_0F_{s}{}^{\beta} - {}_{\mu}F_{\gamma}{}^{\beta}.
\end{aligned}
$$

Using the results of order zero and one, the terms of second order are calculated as follows

$$
\begin{aligned}
{}_{34}E_c{}^b &= -\left({}_0E_c{}^s\,{}_{34}F_s{}^a - \varepsilon^{\mu\nu}\,{}_\mu E_c{}^s\,{}_\nu F_s{}^a + {}_0E_c{}^\sigma\,{}_{34}F_\sigma{}^a + \varepsilon^{\mu\nu}\,{}_\mu E_c{}^\sigma\,{}_\nu F_\sigma{}^a\right)\\
&\quad\cdot{}_0E_a{}^b\\
&= -\Big({}_0E_c{}^s\,{}_{34}F_s{}^a + \varepsilon^{\mu\nu}\,{}_0E_c{}^p\left({}_\mu F_p{}^a + {}_0F_p{}^\tau\,{}_\mu F_\tau{}^a\right){}_0E_a{}^s\,{}_\nu F_s{}^a\\
&\quad - {}_0E_c{}^p\,{}_0F_p{}^\sigma\,{}_{34}F_\sigma{}^a + \varepsilon^{\mu\nu}\,{}_0E_c{}^p\Big(\left({}_\mu F_p{}^a + {}_0F_p{}^\tau\,{}_\mu F_\tau{}^a\right){}_0E_a{}^t\,{}_0F_t{}^\sigma\\
&\quad - {}_\mu F_p{}^\sigma - {}_0F_p{}^\tau\,{}_\mu F_\tau{}^\sigma\Big){}_\nu F_\sigma{}^a\Big){}_0E_a{}^b\\
&= -{}_0E_c{}^s\left({}_{34}F_s{}^a + \varepsilon^{\mu\nu}\left({}_\mu F_s{}^a + {}_0F_s{}^\tau\,{}_\mu F_\tau{}^a\right){}_0E_a{}^t\left({}_\nu F_t{}^a + {}_0F_t{}^\sigma\,{}_\nu F_\sigma{}^a\right)\right.\\
&\quad\left. - {}_0F_s{}^\sigma\,{}_{34}F_\sigma{}^a - \varepsilon^{\mu\nu}\left({}_\mu F_s{}^\sigma + {}_0F_s{}^\tau\,{}_\mu F_\tau{}^\sigma\right){}_\nu F_\sigma{}^a\right){}_0E_a{}^b\\
{}_{34}E_c{}^\alpha &= -{}_0E_c{}^b\,{}_{34}F_b{}^\alpha + \varepsilon^{\mu\nu}\,{}_\mu E_c{}^b\,{}_\nu F_b{}^\alpha - {}_{34}E_c{}^b\,{}_0F_b{}^\alpha - {}_0E_c{}^\beta\,{}_{34}F_\beta{}^\alpha\\
&\quad - \varepsilon^{\mu\nu}\,{}_\mu E_c{}^\beta\,{}_\nu F_\beta{}^\alpha\\
&= -{}_0E_c{}^b\,{}_{34}F_b{}^\alpha - \varepsilon^{\mu\nu}\,{}_0E_c{}^p\left({}_\mu F_p{}^a + {}_0F_p{}^\sigma\,{}_\mu F_\sigma{}^a\right){}_0E_a{}^b\,{}_\nu F_b{}^\alpha\\
&\quad + {}_0E_c{}^s\left({}_{34}F_s{}^a + \varepsilon^{\mu\nu}\left({}_\mu F_s{}^m + {}_0F_s{}^\tau\,{}_\mu F_\tau{}^m\right){}_0E_m{}^t\right.\\
&\quad\cdot\left({}_\nu F_t{}^a + {}_0F_t{}^\sigma\,{}_\nu F_\sigma{}^a\right) - {}_0F_s{}^\sigma\,{}_{34}F_\sigma{}^a\\
&\quad\left. - \varepsilon^{\mu\nu}\left({}_\mu F_s{}^\sigma + {}_0F_s{}^\tau\,{}_\mu F_\tau{}^\sigma\right){}_\nu F_\sigma{}^a\right){}_0E_a{}^b\,{}_0F_b{}^\alpha + {}_0E_c{}^p\,{}_0F_p{}^\beta\,{}_{34}F_\beta{}^\alpha\\
&\quad - \varepsilon^{\mu\nu}\,{}_0E_c{}^p\Big(\left({}_\mu F_p{}^a + {}_0F_p{}^\sigma\,{}_\mu F_\sigma{}^a\right){}_0E_a{}^s\,{}_0F_s{}^\beta\\
&\quad - {}_\mu F_p{}^\beta - {}_0F_p{}^\sigma\,{}_\mu F_\sigma{}^\beta\Big){}_\nu F_\beta{}^\alpha\\
&= {}_0E_c{}^p\Big(\varepsilon^{\mu\nu}\left({}_\mu F_p{}^m + {}_0F_p{}^\tau\,{}_\mu F_\tau{}^m\right){}_0E_m{}^t\\
&\quad\cdot\Big(\left({}_\nu F_t{}^a + {}_0F_t{}^\sigma\,{}_\nu F_\sigma{}^a\right){}_0E_a{}^b\,{}_0F_b{}^\alpha - {}_\nu F_t{}^\alpha - {}_0F_t{}^\beta\,{}_\nu F_\beta{}^\alpha\Big) - {}_{34}F_p{}^\alpha\\
&\quad + \left({}_{34}F_p{}^a - {}_0F_p{}^\sigma\,{}_{34}F_\sigma{}^a\right){}_0E_a{}^b\,{}_0F_b{}^\alpha - \varepsilon^{\mu\nu}\left({}_\mu F_p{}^\sigma + {}_0F_p{}^\tau\,{}_\mu F_\tau{}^\sigma\right)\\
&\quad\cdot\left({}_\nu F_\sigma{}^a\,{}_0E_a{}^b\,{}_0F_b{}^\alpha - {}_\nu F_\sigma{}^\alpha\right) + {}_0F_p{}^\beta\,{}_{34}F_\beta{}^\alpha\Big)\\
{}_{34}E_\gamma{}^b &= -\left({}_0E_\gamma{}^t\,{}_{34}F_t{}^a + \varepsilon^{\mu\nu}\,{}_\mu E_\gamma{}^t\,{}_\nu F_t{}^a + {}_0E_\gamma{}^\tau\,{}_{34}F_\tau{}^a - \varepsilon^{\mu\nu}\,{}_\mu E_\gamma{}^\tau\,{}_\nu F_\tau{}^a\right)\\
&\quad\cdot{}_0E_a{}^b\\
&= \Big(\varepsilon^{\mu\nu}\,{}_\mu F_\gamma{}^s\,{}_0E_s{}^t\,{}_\nu F_t{}^a - {}_{34}F_\gamma{}^a\\
&\quad + \varepsilon^{\mu\nu}\left({}_\mu F_\gamma{}^a\,{}_0E_a{}^s\,{}_0F_s{}^\tau - {}_\mu F_\gamma{}^\tau\right){}_\nu F_\tau{}^a\Big){}_0E_a{}^b
\end{aligned}
$$

$$
\begin{aligned}
{}_{34}{E_\gamma}^\alpha &= -{}_0{E_\gamma}^b\,{}_{34}{F_b}^\alpha - \varepsilon^{\mu\nu}\,{}_\mu{E_\gamma}^b\,{}_\nu{F_b}^\alpha - {}_{34}{E_\gamma}^b\,{}_0{F_b}^\alpha - {}_0{E_\gamma}^\beta\,{}_{34}{F_\beta}^\alpha \\
&\quad + \varepsilon^{\mu\nu}\,{}_\mu{E_\gamma}^\beta\,{}_\nu{F_\beta}^\alpha \\
&= \varepsilon^{\mu\nu}\,{}_\mu{F_\gamma}^a\,{}_0{E_a}^b\,{}_\nu{F_b}^\alpha - \Big(\varepsilon^{\mu\nu}\,{}_\mu{F_\gamma}^s\,{}_0{E_s}^t\,{}_\nu{F_t}^a - {}_{34}{F_\gamma}^a \\
&\quad + \varepsilon^{\mu\nu}\left({}_\mu{F_\gamma}^a\,{}_0{E_a}^s\,{}_0{F_s}^\tau - {}_\mu{F_\gamma}^\tau\right){}_\nu{F_\tau}^a\Big)\,{}_0{E_a}^b\,{}_0{F_b}^\alpha - {}_{34}{F_\gamma}^\alpha \\
&\quad + \varepsilon^{\mu\nu}\left({}_\mu{F_\gamma}^a\,{}_0{E_a}^s\,{}_0{F_s}^\beta - {}_\mu{F_\gamma}^\beta\right){}_\nu{F_\beta}^\alpha
\end{aligned}
$$

□

Remark 13.2.2 The coordinates and $X^A = (x^a, \eta^\alpha)$ are Wess–Zumino-coordinates for F_A and i if and only if $i^\# \eta^\alpha = 0$ and the coefficients of the left dual frame

$$F^B = dX^A {E_A}^B$$

fulfill

$${}_0{E_\alpha}^b = 0, \qquad {}_0{E_\alpha}^\beta = \delta_\alpha^\beta, \qquad \varepsilon^{\mu\alpha}\,{}_\mu{E_\alpha}^C = 0.$$

13.3 Commutators in Wess–Zumino Coordinates

In this section we are assuming that $X^A = (x^a, \eta^\alpha)$ are Wess–Zumino coordinates for the frame F_A and calculate the commutator between the odd frames as expressions of the frame coefficients of F_A.

$$
\begin{aligned}
[F_\alpha, F_\beta] &= [{F_\alpha}^c \partial_{x^c} + {F_\alpha}^\gamma \partial_{\eta^\gamma}, {F_\beta}^d \partial_{x^d} + {F_\beta}^\delta \partial_{\eta^\delta}] \\
&= {F_\alpha}^c \left(\partial_{x^c} {F_\beta}^d\right) \partial_{x^d} + {F_\beta}^d \left(\partial_{x^d} {F_\alpha}^c\right) \partial_{x^c} + {F_\alpha}^c \left(\partial_{x^c} {F_\beta}^\delta\right) \partial_{\eta^\delta} \\
&\quad + {F_\beta}^\delta \left(\partial_{\eta^\delta} {F_\alpha}^c\right) \partial_{x^c} + {F_\alpha}^\gamma \left(\partial_{\eta^\gamma} {F_\beta}^d\right) \partial_{x^d} + {F_\beta}^d \left(\partial_{x^d} {F_\alpha}^\gamma\right) \partial_{\eta^\gamma} \\
&\quad + {F_\alpha}^\gamma \left(\partial_{\eta^\gamma} {F_\beta}^\delta\right) \partial_{\eta^\delta} + {F_\beta}^\delta \left(\partial_{\eta^\delta} {F_\alpha}^\gamma\right) \partial_{\eta^\gamma} \\
&= \Big(2\,{}_\alpha{F_\beta}^t + \eta^\kappa \Big({}_\kappa{F_\alpha}^\mu\,{}_\mu{F_\beta}^t + {}_\kappa{F_\beta}^\mu\,{}_\mu{F_\alpha}^t + \varepsilon_{\alpha\kappa}\,{}_{34}{F_\beta}^t + \varepsilon_{\beta\kappa}\,{}_{34}{F_\alpha}^t\Big) \\
&\quad + \eta^3\eta^4 \Big(\varepsilon^{\mu\nu} \Big({}_\mu{F_\alpha}^c \left(\partial_{x^c}\,{}_\nu{F_\beta}^t\right) + {}_\mu{F_\beta}^d \left(\partial_{x^d}\,{}_\nu{F_\alpha}^t\right)\Big) \\
&\quad - \varepsilon^{\mu\nu}\,{}_\mu{F_\beta}^\delta \varepsilon_{\delta\nu}\,{}_{34}{F_\alpha}^t + {}_{34}{F_\beta}^\delta\,{}_\delta{F_\alpha}^t - \varepsilon^{\mu\nu}\,{}_\mu{F_\alpha}^\gamma \varepsilon_{\gamma\nu}\,{}_{34}{F_\beta}^t \\
&\quad + {}_{34}{F_\alpha}^\gamma\,{}_\gamma{F_\beta}^t\Big)\Big)\partial_{x^t}
\end{aligned}
$$

$$
\begin{aligned}
&+\Big(2\,{}_{\alpha}F_{\beta}{}^{\tau}+\eta^{\kappa}\Big({}_{\kappa}F_{\alpha}{}^{\mu}\,{}_{\mu}F_{\beta}{}^{\tau}+{}_{\kappa}F_{\beta}{}^{\mu}\,{}_{\mu}F_{\alpha}{}^{\tau}+\varepsilon_{\alpha\kappa}\,{}_{34}F_{\beta}{}^{\tau}\\
&+\varepsilon_{\beta\kappa}\,{}_{34}F_{\alpha}{}^{\tau}\Big)+\eta^{3}\eta^{4}\left(\varepsilon^{\mu\nu}\left({}_{\mu}F_{\alpha}{}^{c}\left(\partial_{x^{c}}\,{}_{\nu}F_{\beta}{}^{\tau}\right)+{}_{\mu}F_{\beta}{}^{c}\left(\partial_{x^{c}}\,{}_{\nu}F_{\alpha}{}^{\tau}\right)\right)\right.\\
&-\varepsilon^{\mu\nu}\,{}_{\mu}F_{\beta}{}^{\delta}\varepsilon_{\delta\nu}\,{}_{34}F_{\alpha}{}^{\tau}+{}_{34}F_{\beta}{}^{\delta}\,{}_{\delta}F_{\alpha}{}^{\tau}-\varepsilon^{\mu\nu}\,{}_{\mu}F_{\alpha}{}^{\gamma}\varepsilon_{\gamma\nu}\,{}_{34}F_{\beta}{}^{\tau}\\
&\left.+\,{}_{34}F_{\alpha}{}^{\gamma}\,{}_{\gamma}F_{\beta}{}^{\tau}\right)\Big)\partial_{\eta^{\tau}}\\
=&\left(2\,{}_{\alpha}F_{\beta}{}^{t}+\eta^{\kappa}\left({}_{\kappa}F_{\alpha}{}^{\mu}\,{}_{\mu}F_{\beta}{}^{t}+{}_{\kappa}F_{\beta}{}^{\mu}\,{}_{\mu}F_{\alpha}{}^{t}+\varepsilon_{\alpha\kappa}\,{}_{34}F_{\beta}{}^{t}+\varepsilon_{\beta\kappa}\,{}_{34}F_{\alpha}{}^{t}\right)\right.\\
&\left.+\eta^{3}\eta^{4}\cdots\right)\left(E_{t}{}^{s}F_{s}+E_{t}{}^{\sigma}F_{\sigma}\right)\\
&+\Big(2\,{}_{\alpha}F_{\beta}{}^{\tau}+\eta^{\kappa}\Big({}_{\kappa}F_{\alpha}{}^{\mu}\,{}_{\mu}F_{\beta}{}^{\tau}+{}_{\kappa}F_{\beta}{}^{\mu}\,{}_{\mu}F_{\alpha}{}^{\tau}+\varepsilon_{\alpha\kappa}\,{}_{34}F_{\beta}{}^{\tau}\\
&+\varepsilon_{\beta\kappa}\,{}_{34}F_{\alpha}{}^{\tau}\Big)+\eta^{3}\eta^{4}\cdots\Big)\left(E_{\tau}{}^{s}F_{s}+E_{\tau}{}^{\sigma}F_{\sigma}\right)\\
=&\Big(2\,{}_{\alpha}F_{\beta}{}^{t}\,{}_{0}E_{t}{}^{s}+\eta^{\nu}\Big(2\,{}_{\alpha}F_{\beta}{}^{t}\,{}_{\nu}E_{t}{}^{s}+\Big({}_{\nu}F_{\alpha}{}^{\mu}\,{}_{\mu}F_{\beta}{}^{t}+{}_{\nu}F_{\beta}{}^{\mu}\,{}_{\mu}F_{\alpha}{}^{t}\\
&+\varepsilon_{\alpha\nu}\,{}_{34}F_{\beta}{}^{t}+\varepsilon_{\beta\nu}\,{}_{34}F_{\alpha}{}^{t}\Big)\,{}_{0}E_{t}{}^{s}-2\,{}_{\alpha}F_{\beta}{}^{\tau}\,{}_{\nu}E_{\tau}{}^{s}\Big)+\eta^{3}\eta^{4}\cdots\Big)F_{s}\\
&+\Big(2\,{}_{\alpha}F_{\beta}{}^{t}\,{}_{0}E_{t}{}^{\sigma}+2\,{}_{\alpha}F_{\beta}{}^{\sigma}+\eta^{\nu}\Big(2\,{}_{\alpha}F_{\beta}{}^{t}\,{}_{\nu}E_{t}{}^{\sigma}+\Big({}_{\nu}F_{\alpha}{}^{\mu}\,{}_{\mu}F_{\beta}{}^{t}\\
&+{}_{\nu}F_{\beta}{}^{\mu}\,{}_{\mu}F_{\alpha}{}^{t}+\varepsilon_{\alpha\nu}\,{}_{34}F_{\beta}{}^{t}+\varepsilon_{\beta\nu}\,{}_{34}F_{\alpha}{}^{t}\Big)\,{}_{0}E_{t}{}^{\sigma}-2\,{}_{\alpha}F_{\beta}{}^{\tau}\,{}_{\nu}E_{\tau}{}^{\sigma}\\
&+{}_{\nu}F_{\alpha}{}^{\mu}\,{}_{\mu}F_{\beta}{}^{\sigma}+{}_{\nu}F_{\beta}{}^{\mu}\,{}_{\mu}F_{\alpha}{}^{\sigma}+\varepsilon_{\alpha\nu}\,{}_{34}F_{\beta}{}^{\sigma}+\varepsilon_{\beta\nu}\,{}_{34}F_{\alpha}{}^{\sigma}\Big)\\
&+\eta^{3}\eta^{4}\ldots\Big)F_{\sigma}
\end{aligned}
$$

Using Lemma 13.2.1, the structure coefficients can be expressed with the help of the coefficients $F_{A}{}^{B}$ only. In order to simplify the equations, we keep ${}_{0}E_{a}{}^{b}$ which is the inverse of ${}_{0}F_{a}{}^{b}$. Consequently, the lowest order terms are given by

$$
{}_{0}d^{s}_{\alpha\beta}=2\,{}_{\alpha}F_{\beta}{}^{t}\,{}_{0}E_{t}{}^{s} \tag{13.3.1}
$$

$$
{}_{0}d^{\sigma}_{\alpha\beta}=2\,{}_{\alpha}F_{\beta}{}^{t}\,{}_{0}E_{t}{}^{\sigma}+2\,{}_{\alpha}F_{\beta}{}^{\sigma}=2\left(-\,{}_{\alpha}F_{\beta}{}^{t}\,{}_{0}E_{t}{}^{r}\,{}_{0}F_{r}{}^{\sigma}+{}_{\alpha}F_{\beta}{}^{\sigma}\right). \tag{13.3.2}
$$

The first order terms are given by

$$
\begin{aligned}
{}_{\nu}d^{s}_{\alpha\beta}=&\,2\,{}_{\alpha}F_{\beta}{}^{t}\,{}_{\nu}E_{t}{}^{s}+\Big({}_{\nu}F_{\alpha}{}^{\mu}\,{}_{\mu}F_{\beta}{}^{t}+{}_{\nu}F_{\beta}{}^{\mu}\,{}_{\mu}F_{\alpha}{}^{t}+\varepsilon_{\alpha\nu}\,{}_{34}F_{\beta}{}^{t}\\
&+\varepsilon_{\beta\nu}\,{}_{34}F_{\alpha}{}^{t}\Big)\,{}_{0}E_{t}{}^{s}-2\,{}_{\alpha}F_{\beta}{}^{\tau}\,{}_{\nu}E_{\tau}{}^{s}\\
=&\left({}_{\kappa}F_{\lambda}{}^{t}\,{}_{\nu}E_{t}{}^{s}+\left({}_{\nu}F_{\kappa}{}^{\mu}\,{}_{\mu}F_{\lambda}{}^{t}+\varepsilon_{\kappa\nu}\,{}_{34}F_{\lambda}{}^{t}\right){}_{0}E_{t}{}^{s}-{}_{\kappa}F_{\lambda}{}^{\tau}\,{}_{\nu}E_{\tau}{}^{s}\right)\\
&\cdot\left(\delta^{\kappa}_{\alpha}\delta^{\lambda}_{\beta}+\delta^{\lambda}_{\alpha}\delta^{\kappa}_{\beta}\right)
\end{aligned}
$$

$$= \Big(-{}_\kappa F_\lambda{}^t\,{}_0E_t{}^p\Big({}_\nu F_p{}^q + {}_0F_p{}^\mu\,{}_\nu F_\mu{}^q\Big) + {}_\nu F_\kappa{}^\mu\,{}_\mu F_\lambda{}^q + \varepsilon_{\kappa\nu}\,{}_{34}F_\lambda{}^q$$
$$+ {}_\kappa F_\lambda{}^\mu\,{}_\nu F_\mu{}^q\Big)\,{}_0E_q{}^s\left(\delta_\alpha^\kappa\delta_\beta^\lambda + \delta_\alpha^\lambda\delta_\beta^\kappa\right) \tag{13.3.3}$$

$$\begin{aligned}
{}_\nu d^\sigma_{\alpha\beta} &= 2\,{}_\alpha F_\beta{}^t\,{}_\nu E_t{}^\sigma + \Big({}_\nu F_\alpha{}^\mu\,{}_\mu F_\beta{}^t + {}_\nu F_\beta{}^\mu\,{}_\mu F_\alpha{}^t + \varepsilon_{\alpha\nu}\,{}_{34}F_\beta{}^t \\
&\quad + \varepsilon_{\beta\nu}\,{}_{34}F_\alpha{}^t\Big)\,{}_0E_t{}^\sigma - 2\,{}_\alpha F_\beta{}^\tau\,{}_\nu E_\tau{}^\sigma + {}_\nu F_\alpha{}^\mu\,{}_\mu F_\beta{}^\sigma \qquad (13.3.4)\\
&\quad + {}_\nu F_\beta{}^\mu\,{}_\mu F_\alpha{}^\sigma + \varepsilon_{\alpha\nu}\,{}_{34}F_\beta{}^\sigma + \varepsilon_{\beta\nu}\,{}_{34}F_\alpha{}^\sigma \\
&= \big({}_\kappa F_\lambda{}^t\,{}_\nu E_t{}^\sigma + \big({}_\nu F_\kappa{}^\mu\,{}_\mu F_\lambda{}^t + \varepsilon_{\kappa\nu}\,{}_{34}F_\lambda{}^t\big)\,{}_0E_t{}^\sigma - {}_\kappa F_\lambda{}^\tau\,{}_\nu E_\tau{}^\sigma \\
&\qquad + {}_\nu F_\kappa{}^\mu\,{}_\mu F_\lambda{}^\sigma + \varepsilon_{\kappa\nu}\,{}_{34}F_\lambda{}^\sigma\big)\left(\delta_\alpha^\kappa\delta_\beta^\lambda + \delta_\alpha^\lambda\delta_\beta^\kappa\right) \\
&= \Big({}_\kappa F_\lambda{}^t\,{}_0E_t{}^p\Big(\Big({}_\nu F_p{}^a + {}_0F_p{}^\tau\,{}_\nu F_\tau{}^a\Big)\,{}_0E_a{}^s\,{}_0F_s{}^\sigma - {}_\nu F_p{}^\sigma - {}_0F_p{}^\tau\,{}_\nu F_\tau{}^\sigma\Big) \\
&\quad - \big({}_\nu F_\kappa{}^\mu\,{}_\mu F_\lambda{}^t + \varepsilon_{\kappa\nu}\,{}_{34}F_\lambda{}^t\big)\,{}_0E_t{}^q\,{}_0F_q{}^\sigma \\
&\quad - {}_\kappa F_\lambda{}^\tau\big({}_\nu F_\tau{}^a\,{}_0E_a{}^s\,{}_0F_s{}^\sigma - {}_\nu F_\tau{}^\sigma\big) + {}_\nu F_\kappa{}^\mu\,{}_\mu F_\lambda{}^\sigma + \varepsilon_{\kappa\nu}\,{}_{34}F_\lambda{}^\sigma\Big) \\
&\quad \cdot\left(\delta_\alpha^\kappa\delta_\beta^\lambda + \delta_\alpha^\lambda\delta_\beta^\kappa\right) \\
&= \Big({}_\kappa F_\lambda{}^t\,{}_0E_t{}^p\Big(\Big({}_\nu F_p{}^a + {}_0F_p{}^\tau\,{}_\nu F_\tau{}^a\Big)\,{}_0E_a{}^s\,{}_0F_s{}^\sigma - {}_\nu F_p{}^\sigma - {}_0F_p{}^\tau\,{}_\nu F_\tau{}^\sigma\Big) \\
&\quad + {}_\nu F_\kappa{}^\mu\Big({}_\mu F_\lambda{}^\sigma - {}_\mu F_\lambda{}^t\,{}_0E_t{}^q\,{}_0F_q{}^\sigma\Big) + {}_\kappa F_\lambda{}^\tau\big({}_\nu F_\tau{}^\sigma - {}_\nu F_\tau{}^a\,{}_0E_a{}^s\,{}_0F_s{}^\sigma\big) \\
&\quad + \varepsilon_{\kappa\nu}\Big({}_{34}F_\lambda{}^\sigma - {}_{34}F_\lambda{}^t\,{}_0E_t{}^q\,{}_0F_q{}^\sigma\Big)\Big)\left(\delta_\alpha^\kappa\delta_\beta^\lambda + \delta_\alpha^\lambda\delta_\beta^\kappa\right)
\end{aligned}$$

The degree two part of $d^s_{\alpha\beta}$ equals

$$\begin{aligned}
{}_{34}d^s_{\alpha\beta} &= 2\,{}_\alpha F_\beta{}^t\,{}_{34}E_t{}^s - \varepsilon^{\kappa\lambda}\Big({}_\kappa F_\alpha{}^\mu\,{}_\mu F_\beta{}^t + {}_\kappa F_\beta{}^\mu\,{}_\mu F_\alpha{}^t + \varepsilon_{\alpha\kappa}\,{}_{34}F_\beta{}^t \\
&\quad + \varepsilon_{\beta\kappa}\,{}_{34}F_\alpha{}^t\Big)\,{}_\lambda E_t{}^s + \Big(\varepsilon^{\mu\nu}\Big({}_\mu F_\alpha{}^c\Big(\partial_{x^c}\,{}_\nu F_\beta{}^t\Big) + {}_\mu F_\beta{}^d\left(\partial_{x^d}\,{}_\nu F_\alpha{}^t\right)\Big) \\
&\qquad - \varepsilon^{\mu\nu}\,{}_\mu F_\beta{}^\delta\varepsilon_{\delta\nu}\,{}_{34}F_\alpha{}^t + {}_{34}F_\beta{}^\delta\,{}_\delta F_\alpha{}^t - \varepsilon^{\mu\nu}\,{}_\mu F_\alpha{}^\gamma\varepsilon_{\gamma\nu}\,{}_{34}F_\beta{}^t \qquad (13.3.5)\\
&\qquad + {}_{34}F_\alpha{}^\gamma\,{}_\gamma F_\beta{}^t\Big)\,{}_0E_t{}^s + 2\,{}_\alpha F_\beta{}^\tau\,{}_{34}E_\tau{}^s \\
&\quad + \varepsilon^{\kappa\lambda}\Big({}_\kappa F_\alpha{}^\mu\,{}_\mu F_\beta{}^\tau + {}_\kappa F_\beta{}^\mu\,{}_\mu F_\alpha{}^\tau + \varepsilon_{\alpha\kappa}\,{}_{34}F_\beta{}^\tau + \varepsilon_{\beta\kappa}\,{}_{34}F_\alpha{}^\tau\Big)\,{}_\lambda E_\tau{}^s \\
&= \big({}_\kappa F_\lambda{}^t\,{}_{34}E_t{}^s - \varepsilon^{\sigma\tau}\big({}_\sigma F_\kappa{}^\mu\,{}_\mu F_\lambda{}^t + \varepsilon_{\kappa\sigma}\,{}_{34}F_\lambda{}^t\big)\,{}_\tau E_t{}^s \\
&\quad + \big(\varepsilon^{\sigma\tau}\,{}_\sigma F_\kappa{}^c\left(\partial_{x^c}\,{}_\tau F_\lambda{}^t\right) - {}_\mu F_\lambda{}^\mu\,{}_{34}F_\kappa{}^t + {}_{34}F_\lambda{}^\mu\,{}_\mu F_\kappa{}^t\big)\,{}_0E_t{}^s \\
&\quad + {}_\kappa F_\lambda{}^\mu\,{}_{34}E_\mu{}^s + \varepsilon^{\sigma\tau}\big({}_\sigma F_\kappa{}^\mu\,{}_\mu F_\lambda{}^\rho + \varepsilon_{\kappa\sigma}\,{}_{34}F_\lambda{}^\rho\big)\,{}_\tau E_\rho{}^s\big) \\
&\quad \cdot\left(\delta_\alpha^\kappa\delta_\beta^\lambda + \delta_\alpha^\lambda\delta_\beta^\kappa\right)
\end{aligned}$$

$$
\begin{aligned}
&= \Big(-{}_{\kappa}F_{\lambda}{}^{t}\,{}_{0}E_{t}{}^{p}\Big({}_{34}F_{p}{}^{a} + \varepsilon^{\mu\nu}\Big({}_{\mu}F_{p}{}^{m} + {}_{0}F_{p}{}^{\tau}\,{}_{\mu}F_{\tau}{}^{m}\Big)\,{}_{0}E_{m}{}^{q} \\
&\qquad \cdot\Big({}_{\nu}F_{q}{}^{a} + {}_{0}F_{q}{}^{\sigma}\,{}_{\nu}F_{\sigma}{}^{a}\Big) - {}_{0}F_{p}{}^{\sigma}\,{}_{34}F_{\sigma}{}^{a} \\
&\qquad - \varepsilon^{\mu\nu}\Big({}_{\mu}F_{p}{}^{\sigma} + {}_{0}F_{p}{}^{\tau}\,{}_{\mu}F_{\tau}{}^{\sigma}\Big)\,{}_{\nu}F_{\sigma}{}^{a}\Big) \\
&\quad + \varepsilon^{\sigma\tau}\big({}_{\sigma}F_{\kappa}{}^{\mu}\,{}_{\mu}F_{\lambda}{}^{t} + \varepsilon_{\kappa\sigma}\,{}_{34}F_{\lambda}{}^{t}\big)\,{}_{0}E_{t}{}^{p}\Big({}_{\tau}F_{p}{}^{a} + {}_{0}F_{p}{}^{\nu}\,{}_{\tau}F_{\nu}{}^{a}\Big) \\
&\quad + \big(\varepsilon^{\sigma\tau}\,{}_{\sigma}F_{\kappa}{}^{c}\big(\partial_{x^c}\,{}_{\tau}F_{\lambda}{}^{a}\big) - {}_{\mu}F_{\lambda}{}^{\mu}\,{}_{34}F_{\kappa}{}^{a} + {}_{34}F_{\lambda}{}^{\mu}\,{}_{\mu}F_{\kappa}{}^{a}\big) \\
&\quad + {}_{\kappa}F_{\lambda}{}^{\mu}\Big(\varepsilon^{\sigma\tau}\,{}_{\sigma}F_{\mu}{}^{r}\,{}_{0}E_{r}{}^{t}\,{}_{\tau}F_{t}{}^{a} - {}_{34}F_{\mu}{}^{a} \\
&\qquad + \varepsilon^{\sigma\tau}\Big({}_{\sigma}F_{\mu}{}^{r}\,{}_{0}E_{r}{}^{p}\,{}_{0}F_{p}{}^{\rho} - {}_{\sigma}F_{\mu}{}^{\rho}\Big)\,{}_{\tau}F_{\rho}{}^{a}\Big) \\
&\quad - \varepsilon^{\sigma\tau}\big({}_{\sigma}F_{\kappa}{}^{\mu}\,{}_{\mu}F_{\lambda}{}^{\rho} + \varepsilon_{\kappa\sigma}\,{}_{34}F_{\lambda}{}^{\rho}\big)\,{}_{\tau}F_{\rho}{}^{a}\Big)\,{}_{0}E_{a}{}^{s}\Big(\delta^{\kappa}_{\alpha}\delta^{\lambda}_{\beta} + \delta^{\lambda}_{\alpha}\delta^{\kappa}_{\beta}\Big)
\end{aligned}
$$

The degree two part of $d^{\sigma}_{\alpha\beta}$ equals:

$$
\begin{aligned}
{}_{34}d_{\alpha\beta}{}^{\sigma} &= 2\,{}_{\alpha}F_{\beta}{}^{t}\,{}_{34}E_{t}{}^{\sigma} \\
&\quad - \varepsilon^{\kappa\lambda}\Big({}_{\kappa}F_{\alpha}{}^{\mu}\,{}_{\mu}F_{\beta}{}^{t} + {}_{\kappa}F_{\beta}{}^{\mu}\,{}_{\mu}F_{\alpha}{}^{t} + \varepsilon_{\alpha\kappa}\,{}_{34}F_{\beta}{}^{t} + \varepsilon_{\beta\kappa}\,{}_{34}F_{\alpha}{}^{t}\Big)\,{}_{\lambda}E_{t}{}^{\sigma} \\
&\quad + \Big(\varepsilon^{\mu\nu}\Big({}_{\mu}F_{\alpha}{}^{c}\Big(\partial_{x^c}\,{}_{\nu}F_{\beta}{}^{t}\Big) + {}_{\mu}F_{\beta}{}^{d}\big(\partial_{x^d}\,{}_{\nu}F_{\alpha}{}^{t}\big)\Big) - \varepsilon^{\mu\nu}\,{}_{\mu}F_{\beta}{}^{\delta}\varepsilon_{\delta\nu}\,{}_{34}F_{\alpha}{}^{t} \\
&\qquad + {}_{34}F_{\beta}{}^{\delta}\,{}_{\delta}F_{\alpha}{}^{t} - \varepsilon^{\mu\nu}\,{}_{\mu}F_{\alpha}{}^{\gamma}\varepsilon_{\gamma\nu}\,{}_{34}F_{\beta}{}^{t} + {}_{34}F_{\alpha}{}^{\gamma}\,{}_{\gamma}F_{\beta}{}^{t}\Big)\,{}_{0}E_{t}{}^{\sigma} \\
&\quad + 2\,{}_{\alpha}F_{\beta}{}^{\tau}\,{}_{34}E_{\tau}{}^{\sigma} \\
&\quad + \varepsilon^{\kappa\lambda}\Big({}_{\kappa}F_{\alpha}{}^{\mu}\,{}_{\mu}F_{\beta}{}^{\tau} + {}_{\kappa}F_{\beta}{}^{\mu}\,{}_{\mu}F_{\alpha}{}^{\tau} + \varepsilon_{\alpha\kappa}\,{}_{34}F_{\beta}{}^{\tau} + \varepsilon_{\beta\kappa}\,{}_{34}F_{\alpha}{}^{\tau}\Big)\,{}_{\lambda}E_{\tau}{}^{\sigma} \\
&\quad + \Big(\varepsilon^{\mu\nu}\Big({}_{\mu}F_{\alpha}{}^{c}\Big(\partial_{x^c}\,{}_{\nu}F_{\beta}{}^{\tau}\Big) + {}_{\mu}F_{\beta}{}^{c}\big(\partial_{x^c}\,{}_{\nu}F_{\alpha}{}^{\tau}\big)\Big) - \varepsilon^{\mu\nu}\,{}_{\mu}F_{\beta}{}^{\delta}\varepsilon_{\delta\nu}\,{}_{34}F_{\alpha}{}^{\tau} \\
&\qquad + {}_{34}F_{\beta}{}^{\delta}\,{}_{\delta}F_{\alpha}{}^{\tau} - \varepsilon^{\mu\nu}\,{}_{\mu}F_{\alpha}{}^{\gamma}\varepsilon_{\gamma\nu}\,{}_{34}F_{\beta}{}^{\tau} + {}_{34}F_{\alpha}{}^{\gamma}\,{}_{\gamma}F_{\beta}{}^{\tau}\Big)\,{}_{0}E_{\tau}{}^{\sigma} \\
&= \big({}_{\kappa}F_{\lambda}{}^{t}\,{}_{34}E_{t}{}^{\sigma} - \varepsilon^{\mu\nu}\big({}_{\mu}F_{\kappa}{}^{\rho}\,{}_{\rho}F_{\lambda}{}^{t} + \varepsilon_{\kappa\mu}\,{}_{34}F_{\lambda}{}^{t}\big)\,{}_{\nu}E_{t}{}^{\sigma} \\
&\qquad + \big(\varepsilon^{\mu\nu}\,{}_{\mu}F_{\kappa}{}^{c}\big(\partial_{x^c}\,{}_{\nu}F_{\lambda}{}^{t}\big) + {}_{\mu}F_{\lambda}{}^{\mu}\,{}_{34}F_{\kappa}{}^{t} + {}_{34}F_{\lambda}{}^{\delta}\,{}_{\delta}F_{\kappa}{}^{t}\big)\,{}_{0}E_{t}{}^{\sigma} \\
&\qquad + {}_{\kappa}F_{\lambda}{}^{\tau}\,{}_{34}E_{\tau}{}^{\sigma} + \varepsilon^{\mu\nu}\big({}_{\mu}F_{\kappa}{}^{\rho}\,{}_{\rho}F_{\lambda}{}^{\tau} + \varepsilon_{\kappa\mu}\,{}_{34}F_{\lambda}{}^{\tau}\big)\,{}_{\nu}E_{\tau}{}^{\sigma} \\
&\qquad + \big(\varepsilon^{\mu\nu}\,{}_{\mu}F_{\kappa}{}^{c}\big(\partial_{x^c}\,{}_{\nu}F_{\lambda}{}^{\tau}\big) - {}_{\mu}F_{\lambda}{}^{\mu}\,{}_{34}F_{\kappa}{}^{\tau} + {}_{34}F_{\lambda}{}^{\delta}\,{}_{\delta}F_{\kappa}{}^{\tau}\big)\,{}_{0}E_{\tau}{}^{\sigma}\big) \\
&\quad \cdot\Big(\delta^{\kappa}_{\alpha}\delta^{\lambda}_{\beta} + \delta^{\lambda}_{\alpha}\delta^{\kappa}_{\beta}\Big)
\end{aligned}
$$

$$
\begin{aligned}
&= \Big({}_\kappa F_\lambda{}^t\, {}_0E_t{}^p \Big(\varepsilon^{\mu\nu} \left({}_\mu F_p{}^a + {}_0F_p{}^\tau\, {}_\mu F_\tau{}^a \right) {}_0E_a{}^s \\
&\quad \cdot \Big(\left(\left({}_\nu F_s{}^b + {}_0F_s{}^\rho\, {}_\nu F_\rho{}^b \right) {}_0E_b{}^c\, {}_0F_c{}^\sigma - {}_\nu F_s{}^\sigma - {}_0F_s{}^\beta\, {}_\nu F_\beta{}^\sigma \right) - {}_{34}F_p{}^\sigma \\
&\quad + \left({}_{34}F_p{}^a - {}_0F_p{}^\rho\, {}_{34}F_\rho{}^a \right) {}_0E_a{}^b\, {}_0F_b{}^\sigma - \varepsilon^{\mu\nu} \left({}_\mu F_p{}^\rho + {}_0F_p{}^\tau\, {}_\mu F_\tau{}^\rho \right) \\
&\quad \cdot \left({}_\nu F_\rho{}^a\, {}_0E_a{}^b\, {}_0F_b{}^\sigma - {}_\nu F_\rho{}^\sigma \right) + {}_0F_p{}^\beta\, {}_{34}F_\beta{}^\sigma \Big) \\
&\quad - \varepsilon^{\mu\nu} \left({}_\mu F_\kappa{}^\rho\, {}_\rho F_\lambda{}^t + \varepsilon_{\kappa\mu}\, {}_{34}F_\lambda{}^t \right) {}_0E_t{}^p \\
&\quad \cdot \Big(\left({}_\nu F_p{}^a + {}_0F_p{}^\rho\, {}_\nu F_\rho{}^a \right) {}_0E_a{}^s\, {}_0F_s{}^\sigma - {}_\nu F_p{}^\sigma - {}_0F_p{}^\rho\, {}_\nu F_\rho{}^\sigma \Big) \\
&\quad - \left(\varepsilon^{\mu\nu}\, {}_\mu F_\kappa{}^c \left(\partial_{x^c}\, {}_\nu F_\lambda{}^t \right) + {}_\mu F_\lambda{}^\mu\, {}_{34}F_\kappa{}^t + {}_{34}F_\lambda{}^\delta\, {}_\delta F_\kappa{}^t \right) {}_0E_t{}^q\, {}_0F_q{}^\sigma \\
&\quad + {}_\kappa F_\lambda{}^\tau \Big(\varepsilon^{\mu\nu}\, {}_\mu F_\tau{}^a\, {}_0E_a{}^b\, {}_\nu F_b{}^\sigma - \big(\varepsilon^{\mu\nu}\, {}_\mu F_\tau{}^s\, {}_0E_s{}^t\, {}_\nu F_t{}^a - {}_{34}F_\tau{}^a \\
&\quad + \varepsilon^{\mu\nu} \left({}_\mu F_\tau{}^a\, {}_0E_a{}^s\, {}_0F_s{}^\rho - {}_\mu F_\tau{}^\rho \right) {}_\nu F_\rho{}^a \big) {}_0E_a{}^b\, {}_0F_b{}^\sigma - {}_{34}F_\tau{}^\sigma \\
&\quad + \varepsilon^{\mu\nu} \left({}_\mu F_\tau{}^a\, {}_0E_a{}^s\, {}_0F_s{}^\rho - {}_\mu F_\tau{}^\rho \right) {}_\nu F_\rho{}^\sigma \Big) \\
&\quad + \varepsilon^{\mu\nu} \left({}_\mu F_\kappa{}^\rho\, {}_\rho F_\lambda{}^\tau + \varepsilon_{\kappa\mu}\, {}_{34}F_\lambda{}^\tau \right) \left({}_\nu F_\tau{}^a\, {}_0E_a{}^s\, {}_0F_s{}^\sigma - {}_\nu F_\tau{}^\sigma \right) \\
&\quad + \varepsilon^{\mu\nu}\, {}_\mu F_\kappa{}^c \left(\partial_{x^c}\, {}_\nu F_\lambda{}^\sigma \right) - {}_\mu F_\lambda{}^\mu\, {}_{34}F_\kappa{}^\sigma + {}_{34}F_\lambda{}^\delta\, {}_\delta F_\kappa{}^\sigma \Big) \\
&\quad \cdot \left(\delta^\kappa_\alpha \delta^\lambda_\beta + \delta^\lambda_\alpha \delta^\kappa_\beta \right)
\end{aligned}
$$

13.4 Wess–Zumino Pairs

Let F_A be a Wess–Zumino frame with respect to $i\colon |M| \to M$ and $X^A = (x^a, \eta^\alpha)$ be corresponding Wess–Zumino coordinates. In the following we will call a pair consisting of F_A and X^A a Wess–Zumino pair.

Lemma 13.4.1 *The integrability conditions and commutator conditions for a Wess–Zumino pair are equivalent to*

$$2\Gamma^c_{\alpha\beta} = d^c_{\alpha\beta}, \tag{13.4.2a}$$

$$0 = \Gamma^r_{\kappa\lambda} \varepsilon^{\lambda\alpha} \varepsilon^{\beta\kappa} d^\sigma_{\alpha\beta}, \tag{13.4.2b}$$

$$0 = {}_0d^\sigma_{\alpha\beta}, \tag{13.4.2c}$$

$$0 = \delta^{\alpha\beta} \delta^\nu_\sigma\, {}_\nu d^\sigma_{\alpha\beta}, \tag{13.4.2d}$$

$$0 = \delta^{\alpha\beta} \varepsilon_{\sigma\tau} \delta^{\tau\nu}\, {}_\nu d^\sigma_{\alpha\beta}, \tag{13.4.2e}$$

Proof The proof relies on the results of Sect. 7.3 that relate real and complex commutators. Recall from Lemma 9.2.8 that the integrability conditions are given by $d^{\bar{z}}_{++} = d^{z}_{+-} = d^{-}_{++} = 0$ and $d^{z}_{++} = 2$, whereas the commutator conditions for a Wess–Zumino frame are given by $d^{+}_{++} = i^{\#}d^{+}_{+-} = i^{\#}F_{+}d^{+}_{+-} = 0$, see Eq. (11.3.2). We have already used in Proposition 9.2.6 that $d^{\bar{z}}_{++} = d^{z}_{+-} = 0$ and $d^{z}_{++} = 2$ are equivalent to $d^{c}_{\alpha\beta} = 2\Gamma^{c}_{\alpha\beta}$, proving Eq. (13.4.2a).

In Lemma 7.3.4 we have proven that

$$4d^{+}_{++} = d^{3}_{33} - d^{3}_{44} + 2d^{4}_{34} + \mathrm{i}\left(d^{4}_{33} - d^{4}_{44} - 2d^{3}_{34}\right)$$

$$4d^{-}_{++} = d^{3}_{33} - d^{3}_{44} - 2d^{4}_{34} + \mathrm{i}\left(-d^{4}_{33} + d^{4}_{44} - 2d^{3}_{34}\right)$$

$$4d^{+}_{+-} = d^{3}_{33} + d^{3}_{44} + \mathrm{i}\left(d^{4}_{33} + d^{4}_{44}\right)$$

Thus the $d^{-}_{++} = 0$ from the integrability conditions and $d^{+}_{++} = 0$ from the conditions for a Wess–Zumino pair imply

$$0 = d^{\sigma}_{33} - d^{\sigma}_{44} = \Gamma^{1}_{\kappa\lambda}\varepsilon^{\lambda\alpha}\varepsilon^{\beta\kappa}d^{\sigma}_{\alpha\beta}$$

$$0 = 2d^{\sigma}_{34} = \Gamma^{2}_{\kappa\lambda}\varepsilon^{\lambda\alpha}\varepsilon^{\beta\kappa}d^{\sigma}_{\alpha\beta}$$

showing Eq. (13.4.2b). Furthermore ${}_{0}d^{+}_{+-} = 0$ implies then Eq. (13.4.2c) and

$$\begin{aligned} 0 = 8i^{*}F_{+}d^{+}_{+-} &= (F_3 - \mathrm{i}F_4)\left(d^{3}_{33} + d^{3}_{44} + \mathrm{i}\left(d^{4}_{33} + d^{4}_{44}\right)\right) \\ &= {}_{3}d^{3}_{33} + {}_{3}d^{3}_{44} + {}_{4}d^{4}_{33} + {}_{4}d^{4}_{44} + \mathrm{i}\left({}_{3}d^{4}_{33} + {}_{3}d^{4}_{44} - {}_{4}d^{3}_{33} - {}_{4}d^{3}_{44}\right) \\ &= \delta^{\alpha\beta}\delta_{\sigma}^{\ \nu}\,{}_{\nu}d^{\sigma}_{\alpha\beta} + \mathrm{i}\delta^{\alpha\beta}\varepsilon_{\sigma\tau}\delta^{\tau\nu}\,{}_{\nu}d^{\sigma}_{\alpha\beta} \end{aligned}$$

proves Eqs. (13.4.2d) and (13.4.2e). □

In the remaining part of this section we are going to show

Lemma 11.3.4 *Let $i\colon |M| \to M$ be an underlying even manifold for a super Riemann surface M, F_A a local Wess–Zumino frame and $X^A = (x^a, \eta^\alpha)$ Wess–Zumino coordinates for F_A. We denote the frame coefficients by ${F_A}^B$, that is $F_A = {F_A}^B\partial_{X^B}$, and their expansion in orders of η as follows:*

$$ {F_A}^B = {}_{0}{F_A}^B + \eta^\mu\,{}_{\mu}{F_A}^B + \eta^3\eta^4\,{}_{34}{F_A}^B $$

All frame coefficients of F_A are completely determined by its independent components ${}_{0}{F_a}^b$ and ${}_{0}{F_a}^\beta$. Recall from Corollary 11.2.10 that $f_a = {f_a}^b\partial_{x^b} = {}_{0}{F_a}^b\partial_{x^b}$,

$s_\alpha = i^* F_\alpha$ *and* $\chi(f_a) = \chi_a{}^\beta s_\beta = {}_0F_a{}^\beta s_\beta$. *Then*

$$\begin{aligned} F_a &= \left(\delta_a^s + \eta^\mu \Gamma_{\mu\nu}^s \chi_a{}^\nu + \eta^3\eta^4\, {}_{34}\overline{F}_a{}^s\right) f_s{}^b \partial_{x^b} \\ &\quad + \left(\chi_a{}^\beta + \eta^\mu\, {}_\mu F_a{}^\beta + \eta^3\eta^4\, {}_{34}F_a{}^\beta\right) \partial_{\eta^\beta}, \\ F_\alpha &= \left(\eta^\mu \Gamma_{\mu\alpha}^s + \eta^3\eta^4 \gamma^t{}_\alpha{}^\lambda \Gamma_{\lambda\tau}^s \chi_t{}^\tau\right) f_s{}^b \partial_{x^b} \\ &\quad + \left(\delta_\alpha^\beta + \eta^\mu \Gamma_{\mu\alpha}^s \chi_s{}^\beta + \eta^3\eta^4\, {}_{34}F_\alpha{}^\beta\right) \partial_{\eta^\beta}. \end{aligned}$$

The remaining coefficients are given by

$$\begin{aligned} {}_{34}\overline{F}_a{}^s &= -\left(\|Q\chi\|^2 \delta_a^s + 2g_S\left((P\chi)_a, (Q\chi)_t\right)\delta^{ts}\right), \\ {}_\mu F_a{}^\beta &= \Gamma_{\mu\tau}^s \chi_a{}^\tau \chi_s{}^\beta + \frac{1}{4}\gamma_{a\mu}{}^\lambda \gamma^t{}_\lambda{}^\beta \left(\mathrm{I}_t{}^a\, \omega_a^{LC} + 2g_S\left(\delta_\gamma\chi, \chi_t\right)\right), \\ {}_{34}F_a{}^\beta &= \|Q\chi\|^2 (Q\chi)_a{}^\beta - 4g_S\left((P\chi)_a, (Q\chi)_t\right)\delta^{tr}\chi_r{}^\beta \\ &\quad - \frac{1}{2}(P\chi)_a{}^\kappa \gamma^t{}_\kappa{}^\beta \mathrm{I}_t{}^r\, \omega_r^{LC}, \\ {}_{34}F_\alpha{}^\beta &= -\left(\|Q\chi\|^2 \delta_\alpha^\beta + \frac{1}{2}\gamma^t{}_\alpha{}^\mu \mathrm{I}_\mu{}^\beta\, \omega_t^{LC}\right). \end{aligned}$$

Here, we denote by $Q\chi \in \Gamma\left(T^\vee|M| \otimes S\right)$ *the* $\frac{3}{2}$*-part of the gravitino, see Appendix A.3. That is,*

$$Q\chi = (Q\chi)_t f^t = (Q\chi)_t{}^\tau f^t \otimes s_\tau = \frac{1}{2}\chi_s{}^\sigma \gamma_{t\sigma}{}^\mu \gamma^s{}_\mu{}^\tau f^t \otimes s_\tau,$$

and similarly for the $\frac{1}{2}$*-part of the gravitino,* $P\chi = \chi - Q\chi$. *The term* $\delta_\gamma\chi = \chi_k{}^\kappa \gamma^k{}_\kappa{}^\lambda s_\lambda \in \Gamma(S)$ *is also called* γ*-trace of the gravitino. Furthermore, we denote by* $\omega_a^{LC} = -\varepsilon^{bd} f_b{}^c \left(\partial_{x^c} f_d{}^m\right) {}_0E_m{}^n \delta_{na}$, *such that* $\nabla_{f_a} f_b = \omega_a^{LC}\, \mathrm{I}\, f_b$ *denotes the Levi-Civita covariant derivative with respect to the metric defined by the orthonormal frame* f_a.

Proof The proof proceeds by applying the conditions (13.4.2) to the commutator expressions from Sect. 13.3. We will proceed step by step with ascending order of η.

The conditions of lowest order in η are

$$2\Gamma_{\alpha\beta}^s = {}_0d_{\alpha\beta}^s = 2\, {}_\alpha F_\beta{}^t\, {}_0E_t{}^s$$

and

$$0 = {}_0d_{\alpha\beta}^\sigma = 2\left(-\, {}_\alpha F_\beta{}^t\, {}_0E_t{}^r\, {}_0F_r{}^\sigma + {}_\alpha F_\beta{}^\sigma\right).$$

Here we have combined Eq. (13.3.1) with (13.4.2a) and Eq. (13.3.2) with (13.4.2c) respectively. Reordering yields

$$ {}_\alpha F_\beta{}^t = \Gamma^s_{\alpha\beta} f_s{}^t, \tag{13.4.3} $$

$$ {}_\alpha F_\beta{}^\sigma = \Gamma^s_{\alpha\beta} \chi_s{}^\sigma. \tag{13.4.4} $$

Now we come to the terms of first order in η. With the help of Eqs. (13.4.3) and (13.4.4), we can continue the calculation of ${}_\nu d^s_{\alpha\beta}$ from Eq. (13.3.3):

$$ \begin{aligned} {}_\nu d^s_{\alpha\beta} &= \Big(- {}_\kappa F_\lambda{}^t\, {}_0E_t{}^p \left({}_\nu F_p{}^q + {}_0F_p{}^\mu\, {}_\nu F_\mu{}^q \right) + {}_\nu F_\kappa{}^\mu\, {}_\mu F_\lambda{}^q + \varepsilon_{\kappa\nu}\, {}_{34}F_\lambda{}^q \\ &\quad + {}_\kappa F_\lambda{}^\mu\, {}_\nu F_\mu{}^q \Big)\, {}_0E_q{}^s \left(\delta^\kappa_\alpha \delta^\lambda_\beta + \delta^\lambda_\alpha \delta^\kappa_\beta \right) \\ &= \left(-\Gamma^p_{\kappa\lambda}\, {}_\nu F_p{}^q\, {}_0E_q{}^s + \Gamma^m_{\nu\kappa} \chi_m{}^\mu \Gamma^s_{\mu\lambda} + \varepsilon_{\kappa\nu}\, {}_{34}F_\lambda{}^q\, {}_0E_q{}^s \right) \left(\delta^\kappa_\alpha \delta^\lambda_\beta + \delta^\lambda_\alpha \delta^\kappa_\beta \right) \end{aligned} \tag{13.4.5} $$

Multiplying with $\varepsilon^{\beta\sigma} \Gamma^k_{\sigma\tau} \varepsilon^{\tau\alpha} f_s{}^b$ yields

$$ \begin{aligned} 0 &= \left(\varepsilon^{\beta\sigma} \Gamma^k_{\sigma\tau} \varepsilon^{\tau\alpha}\, {}_\nu d^s_{\alpha\beta} \right) f_s{}^b \\ &= -4\delta^{kp}\, {}_\nu F_p{}^b + 2\Gamma^m_{\nu\alpha} \varepsilon^{\alpha\tau} \Gamma^k_{\tau\sigma} \varepsilon^{\sigma\beta} \Gamma^s_{\beta\mu} \chi_m{}^\mu f_s{}^b + 2\Gamma^k_{\nu\sigma} \varepsilon^{\sigma\beta}\, {}_{34}F_\beta{}^b \\ &= -4\delta^{kp}\, {}_\nu F_p{}^b + 2\gamma^m{}_\nu{}^\tau \gamma^k{}_\tau{}^\beta \Gamma^s_{\beta\mu} \chi_m{}^\mu f_s{}^b + 2\gamma^k{}_\nu{}^\beta\, {}_{34}F_\beta{}^b, \end{aligned} $$

and hence

$$ {}_\nu F_p{}^b = \frac{1}{2} \left(\gamma^m{}_\nu{}^\tau \gamma_{p\tau}{}^\beta \Gamma^s_{\beta\mu} \chi_m{}^\mu f_s{}^b + \gamma_{p\nu}{}^\beta\, {}_{34}F_\beta{}^b \right). $$

Inserting into Eq. (13.4.5) gives

$$ \begin{aligned} {}_\nu d^s_{\alpha\beta} &= \frac{1}{2} \Big(-\Gamma^p_{\kappa\lambda} \left(\gamma^m{}_\nu{}^\tau \gamma_{p\tau}{}^\sigma \Gamma^s_{\sigma\mu} \chi_m{}^\mu + \gamma_{p\nu}{}^\sigma\, {}_{34}F_\sigma{}^q\, {}_0E_q{}^s \right) \\ &\quad + 2\Gamma^m_{\nu\kappa} \chi_m{}^\mu \Gamma^s_{\mu\lambda} + 2\varepsilon_{\kappa\nu}\, {}_{34}F_\lambda{}^q\, {}_0E_q{}^s \Big) \left(\delta^\kappa_\alpha \delta^\lambda_\beta + \delta^\lambda_\alpha \delta^\kappa_\beta \right). \end{aligned} $$

Consequently, we obtain

$$ 0 = \varepsilon^{\nu\alpha}\, {}_\nu d^s_{\alpha\beta} f_s{}^b = 2\, {}_{34}F_\beta{}^b + \gamma^m{}_\beta{}^\lambda \Gamma^s_{\lambda\mu} \chi_m{}^\mu f_s{}^b - 3\, {}_{34}F_\beta{}^b, $$

yielding

$$ {}_{34}F_\beta{}^b = \gamma^m{}_\beta{}^\lambda \Gamma^s_{\lambda\mu} \chi_m{}^\mu f_s{}^b. $$

Now we can simplify the expression for ${}_{\nu}F_{p}{}^{b}$ to obtain

$$
{}_{\nu}F_{p}{}^{b} = \frac{1}{2}\left(\gamma^{m}{}_{\nu}{}^{\tau}\gamma_{p\tau}{}^{\beta}\Gamma^{s}_{\beta\mu}\chi_{m}{}^{\mu}f_{s}{}^{b} + \gamma_{p\nu}{}^{\beta}\gamma^{t}{}_{\beta}{}^{\lambda}\Gamma^{s}_{\lambda\mu}\chi_{m}{}^{\mu}f_{s}{}^{b}\right) = \Gamma^{s}_{\nu\mu}\chi_{p}{}^{\mu}f_{s}{}^{b}.
$$

We have now exploited the condition ${}_{\nu}d^{s}_{\alpha\beta} = 0$ completely and return to Eq. (13.3.4):

$$
\begin{aligned}
{}_{\nu}d^{\sigma}_{\alpha\beta} &= \Big({}_{\kappa}F_{\lambda}{}^{t}\,{}_{0}E_{t}{}^{p}\left(\left({}_{\nu}F_{p}{}^{a} + {}_{0}F_{p}{}^{\tau}\,{}_{\nu}F_{\tau}{}^{a}\right){}_{0}E_{a}{}^{s}\,{}_{0}F_{s}{}^{\sigma} - {}_{\nu}F_{p}{}^{\sigma} - {}_{0}F_{p}{}^{\tau}\,{}_{\nu}F_{\tau}{}^{\sigma}\right) \\
&\quad + {}_{\nu}F_{\kappa}{}^{\mu}\left({}_{\mu}F_{\lambda}{}^{\sigma} - {}_{\mu}F_{\lambda}{}^{t}\,{}_{0}E_{t}{}^{q}\,{}_{0}F_{q}{}^{\sigma}\right) + {}_{\kappa}F_{\lambda}{}^{\tau}\left({}_{\nu}F_{\tau}{}^{\sigma} - {}_{\nu}F_{\tau}{}^{a}\,{}_{0}E_{a}{}^{s}\,{}_{0}F_{s}{}^{\sigma}\right) \\
&\quad + \varepsilon_{\kappa\nu}\left({}_{34}F_{\lambda}{}^{\sigma} - {}_{34}F_{\lambda}{}^{t}\,{}_{0}E_{t}{}^{q}\,{}_{0}F_{q}{}^{\sigma}\right)\Big)\left(\delta^{\kappa}_{\alpha}\delta^{\lambda}_{\beta} + \delta^{\lambda}_{\alpha}\delta^{\kappa}_{\beta}\right) \\
&= \left(\Gamma^{p}_{\kappa\lambda}\left(\Gamma^{s}_{\nu\mu}\chi_{p}{}^{\mu}\chi_{s}{}^{\sigma} - {}_{\nu}F_{p}{}^{\sigma}\right) + \varepsilon_{\kappa\nu}\left({}_{34}F_{\lambda}{}^{\sigma} - \gamma^{m}{}_{\lambda}{}^{\tau}\Gamma^{q}_{\tau\mu}\chi_{m}{}^{\mu}\chi_{q}{}^{\sigma}\right)\right) \\
&\quad \cdot\left(\delta^{\kappa}_{\alpha}\delta^{\lambda}_{\beta} + \delta^{\lambda}_{\alpha}\delta^{\kappa}_{\beta}\right)
\end{aligned}
\tag{13.4.6}
$$

Let us decompose

$$
{}_{34}F_{\lambda}{}^{\sigma} = A\delta^{\sigma}_{\lambda} + B_{t}\gamma^{t}{}_{\lambda}{}^{\sigma} + C\,\mathrm{I}_{\lambda}{}^{\sigma}.
$$

Then the conditions (13.4.2d) and (13.4.2e) give

$$
\begin{aligned}
0 &= \delta^{\nu}_{\sigma}\delta^{\alpha\beta}\,{}_{\nu}d^{\sigma}_{\alpha\beta} = 2\delta^{\kappa\lambda}\varepsilon_{\kappa\sigma}\left({}_{34}F_{\lambda}{}^{\sigma} - \gamma^{m}{}_{\lambda}{}^{\tau}\Gamma^{q}_{\tau\mu}\chi_{m}{}^{\mu}\chi_{q}{}^{\sigma}\right) = 4C, \\
0 &= \varepsilon_{\sigma\tau}\delta^{\tau\nu}\delta^{\alpha\beta}\,{}_{\nu}d^{\sigma}_{\alpha\beta} = 2\delta^{\lambda}_{\sigma}\left({}_{34}F_{\lambda}{}^{\sigma} - \gamma^{m}{}_{\lambda}{}^{\tau}\Gamma^{q}_{\tau\mu}\chi_{m}{}^{\mu}\chi_{q}{}^{\sigma}\right) \\
&= 4A - 2\chi_{m}{}^{\mu}\gamma^{q}{}_{\mu}{}^{\tau}\gamma^{m}{}_{\tau}{}^{\kappa}\varepsilon_{\kappa\sigma}\chi_{q}{}^{\sigma} = 4A + 4\|Q\chi\|^{2},
\end{aligned}
$$

whereas the condition (13.4.2b) yields

$$
\begin{aligned}
0 &= \Gamma^{r}_{\pi\rho}\varepsilon^{\rho\alpha}\varepsilon^{\beta\pi}\,{}_{\nu}d^{\sigma}_{\alpha\beta} \\
&= 4\delta^{rp}\left(\Gamma^{s}_{\nu\mu}\chi_{p}{}^{\mu}\chi_{s}{}^{\sigma} - {}_{\nu}F_{p}{}^{\sigma}\right) + 2\gamma^{r}{}_{\nu}{}^{\lambda}\left({}_{34}F_{\lambda}{}^{\sigma} - \gamma^{m}{}_{\lambda}{}^{\tau}\Gamma^{q}_{\tau\mu}\chi_{m}{}^{\mu}\chi_{q}{}^{\sigma}\right) \\
&= -4\delta^{rp}\,{}_{\nu}F_{p}{}^{\sigma} + 4\delta^{rt}\Gamma^{s}_{\nu\tau}\chi_{t}{}^{\tau}\chi_{s}{}^{\sigma} \\
&\quad - 2\gamma^{r}{}_{\nu}{}^{\lambda}\left(\|Q\chi\|^{2}\delta^{\sigma}_{\lambda} + \gamma^{t}{}_{\lambda}{}^{\mu}\left(-\delta^{\sigma}_{\mu}B_{t} + \Gamma^{s}_{\mu\tau}\chi_{t}{}^{\tau}\chi_{s}{}^{\sigma}\right)\right).
\end{aligned}
$$

We use now the following Fierz-identity, that is, any endomorphism of spinors can be written as linear combination of elements of the Clifford algebra and the fact that the coefficients of the gravitino are odd:

$$\begin{aligned}\gamma^{t}_{\lambda}{}^{\mu}\Gamma^{s}_{\mu\tau}\chi_t{}^{\tau}\chi_s{}^{\sigma} &= \frac{1}{2}\gamma^{t}_{\beta}{}^{\mu}\Gamma^{s}_{\mu\tau}\chi_t{}^{\tau}\chi_s{}^{\alpha}\left(\delta^{\sigma}_{\lambda}\delta^{\beta}_{\alpha} + \gamma^{l}_{\lambda}{}^{\sigma}\gamma_{l\alpha}{}^{\beta} - \mathrm{I}_{\lambda}{}^{\sigma}\,\mathrm{I}_{\alpha}{}^{\beta}\right)\\ &= -\|Q\chi\|^2\delta^{\sigma}_{\lambda} - gs\left(\delta_{\gamma}\chi,\chi_l\right)\gamma^{l}_{\lambda}{}^{\sigma}.\end{aligned} \tag{13.4.7}$$

Hence,

$${}_{\nu}F_{p}{}^{\sigma} = \Gamma^{s}_{\nu\tau}\chi_p{}^{\tau}\chi_s{}^{\sigma} + \frac{1}{2}\gamma_{p\nu}{}^{\lambda}\gamma^{t}_{\lambda}{}^{\sigma}\left(B_t + gs\left(\delta_{\gamma}\chi,\chi_t\right)\right).$$

We now turn to the conditions of order two in η. Here we simplify Eq. (13.3.5) with what we know:

$$\begin{aligned}{}_{34}d^{s}_{\alpha\beta} = \Bigg(&-{}_{\kappa}F_{\lambda}{}^{t}\,{}_{0}E_{t}{}^{p}\Big({}_{34}F_{p}{}^{a} + \varepsilon^{\mu\nu}\Big({}_{\mu}F_{p}{}^{m} + {}_{0}F_{p}{}^{\tau}\,{}_{\mu}F_{\tau}{}^{m}\Big){}_{0}E_{m}{}^{q}\\
&\cdot\Big({}_{\nu}F_{q}{}^{a} + {}_{0}F_{q}{}^{\sigma}\,{}_{\nu}F_{\sigma}{}^{a}\Big) - {}_{0}F_{p}{}^{\sigma}\,{}_{34}F_{\sigma}{}^{a}\\
&-\varepsilon^{\mu\nu}\Big({}_{\mu}F_{p}{}^{\sigma} + {}_{0}F_{p}{}^{\tau}\,{}_{\mu}F_{\tau}{}^{\sigma}\Big){}_{\nu}F_{\sigma}{}^{a}\Big)\\
&+\varepsilon^{\sigma\tau}\left({}_{\sigma}F_{\kappa}{}^{\mu}\,{}_{\mu}F_{\lambda}{}^{t} + \varepsilon_{\kappa\sigma}\,{}_{34}F_{\lambda}{}^{t}\right){}_{0}E_{t}{}^{p}\Big({}_{\tau}F_{p}{}^{a} + {}_{0}F_{p}{}^{\nu}\,{}_{\tau}F_{\nu}{}^{a}\Big)\\
&+\left(\varepsilon^{\sigma\tau}\,{}_{\sigma}F_{\kappa}{}^{c}\left(\partial_{x^c}\,{}_{\tau}F_{\lambda}{}^{a}\right) - {}_{\mu}F_{\lambda}{}^{\mu}\,{}_{34}F_{\kappa}{}^{a} + {}_{34}F_{\lambda}{}^{\mu}\,{}_{\mu}F_{\kappa}{}^{a}\right)\\
&+{}_{\kappa}F_{\lambda}{}^{\mu}\Big(\varepsilon^{\sigma\tau}\,{}_{\sigma}F_{\mu}{}^{r}\,{}_{0}E_{r}{}^{t}\,{}_{\tau}F_{t}{}^{a} - {}_{34}F_{\mu}{}^{a}\\
&\quad+\varepsilon^{\sigma\tau}\Big({}_{\sigma}F_{\mu}{}^{r}\,{}_{0}E_{r}{}^{p}\,{}_{0}F_{p}{}^{\rho} - {}_{\sigma}F_{\mu}{}^{\rho}\Big){}_{\tau}F_{\rho}{}^{a}\Big)\\
&-\varepsilon^{\sigma\tau}\left({}_{\sigma}F_{\kappa}{}^{\mu}\,{}_{\mu}F_{\lambda}{}^{\rho} + \varepsilon_{\kappa\sigma}\,{}_{34}F_{\lambda}{}^{\rho}\right){}_{\tau}F_{\rho}{}^{a}\Bigg){}_{0}E_{a}{}^{s}\left(\delta^{\kappa}_{\alpha}\delta^{\lambda}_{\beta} + \delta^{\lambda}_{\alpha}\delta^{\kappa}_{\beta}\right)\\
= \Bigg(&-\Gamma^{p}_{\kappa\lambda}\Bigg({}_{34}F_{p}{}^{a}\,{}_{0}E_{a}{}^{s} + 4\chi_p{}^{\tau}\gamma^{q}_{\tau}{}^{\nu}\Gamma^{s}_{\nu\sigma}\chi_q{}^{\sigma} - \chi_p{}^{\sigma}\gamma^{q}_{\sigma}{}^{\nu}\Gamma^{s}_{\nu\tau}\chi_q{}^{\tau}\\
&+\left(\Gamma^{q}_{\mu\tau}\chi_p{}^{\tau}\chi_q{}^{\sigma} + \frac{1}{2}\gamma_{p\mu}{}^{\lambda}\gamma^{t}_{\lambda}{}^{\sigma}\left(B_t + gs\left(\delta_{\gamma}\chi,\chi_t\right)\right) + \chi_p{}^{\tau}\Gamma^{q}_{\mu\tau}\chi_q{}^{\sigma}\right)\\
&\cdot\gamma^{s}_{\sigma}{}^{\mu}\Bigg) + 2\varepsilon^{\sigma\tau}\left(\Gamma^{q}_{\sigma\kappa}\chi_q{}^{\mu}\Gamma^{p}_{\mu\lambda} + \varepsilon_{\kappa\sigma}\gamma^{q}_{\lambda}{}^{\pi}\Gamma^{p}_{\pi\rho}\chi_q{}^{\rho}\right)\chi_p{}^{\nu}\Gamma^{s}_{\tau\nu}\\
&+\varepsilon^{\sigma\tau}\Gamma^{b}_{\sigma\kappa}\Gamma^{d}_{\tau\lambda}f_b{}^{c}\left(\partial_{x^c}f_d{}^{a}\right){}_{0}E_{a}{}^{s} - \Gamma^{p}_{\mu\lambda}\chi_p{}^{\mu}\gamma^{q}_{\kappa}{}^{\pi}\Gamma^{s}_{\pi\rho}\chi_q{}^{\rho}\\
&+2\left(-\|Q\chi\|^2\delta^{\mu}_{\lambda} + B_t\gamma^{t}_{\lambda}{}^{\mu}\right)\Gamma^{s}_{\mu\kappa} - \varepsilon^{\sigma\tau}\Gamma^{p}_{\sigma\kappa}\chi_p{}^{\mu}\Gamma^{q}_{\mu\lambda}\chi_q{}^{\rho}\Gamma^{s}_{\tau\rho}\Bigg)\\
&\cdot\left(\delta^{\kappa}_{\alpha}\delta^{\lambda}_{\beta} + \delta^{\lambda}_{\alpha}\delta^{\kappa}_{\beta}\right)\end{aligned}$$

$$
\begin{aligned}
&= \left(-\Gamma^p_{\kappa\lambda}\left({}_{34}F_p{}^a{}_0E_a{}^s + \chi_p{}^\pi\gamma^q{}_\tau{}^\nu\Gamma^s_{\nu\rho}\chi_q{}^\rho - 2\|Q\chi\|^2\delta^s_p\right)\right.\\
&\quad + \chi_p{}^\pi\chi_q{}^\rho\left(\gamma^p{}_\kappa{}^\tau\Gamma^s_{\tau\rho}\Gamma^q_{\pi\lambda} - 2\gamma^p{}_\lambda{}^\nu\Gamma^q_{\nu\pi}\Gamma^s_{\kappa\rho} - \Gamma^p_{\pi\lambda}\gamma^q{}_\kappa{}^\sigma\Gamma^s_{\sigma\rho}\right)\\
&\quad \left.+ \gamma^b{}_\kappa{}^\tau\Gamma^d_{\tau\lambda}f_b{}^c\left(\partial_{x^c}f_d{}^a\right){}_0E_a{}^s + 2B_t\gamma^t{}_\lambda{}^\mu\Gamma^s_{\mu\kappa}\right)\left(\delta^\kappa_\alpha\delta^\lambda_\beta + \delta^\lambda_\alpha\delta^\kappa_\beta\right)
\end{aligned}
$$

We decompose the equation ${}_{34}d^s_{\alpha\beta} = 0$ in two parts fixing B_t and ${}_{34}F_p{}^a$ respectively. Let us start with B_t:

$$
\begin{aligned}
0 &= \frac{1}{2}\delta^{\alpha\beta}\,{}_{34}d^s_{\alpha\beta}\\
&= \chi_p{}^\pi\chi_q{}^\rho\left(\gamma^p{}_\kappa{}^\tau\Gamma^s_{\tau\rho}\Gamma^q_{\pi\lambda} - 2\gamma^p{}_\lambda{}^\nu\Gamma^q_{\nu\pi}\Gamma^s_{\kappa\rho} - \Gamma^p_{\pi\lambda}\gamma^q{}_\kappa{}^\sigma\Gamma^s_{\sigma\rho}\right)\delta^{\kappa\lambda}\\
&\quad + \delta^{\kappa\lambda}\gamma^b{}_\kappa{}^\tau\Gamma^d_{\tau\lambda}f_b{}^c\left(\partial_{x^c}f_d{}^a\right){}_0E_a{}^s + 2B_t\gamma^t{}_\lambda{}^\mu\Gamma^s_{\mu\kappa}\delta^{\kappa\lambda}\\
&= 2\varepsilon^{bd}f_b{}^c\left(\partial_{x^c}f_d{}^a\right){}_0E_a{}^s + 4B_t\varepsilon^{ts}
\end{aligned}
$$

If we now write the Levi-Civita covariant derivative on $|M|$ that is defined by f_a as $\nabla^{LC}_{f_a}f_b = \omega^{LC}_a\,\mathrm{I}\,f_b$, we obtain that

$$
B_t = \frac{1}{2}\mathrm{I}_t{}^a\,\omega^{LC}_a.
$$

Since $\gamma^t{}_\lambda{}^\sigma\,\mathrm{I}_t{}^a = -\gamma^a{}_\lambda{}^\mu\,\mathrm{I}_\mu{}^\sigma$, we obtain

$$
\begin{aligned}
{}_{34}F_\lambda{}^\sigma &= -\left(\|Q\chi\|^2\delta^\sigma_\lambda + \frac{1}{2}\gamma^t{}_\lambda{}^\mu\,\mathrm{I}_\mu{}^\sigma\,\omega^{LC}_t\right),\\
{}_\nu F_p{}^\sigma &= \Gamma^s_{\nu\tau}\chi_p{}^\tau\chi_s{}^\sigma + \frac{1}{4}\gamma_{p\nu}{}^\lambda\gamma^t{}_\lambda{}^\sigma\left(\mathrm{I}_t{}^a\,\omega^{LC}_a + 2g_S\left(\delta_\gamma\chi,\chi_t\right)\right).
\end{aligned}
$$

The remaining part of ${}_{34}d^s_{\alpha\beta} = 0$ is given by

$$
\begin{aligned}
0 &= \frac{1}{2}\varepsilon^{\beta\gamma}\Gamma^r_{\gamma\delta}\varepsilon^{\delta\alpha}\,{}_{34}d^s_{\alpha\beta}\\
&= -2\delta^{rp}\left({}_{34}F_p{}^a{}_0E_a{}^s + \chi_p{}^\pi\gamma^q{}_\tau{}^\nu\Gamma^s_{\nu\rho}\chi_q{}^\rho\right)\\
&\quad + \chi_t{}^\pi\chi_q{}^\rho\left(\gamma^t{}_\kappa{}^\tau\Gamma^s_{\tau\rho}\Gamma^q_{\pi\lambda}\varepsilon^{\lambda\gamma}\Gamma^r_{\gamma\delta}\varepsilon^{\delta\kappa}\right.\\
&\quad \left.- 2\gamma^t{}_\lambda{}^\nu\Gamma^q_{\nu\pi}\varepsilon^{\lambda\gamma}\Gamma^r_{\gamma\delta}\varepsilon^{\delta\kappa}\Gamma^s_{\kappa\rho} - \Gamma^t_{\pi\lambda}\varepsilon^{\lambda\gamma}\Gamma^r_{\gamma\delta}\varepsilon^{\delta\kappa}\gamma^q{}_\kappa{}^\sigma\Gamma^s_{\sigma\rho}\right)\\
&= -2\delta^{rp}\,{}_{34}F_p{}^a{}_0E_a{}^s - 4\|Q\chi\|^2\delta^{rs} + \chi_t{}^\pi\chi_q{}^\rho\left(-2\delta^{rt}\gamma^q{}_\pi{}^\nu\Gamma^s_{\nu\rho}\right.\\
&\quad \left.+ \gamma^q{}_\pi{}^\gamma\gamma^r{}_\gamma{}^\kappa\gamma^t{}_\kappa{}^\tau\Gamma^s_{\tau\rho} + 2\gamma^q{}_\pi{}^\lambda\gamma^t{}_\lambda{}^\nu\gamma^r{}_\nu{}^\kappa\Gamma^s_{\kappa\rho} - \gamma^t{}_\pi{}^\gamma\gamma^r{}_\gamma{}^\kappa\gamma^q{}_\kappa{}^\sigma\Gamma^s_{\sigma\rho}\right)\\
&= -2\delta^{rp}\,{}_{34}F_p{}^a{}_0E_a{}^s - 2\|Q\chi\|^2\delta^{rs} - \chi_t{}^\pi\chi_q{}^\rho\left(\gamma^t{}_\pi{}^\gamma\gamma^r{}_\gamma{}^\kappa\gamma^q{}_\kappa{}^\sigma\Gamma^s_{\sigma\rho}\right)
\end{aligned}
$$

and hence

$$ {}_{34}F_p{}^a = -\left(\|Q\chi\|^2\delta_p^s + 2gs\left(\left(P\chi\right)_p, \left(Q\chi\right)_t\right)\delta^{ts}\right) f_s{}^a. $$

The only remaining equation to check is

$$
\begin{aligned}
0 &= \delta_{nr}\frac{1}{2}\Gamma^r_{\xi\zeta}\varepsilon^{\zeta\alpha}\varepsilon^{\beta\xi}\ {}_{34}d_{\alpha\beta}{}^\sigma \\
&= \delta_{nr}\frac{1}{2}\Gamma^r_{\xi\zeta}\varepsilon^{\zeta\alpha}\varepsilon^{\beta\xi}\left({}_\kappa F_\lambda{}^t\ {}_0E_t{}^p\left(\varepsilon^{\mu\nu}\left({}_\mu F_p{}^a + {}_0F_p{}^\tau\ {}_\mu F_\tau{}^a\right){}_0E_a{}^s\right.\right. \\
&\qquad \cdot\left(\left({}_\nu F_s{}^b + {}_0F_s{}^\rho\ {}_\nu F_\rho{}^b\right){}_0E_b{}^c\ {}_0F_c{}^\sigma - {}_\nu F_s{}^\sigma - {}_0F_s{}^\beta\ {}_\nu F_\beta{}^\sigma\right) - {}_{34}F_p{}^\sigma \\
&\qquad + \left({}_{34}F_p{}^a - {}_0F_p{}^\rho\ {}_{34}F_\rho{}^a\right){}_0E_a{}^b\ {}_0F_b{}^\sigma - \varepsilon^{\mu\nu}\left({}_\mu F_p{}^\rho + {}_0F_p{}^\tau\ {}_\mu F_\tau{}^\rho\right) \\
&\qquad \left.\cdot\left({}_\nu F_\rho{}^a\ {}_0E_a{}^b\ {}_0F_b{}^\sigma - {}_\nu F_\rho{}^\sigma\right) + {}_0F_p{}^\beta\ {}_{34}F_\beta{}^\sigma\right) \\
&\quad - \varepsilon^{\mu\nu}\left({}_\mu F_\kappa{}^\rho\ {}_\rho F_\lambda{}^t + \varepsilon_{\kappa\mu}\ {}_{34}F_\lambda{}^t\right){}_0E_t{}^p \\
&\qquad \cdot\left(\left({}_\nu F_p{}^a + {}_0F_p{}^\rho\ {}_\nu F_\rho{}^a\right){}_0E_a{}^s\ {}_0F_s{}^\sigma - {}_\nu F_p{}^\sigma - {}_0F_p{}^\rho\ {}_\nu F_\rho{}^\sigma\right) \\
&\quad - \left(\varepsilon^{\mu\nu}\ {}_\mu F_\kappa{}^c\left(\partial_{x^c}\ {}_\nu F_\lambda{}^t\right) + {}_\mu F_\lambda{}^\mu\ {}_{34}F_\kappa{}^t + {}_{34}F_\lambda{}^\delta\ {}_\delta F_\kappa{}^t\right){}_0E_t{}^q\ {}_0F_q{}^\sigma \\
&\quad + {}_\kappa F_\lambda{}^\tau\left(\varepsilon^{\mu\nu}\ {}_\mu F_\tau{}^a\ {}_0E_a{}^b\ {}_\nu F_b{}^\sigma - \left(\varepsilon^{\mu\nu}\ {}_\mu F_\tau{}^s\ {}_0E_s{}^t\ {}_\nu F_t{}^a - {}_{34}F_\tau{}^a\right.\right. \\
&\qquad \left.+ \varepsilon^{\mu\nu}\left({}_\mu F_\tau{}^a\ {}_0E_a{}^s\ {}_0F_s{}^\rho - {}_\mu F_\tau{}^\rho\right){}_\nu F_\rho{}^a\right){}_0E_a{}^b\ {}_0F_b{}^\sigma - {}_{34}F_\tau{}^\sigma \\
&\qquad \left.+ \varepsilon^{\mu\nu}\left({}_\mu F_\tau{}^a\ {}_0E_a{}^s\ {}_0F_s{}^\rho - {}_\mu F_\tau{}^\rho\right){}_\nu F_\rho{}^\sigma\right) \\
&\quad + \varepsilon^{\mu\nu}\left({}_\mu F_\kappa{}^\rho\ {}_\rho F_\lambda{}^\tau + \varepsilon_{\kappa\mu}\ {}_{34}F_\lambda{}^\tau\right)\left({}_\nu F_\tau{}^a\ {}_0E_a{}^s\ {}_0F_s{}^\sigma - {}_\nu F_\tau{}^\sigma\right) \\
&\quad \left.+ \varepsilon^{\mu\nu}\ {}_\mu F_\kappa{}^c\left(\partial_{x^c}\ {}_\nu F_\lambda{}^\sigma\right) - {}_\mu F_\lambda{}^\mu\ {}_{34}F_\kappa{}^\sigma + {}_{34}F_\lambda{}^\delta\ {}_\delta F_\kappa{}^\sigma\right) \\
&\quad \cdot\left(\delta^\kappa_\alpha\delta^\lambda_\beta + \delta^\lambda_\alpha\delta^\kappa_\beta\right) \\
&= 4\chi_n{}^\tau\gamma^s_\tau{}^\nu\left(\Gamma^c_{\nu\rho}\chi_s{}^\rho\chi_c{}^\sigma - {}_\nu F_s{}^\sigma\right) - 2\ {}_{34}F_n{}^\sigma \\
&\qquad + 2\left({}_{34}F_n{}^a\ {}_0E_a{}^b - \chi_n{}^\rho\gamma^t_\rho{}^\tau\Gamma^b_{\tau\lambda}\chi_t{}^\lambda\right)\chi_b{}^\sigma + 2\chi_n{}^\beta\ {}_{34}F_\beta{}^\sigma \\
&\quad - \chi_s{}^\rho\left(\gamma^p_\rho{}^\beta\gamma_{n\beta}{}^\alpha\gamma^s_\alpha{}^\nu + \gamma^p_\rho{}^\tau\gamma^s_\tau{}^\beta\gamma_{n\beta}{}^\nu\right)\left(\Gamma^s_{\nu\xi}\chi_p{}^\xi\chi_s{}^\sigma - {}_\nu F_p{}^\sigma\right) \\
&\quad - \left(\chi_s{}^\mu\gamma^s_\mu{}^\xi\gamma_{n\xi}{}^\kappa\gamma^l_\kappa{}^\lambda\Gamma^q_{\lambda\tau}\chi_l{}^\tau + \gamma^q_\delta{}^\alpha\gamma_{n\alpha}{}^\lambda\ {}_{34}F_\lambda{}^\delta\right)\chi_q{}^\sigma \\
&\quad + 2\chi_n{}^\tau\left(\gamma^c_\tau{}^\nu\left({}_\nu F_c{}^\sigma - \Gamma^a_{\nu\xi}\chi_a{}^\xi\chi_b{}^\sigma\right) + \gamma^s_\tau{}^\kappa\Gamma^b_{\kappa\lambda}\chi_s{}^\lambda\chi_b{}^\sigma - {}_{34}F_\tau{}^\sigma\right) \\
&\quad - \chi_s{}^\mu\gamma^s_\mu{}^\xi\gamma_{n\xi}{}^\kappa\ {}_{34}F_\kappa{}^\sigma + \gamma^s_\delta{}^\alpha\gamma_{n\alpha}{}^\lambda\ {}_{34}F_\lambda{}^\delta\chi_s{}^\sigma
\end{aligned}
$$

$$
\begin{aligned}
&= -2\,{}_{34}F_n{}^{\sigma} + 2\,{}_{34}F_n{}^{a}\,{}_0E_a{}^{b}\chi_b{}^{\sigma} \\
&\quad - \chi_s{}^{\mu}\gamma^{s}{}_{\mu}{}^{\xi}\gamma_{n\xi}{}^{\kappa}\gamma^{l}{}_{\kappa}{}^{\lambda}\Gamma^{q}_{\lambda\tau}\chi_l{}^{\tau}\chi_q{}^{\sigma} - \chi_s{}^{\mu}\gamma^{s}{}_{\mu}{}^{\xi}\gamma_{n\xi}{}^{\kappa}\,{}_{34}F_{\kappa}{}^{\sigma} \\
&= -2\,{}_{34}F_n{}^{\sigma} - 2\left(\|Q\chi\|^2\delta^b_n + 2gs\left((P\chi)_n, (Q\chi)_t\right)\delta^{tb}\right)\chi_b{}^{\sigma} \\
&\quad - 4gs\left((P\chi)_n, (Q\chi)_t\right)\delta^{tq}\chi_q{}^{\sigma} \\
&\quad - 2\,(P\chi)_n{}^{\kappa}\left(\|Q\chi\|^2\delta^{\sigma}_{\kappa} + \frac{1}{2}\gamma^{t}{}_{\kappa}{}^{\sigma}\,\mathrm{I}_t{}^{a}\,\omega^{LC}_a\right) \\
&= -2\,{}_{34}F_n{}^{\sigma} - 2\|Q\chi\|^2\left(\chi_n{}^{\sigma} + (P\chi)_n{}^{\sigma}\right) - 8gs\left((P\chi)_n, (Q\chi)_t\right)\delta^{tq}\chi_q{}^{\sigma} \\
&\quad - (P\chi)_n{}^{\kappa}\,\gamma^{t}{}_{\kappa}{}^{\sigma}\,\mathrm{I}_t{}^{a}\,\omega^{LC}_a
\end{aligned}
$$

That is,

$$
\begin{aligned}
{}_{34}F_n{}^{\sigma} &= -\|Q\chi\|^2\left(\chi_n{}^{\sigma} + (P\chi)_n{}^{\sigma}\right) - 4gs\left((P\chi)_n, (Q\chi)_t\right)\delta^{tq}\chi_q{}^{\sigma} \\
&\quad - \frac{1}{2}(P\chi)_n{}^{\kappa}\,\gamma^{t}{}_{\kappa}{}^{\sigma}\,\mathrm{I}_t{}^{a}\,\omega^{LC}_a.
\end{aligned}
$$

This finishes the proof of Lemma 11.3.4. □

For the remainder of the calculation we need certain of the remaining non-zero commutators:

Lemma 13.4.8 *It holds*

$$
{}_{\nu}d^{\sigma}_{\alpha\beta} = \frac{1}{2}\left(\varepsilon_{\alpha\nu}\delta^{\rho}_{\beta} + \varepsilon_{\beta\nu}\delta^{\rho}_{\alpha} - \Gamma^{p}_{\alpha\beta}\gamma_{p\nu}{}^{\rho}\right)\gamma^{s}{}_{\rho}{}^{\sigma}\left(\mathrm{I}_s{}^{t}\,\omega^{LC}_t + 2gs(\delta_{\gamma}\chi, \chi_s)\right).
$$

Proof Inserting the results of Lemma 11.3.4 into Eq. (13.4.6) yields

$$
\begin{aligned}
{}_{\nu}d^{\sigma}_{\alpha\beta} &= \left(\Gamma^{p}_{\kappa\lambda}\left(\Gamma^{s}_{\nu\mu}\chi_p{}^{\mu}\chi_s{}^{\sigma} - {}_{\nu}F_p{}^{\sigma}\right) + \varepsilon_{\kappa\nu}\left({}_{34}F_{\lambda}{}^{\sigma} - \gamma^{m}{}_{\lambda}{}^{\tau}\Gamma^{q}_{\tau\mu}\chi_m{}^{\mu}\chi_q{}^{\sigma}\right)\right) \\
&\quad \cdot\left(\delta^{\kappa}_{\alpha}\delta^{\lambda}_{\beta} + \delta^{\lambda}_{\alpha}\delta^{\kappa}_{\beta}\right) \\
&= \left(-\frac{1}{4}\Gamma^{p}_{\kappa\lambda}\gamma_{p\nu}{}^{\rho}\gamma^{t}{}_{\rho}{}^{\sigma}\left(\mathrm{I}_t{}^{a}\,\omega^{LC}_a + 2gs\left(\delta_{\gamma}\chi, \chi_t\right)\right)\right. \\
&\qquad \left. - \varepsilon_{\kappa\nu}\left(\|Q\chi\|^2\delta^{\sigma}_{\lambda} + \frac{1}{2}\gamma^{t}{}_{\lambda}{}^{\mu}\,\mathrm{I}_{\mu}{}^{\sigma}\,\omega^{LC}_t + \gamma^{m}{}_{\lambda}{}^{\tau}\Gamma^{q}_{\tau\mu}\chi_m{}^{\mu}\chi_q{}^{\sigma}\right)\right) \\
&\quad \cdot\left(\delta^{\kappa}_{\alpha}\delta^{\lambda}_{\beta} + \delta^{\lambda}_{\alpha}\delta^{\kappa}_{\beta}\right) \\
&= \frac{1}{2}\left(\varepsilon_{\alpha\nu}\delta^{\rho}_{\beta} + \varepsilon_{\beta\nu}\delta^{\rho}_{\alpha} - \Gamma^{p}_{\alpha\beta}\gamma_{p\nu}{}^{\rho}\right)\gamma^{t}{}_{\rho}{}^{\sigma}\left(\mathrm{I}_t{}^{a}\,\omega^{LC}_a + 2gs(\delta_{\gamma}\chi, \chi_t)\right).
\end{aligned}
$$

For the last step we use Eq. (13.4.7). □

Lemma 13.4.9 *For the commutators of a Wess–Zumino frame F_A in Wess–Zumino coordinates it holds that*

$$i^*[F_\alpha, F_\beta] = 2\Gamma^s_{\alpha\beta}\left(dif_s + \chi_s{}^\tau i^* F_\tau\right), \tag{13.4.10}$$

$$i^*[F_a, F_\beta] = -\frac{1}{4}\gamma_{a\beta}{}^\rho \gamma^t{}_\rho{}^\sigma \left(\mathrm{I}_t{}^a\, \omega^{LC}_a + 2g_S(\delta_\gamma\chi, \chi_t)\right) i^* F_\sigma, \tag{13.4.11}$$

$$i^*\varepsilon^{\beta\gamma}[[F_\alpha, F_\beta], F_\gamma] = -\frac{3}{2}\gamma^s{}_\alpha{}^\sigma \left(\mathrm{I}_s{}^t\, \omega^{LC}_t + 2g_S(\delta_\gamma\chi, \chi_s)\right) i^* F_\sigma \tag{13.4.12}$$

$$i^*\varepsilon^{\mu\nu}\varepsilon^{\alpha\beta}[[F_\mu, F_\alpha], [F_\nu, F_\beta]] = 0. \tag{13.4.13}$$

Proof By definition of a Wess–Zumino frame the zero degree part of $d^\sigma_{\alpha\beta}$ vanishes and hence by definition of the gravitino:

$$i^*[F_\alpha, F_\beta] = 2\Gamma^s_{\alpha\beta} i^* F_s = 2\Gamma^s_{\alpha\beta}\left(dif_s + \chi_s{}^\tau i^* F_\tau\right).$$

This proves Eq. (13.4.10). To show Eq. (13.4.11), we use Lemma 9.2.11 and specialize to a Wess–Zumino pair:

$$i^*d^z_{z+} = -i^*d^+_{++} = 0 \qquad i^*d^z_{z-} = 2i^*d^+_{+-} = 0$$
$$i^*d^{\overline{z}}_{z+} = 0 \qquad i^*d^{\overline{z}}_{z-} = 0$$

Hence ${}_0d^s_{a\beta} = 0$. Similarly,

$$\begin{aligned}
i^*d^+_{z+} &= \frac{1}{2} i^* F_+ d^+_{++} = 0,\\
i^*d^+_{z-} &= i^*\left(F_+ d^+_{+-} + \frac{1}{2}\left(F_- d^+_{++} - d^+_{++}\overline{d^+_{+-}}\right)\right) = 0,\\
i^*d^-_{z+} &= 0,\\
i^*d^-_{z-} &= i^*\left(\overline{F_- d^+_{+-}} - \frac{1}{2} d^+_{++}\overline{d^+_{+-}}\right) = i^*\overline{F_- d^+_{+-}}.
\end{aligned}$$

Lemma 7.3.4 together with Eq. (13.4.2) implies

$$\begin{aligned}
8i^*d^-_{z-} &= i^*\left(F_3 - \mathrm{i}F_4\right)\left(d^3_{33} + d^3_{44} - \mathrm{i}\left(d^4_{33} + d^4_{44}\right)\right)\\
&= \left({}_3d^3_{33} + {}_3d^3_{44} - {}_4d^4_{33} - {}_4d^4_{44}\right) - \mathrm{i}\left({}_4d^3_{33} + {}_4d^3_{44} + {}_3d^4_{33} + {}_3d^4_{44}\right).
\end{aligned}$$

Again by Lemma 7.3.4 and Eqs. (13.4.2d) and (13.4.2e), we have

$$
\begin{aligned}
4d^3_{13} &= \operatorname{Re} d^-_{z-} = {}_3d^3_{33} + {}_3d^3_{44}, & 4d^4_{13} &= -\operatorname{Im} d^-_{z-} = {}_3d^4_{33} + {}_3d^4_{44},\\
4d^3_{23} &= -\operatorname{Im} d^-_{z-} = {}_4d^3_{33} + {}_4d^3_{44}, & 4d^4_{23} &= -\operatorname{Re} d^-_{z-} = {}_4d^4_{33} + {}_4d^4_{44},\\
4d^3_{14} &= \operatorname{Im} d^-_{z-} = -{}_4d^3_{33} - {}_4d^3_{44}, & 4d^4_{14} &= \operatorname{Re} d^-_{z-} = -{}_4d^4_{33} - {}_4d^4_{44},\\
4d^3_{24} &= \operatorname{Re} d^-_{z-} = d^3_{33} + d^3_{44}, & 4d^4_{24} &= -\operatorname{Im} d^-_{z-} = {}_3d^4_{33} + {}_3d^4_{44}.
\end{aligned}
$$

Hence, $4d^\sigma_{p\alpha} = \delta_{pq}\Gamma^q_{\alpha\mu}\delta^{\mu\nu}\ {}_\nu d^\sigma_{\kappa\lambda}\delta^{\kappa\lambda}$, and consequently

$$
\begin{aligned}
{}_0d^\sigma_{p\alpha} &= \frac{1}{4}\delta_{pq}\Gamma^q_{\alpha\mu}\delta^{\mu\nu}\delta^{\kappa\lambda}\frac{1}{2}\left(\varepsilon_{\kappa\nu}\delta^\rho_\lambda + \varepsilon_{\lambda\nu}\delta^\rho_\kappa - \Gamma^p_{\kappa\lambda}\gamma_{p\nu}{}^\rho\right)\gamma^t{}_\rho{}^\sigma\\
&\quad\cdot\left(\mathrm{I}_t{}^a\,\omega^{LC}_a + 2gs(\delta_\gamma\chi,\chi_t)\right)\\
&= -\frac{1}{4}\gamma_{p\alpha}{}^\rho\gamma^t{}_\rho{}^\sigma\left(\mathrm{I}_t{}^a\,\omega^{LC}_a + 2gs(\delta_\gamma\chi,\chi_t)\right).
\end{aligned}
$$

For Eq. (13.4.12) verify

$$
\begin{aligned}
i^*\varepsilon^{\beta\gamma}[[F_\alpha,F_\beta],F_\gamma] &= i^*\varepsilon^{\beta\gamma}[2\Gamma^s_{\alpha\beta}F_s + d^\sigma_{\alpha\beta}F_\sigma, F_\gamma]\\
&= i^*\left(2\Gamma^s_{\alpha\beta}\varepsilon^{\beta\gamma}\left(d^t_{s\gamma}F_t + d^\tau_{s\gamma}F_\tau\right) - \varepsilon^{\beta\gamma}\left(F_\gamma d^\sigma_{\alpha\beta}\right)F_\sigma + \varepsilon^{\beta\gamma}d^\sigma_{\alpha\beta}[F_\sigma,F_\gamma]\right)\\
&= \left(2\gamma^s{}_\alpha{}^\gamma\ {}_0d^\sigma_{s\gamma} - \varepsilon^{\beta\gamma}\ {}_\gamma d^\sigma_{\alpha\beta}\right)i^*F_\sigma\\
&= -\left(\mathrm{I}_\alpha{}^\nu\,\delta^{\kappa\lambda} + \varepsilon^{\lambda\nu}\delta_\alpha{}^\kappa\right){}_\nu d^\sigma_{\kappa\lambda}i^*F_\sigma\\
&= -\left(\mathrm{I}_\alpha{}^\nu\,\delta^{\kappa\lambda} + \varepsilon^{\lambda\nu}\delta_\alpha{}^\kappa\right)\frac{1}{2}\left(\varepsilon_{\kappa\nu}\delta^\rho_\lambda + \varepsilon_{\lambda\nu}\delta^\rho_\kappa - \Gamma^p_{\kappa\lambda}\gamma_{p\nu}{}^\rho\right)\gamma^t{}_\rho{}^\sigma\\
&\quad\cdot\left(\mathrm{I}_t{}^a\,\omega^{LC}_a + 2gs(\delta_\gamma\chi,\chi_t)\right)i^*F_\sigma\\
&= -\frac{3}{2}\gamma^t{}_\alpha{}^\sigma\left(\mathrm{I}_t{}^a\,\omega^{LC}_a + 2gs(\delta_\gamma\chi,\chi_t)\right)i^*F_\sigma.
\end{aligned}
$$

For the term of order four, Eq. (13.4.13),

$$
\begin{aligned}
i^*\varepsilon^{\mu\nu}\varepsilon^{\alpha\beta}[[F_\mu,F_\alpha],[F_\nu,F_\beta]] &= i^*\varepsilon^{\mu\nu}\varepsilon^{\alpha\beta}[2\Gamma^k_{\mu\alpha}F_k + d^\kappa_{\mu\alpha}F_\kappa, 2\Gamma^l_{\nu\beta}F_l + d_{\nu\beta}{}^\lambda F_\lambda]\\
&= i^*8\delta^{kl}[F_k,F_l] = 0. \qquad \square
\end{aligned}
$$

13.5 Covariant Derivative Identities

In this section we derive expressions that involve the pullback of the Levi-Civita connection on the Riemannian target manifold N with metric n. The formulas will ultimately be needed for the calculation of the component action in Sect. 13.9 and the supersymmetry of the component fields in Sect. 12.2. The formulas can be seen as a version of Deligne and Freed (1999b, (4.10)–(4.11)) specialized to super Riemann surfaces and for curved targets. For the rest of the section we denote by ∇ the pullback of the Levi-Civita connection on TN along $\Phi\colon M \to N$ and by R^N its curvature.

Lemma 13.5.1 *For any frame F_α of $\mathcal{D}$, we have the following relations*

$$\nabla_{F_\alpha} F_\beta \Phi = \frac{1}{2}\left([F_\alpha, F_\beta]\Phi + \varepsilon_{\alpha\beta}\varepsilon^{\gamma\delta}\nabla_{F_\gamma} F_\delta \Phi\right), \tag{13.5.2}$$

$$\begin{aligned}\varepsilon^{\alpha\beta}\nabla_{F_\alpha}\nabla_{F_\beta} F_\gamma \Phi = & -\varepsilon^{\alpha\beta}\left(\left(\nabla_{[F_\gamma, F_\alpha]} + \frac{1}{3}R^N(F_\gamma\Phi, F_\alpha\Phi)\right)F_\beta\Phi\right.\\ & \left. - \frac{2}{3}[[F_\gamma, F_\alpha], F_\beta]\Phi\right),\end{aligned} \tag{13.5.3}$$

$$\begin{aligned}\varepsilon^{\alpha\beta}\nabla_{F_\gamma}\nabla_{F_\alpha} F_\beta \Phi = & \ \varepsilon^{\alpha\beta}\left(\left(\nabla_{[F_\gamma, F_\alpha]} + \frac{2}{3}R^N(F_\gamma\Phi, F_\alpha\Phi)\right)F_\beta\Phi\right.\\ & \left. - \frac{1}{3}[[F_\gamma, F_\alpha], F_\beta]\Phi\right),\end{aligned} \tag{13.5.4}$$

$$\begin{aligned}\varepsilon^{\mu\nu}\varepsilon^{\alpha\beta}\nabla_{F_\mu}\nabla_{F_\nu}\nabla_{F_\alpha} F_\beta\Phi = & \ \varepsilon^{\mu\nu}\varepsilon^{\alpha\beta}\left(\frac{1}{2}\nabla_{[F_\nu, F_\alpha]}[F_\mu, F_\beta]\Phi\right.\\ & + \frac{2}{3}\nabla_{[[F_\alpha, F_\mu], F_\nu,]}F_\beta\Phi - \frac{1}{6}[[F_\mu, F_\alpha], [F_\nu, F_\beta]]\Phi\\ & \left. + \frac{2}{3}\left(\nabla_{F_\mu}R^N\right)(F_\nu\Phi, F_\alpha\Phi)F_\beta\Phi - R^N\left(F_\mu\Phi, F_\alpha\Phi\right)[F_\nu, F_\beta]\Phi\right).\end{aligned} \tag{13.5.5}$$

Proof This lemma follows from the fact that $\mathcal{D}$ has only two odd real directions and the fact that the Levi-Civita connection ∇ on the target N is torsion free and compatible with the metric. The proof of Eq. (13.5.2) uses only torsion freeness:

$$\begin{aligned}2\nabla_{F_\alpha} F_\beta\Phi &= \nabla_{F_\alpha} F_\beta\Phi - \nabla_{F_\beta} F_\alpha\Phi + [F_\alpha, F_\beta]\Phi\\ &= [F_\alpha, F_\beta]\Phi + \varepsilon_{\alpha\beta}\varepsilon^{\gamma\delta}\nabla_{F_\gamma} F_\delta\Phi\end{aligned}$$

For Eq. (13.5.3) we need the curvature equation and need to proceed for $\gamma = 3, 4$ separately:

$$\begin{aligned}
\varepsilon^{\alpha\beta}\nabla_{F_\alpha}\nabla_{F_\beta}F_3\Phi &= \nabla_{F_3}\nabla_{F_4}F_3\Phi - \nabla_{F_4}\nabla_{F_3}F_3\Phi \\
&= \left(R^N(F_3\Phi, F_4\Phi)F_3\Phi - \nabla_{F_4}\nabla_{F_3}F_3\Phi + \nabla_{[F_3,F_4]}F_3\Phi\right) - \nabla_{F_4}\nabla_{F_3}F_3\Phi \\
&= \left(\frac{1}{3} + \frac{2}{3}\right) R^N(F_3\Phi, F_4\Phi)F_3\Phi + \nabla_{[F_3,F_4]}F_3\Phi - \nabla_{F_4}[F_3, F_3]\Phi \\
&= \frac{1}{3}\left(R^N(F_3\Phi, F_4\Phi)F_3\Phi - R^N(F_3\Phi, F_3\Phi)F_4\Phi\right) + \nabla_{[F_3,F_4]}F_3\Phi \\
&\quad - \nabla_{[F_3,F_3]}F_4\Phi - [F_4, [F_3, F_3]]\Phi \\
&= \frac{1}{3}\left(R^N(F_3\Phi, F_4\Phi)F_3\Phi - R^N(F_3\Phi, F_3\Phi)F_4\Phi\right) + \nabla_{[F_3,F_4]}F_3\Phi \\
&\quad - \nabla_{[F_3,F_3]}F_4\Phi + \frac{2}{3}\left([[F_3, F_3], F_4] - [[F_3, F_4], F_3]\right)\Phi \\
&= -\varepsilon^{\alpha\beta}\left(\left(\nabla_{[F_3,F_\alpha]} + \frac{1}{3}R^N(F_3\Phi, F_\alpha\Phi)\right)F_\beta\Phi - \frac{2}{3}[[F_3, F_\alpha], F_\beta]\Phi\right)
\end{aligned}$$

We have used the first Bianchi identity:

$$\begin{aligned}
0 &= R^N(F_\mu, F_\nu)F_\mu + R^N(F_\nu, F_\mu)F_\mu + R^N(F_\mu, F_\mu)F_\nu \\
&= 2R^N(F_\mu, F_\nu)F_\mu + R^N(F_\mu, F_\mu)F_\nu
\end{aligned}$$

and the Jacobi identity

$$\begin{aligned}
0 &= [F_\mu, [F_\mu, F_\nu]] + [F_\mu, [F_\nu, F_\mu]] + [F_\nu, [F_\mu, F_\mu]] \\
&= 2[F_\mu, [F_\nu, F_\mu]] + [F_\nu, [F_\mu, F_\mu]].
\end{aligned}$$

The case $\gamma = 4$ follows analogously. Equation (13.5.4) is derived from Eq. (13.5.3) as follows:

$$\begin{aligned}
\varepsilon^{\alpha\beta}\nabla_{F_\gamma}\nabla_{F_\alpha}F_\beta\Phi &= \varepsilon^{\alpha\beta}\left(R^N(F_\gamma\Phi, F_\alpha\Phi)F_\beta\Phi - \nabla_{F_\alpha}\nabla_{F_\gamma}F_\beta\Phi + \nabla_{[F_\gamma,F_\alpha]}F_\beta\Phi\right) \\
&= \varepsilon^{\alpha\beta}\left(R^N(F_\gamma\Phi, F_\alpha\Phi)F_\beta\Phi + \nabla_{F_\alpha}\nabla_{F_\beta}F_\gamma\Phi - \nabla_{F_\alpha}[F_\gamma, F_\beta]\Phi\right. \\
&\quad \left. + \nabla_{[F_\gamma,F_\alpha]}F_\beta\Phi\right)
\end{aligned}$$

$$= \varepsilon^{\alpha\beta} \left(R^N(F_\gamma \Phi, F_\alpha \Phi) F_\beta \Phi - \left(\nabla_{[F_\gamma, F_\alpha]} + \frac{1}{3} R^N(F_\gamma \Phi, F_\alpha \Phi) \right) F_\beta \Phi \right.$$
$$\left. + \frac{2}{3}[[F_\gamma, F_\alpha], F_\beta]\Phi - [F_\alpha, [F_\gamma, F_\beta]]\Phi - \nabla_{[F_\gamma, F_\beta]} F_\alpha \Phi + \nabla_{[F_\gamma, F_\alpha]} F_\beta \Phi \right)$$
$$= \varepsilon^{\alpha\beta} \left(\left(\nabla_{[F_\gamma, F_\alpha]} + \frac{2}{3} R^N(F_\gamma \Phi, F_\alpha \Phi) \right) F_\beta \Phi - \frac{1}{3}[[F_\gamma, F_\alpha], F_\beta]\Phi \right)$$

Equation (13.5.5) follows from Eq. (13.5.4) by further differentiation:

$$\varepsilon^{\mu\nu}\varepsilon^{\alpha\beta} \nabla_{F_\mu} \nabla_{F_\nu} \nabla_{F_\alpha} F_\beta \Phi$$
$$= \varepsilon^{\mu\nu}\varepsilon^{\alpha\beta} \nabla_{F_\mu} \left(\left(\nabla_{[F_\nu, F_\alpha]} + \frac{2}{3} R^N(F_\nu \Phi, F_\alpha \Phi) \right) F_\beta \Phi - \frac{1}{3}[[F_\nu, F_\alpha], F_\beta]\Phi \right)$$
$$= \varepsilon^{\mu\nu}\varepsilon^{\alpha\beta} \left(\nabla_{[F_\nu, F_\alpha]} \nabla_{F_\mu} F_\beta \Phi + \nabla_{[F_\mu, [F_\nu, F_\alpha]]} F_\beta \Phi \right.$$
$$+ R^N\left(F_\mu \Phi, [F_\nu, F_\alpha]\Phi\right) F_\beta \Phi + \frac{1}{3}\left(2\left(\nabla_{F_\mu} R^N\right)(F_\nu \Phi, F_\alpha \Phi) F_\beta \Phi \right.$$
$$\left. - R^N\left(F_\nu \Phi, [F_\mu, F_\alpha]\Phi\right) F_\beta \Phi + R^N(F_\nu \Phi, F_\alpha \Phi)[F_\mu, F_\beta]\Phi \right)$$
$$\left. - \frac{1}{3}\left([F_\mu, [[F_\nu, F_\alpha], F_\beta]]\Phi - \nabla_{[[F_\nu, F_\alpha], F_\beta]} F_\mu \Phi\right) \right)$$
$$= \varepsilon^{\mu\nu}\varepsilon^{\alpha\beta} \left(\frac{1}{2} \nabla_{[F_\nu, F_\alpha]}[F_\mu, F_\beta]\Phi + \frac{2}{3} \nabla_{[[F_\alpha, F_\mu], F_\nu,]} F_\beta \Phi \right.$$
$$- \frac{1}{6}[[F_\mu, F_\alpha], [F_\nu, F_\beta]]\Phi + \frac{2}{3}\left(\nabla_{F_\mu} R^N\right)(F_\nu \Phi, F_\alpha \Phi) F_\beta \Phi$$
$$\left. - R^N\left(F_\mu \Phi, F_\alpha \Phi\right)[F_\nu, F_\beta]\Phi \right)$$

Here we have used the Jacobi identity for the commutators

$$0 = \varepsilon^{\mu\nu}\varepsilon^{\alpha\beta} \left([F_\mu, [[F_\nu, F_\alpha], F_\beta]] - [F_\beta, [F_\mu, [F_\nu, F_\alpha]]] - [[F_\nu, F_\alpha], [F_\beta, F_\mu]]\right)$$
$$= \varepsilon^{\mu\nu}\varepsilon^{\alpha\beta} \left(2[F_\mu, [[F_\nu, F_\alpha], F_\beta]] + [[F_\mu, F_\alpha], [F_\nu, F_\beta]]\right),$$

and the Bianchi-identity:

$$0 = \varepsilon^{\mu\nu}\varepsilon^{\alpha\beta} \left(R^N\left(F_\mu, [F_\nu, F_\alpha]\right) F_\beta - R^N\left(F_\beta, F_\mu\right)[F_\nu, F_\alpha] \right.$$
$$\left. - R^N\left([F_\nu, F_\alpha], F_\beta\right) F_\mu \right)$$
$$= \varepsilon^{\mu\nu}\varepsilon^{\alpha\beta} \left(2R^N\left(F_\mu, [F_\nu, F_\alpha]\right) F_\beta + R^N\left(F_\mu, F_\alpha\right)[F_\nu, F_\beta] \right). \qquad \square$$

13.6 Covariant Derivative Identities in Component Fields

In this section we calculate the pullback of the formulas of Lemma 13.5.1 along i. We obtain expressions in terms of the component fields of $\Phi\colon M \to N$ and its derivatives. In particular we obtain terms involving the twisted Dirac operator:

Definition 13.6.1 (Twisted Dirac Operator) Let $\nabla^{S^\vee}$ be a U(1)-covariant derivative on $S^\vee$. Denote by ∇^{φ^*TN} be the pullback of the Levi-Civita connection on (N,n) along $\varphi\colon |M| \to N$. Let $\psi = s^\alpha \otimes {}_\alpha\psi$ be a twisted spinor, that is, a section of $S^\vee \otimes \varphi^*TN$. Then the twisted Dirac operator acts on ψ as

$$\not{D}\psi = \gamma^k \nabla_k^{S^\vee \otimes \varphi^*TN} \psi = \left(\gamma^k \nabla_k^{S^\vee} s^\alpha\right) \otimes {}_\alpha\psi + \gamma^k s^\alpha \otimes \nabla_k^{\varphi^*TN}{}_\alpha\psi$$

Notice that γ^k acts on the dual basis s^α as $\gamma^k s^\alpha = -s^\beta \gamma^k{}_\beta{}^\alpha$ as we identify $S^\vee$ and S with the help of the spinor metric g_S. Let us write as before $\nabla^{T|M|} f_k = (\omega^{LC} + A)\,\mathrm{I}\, f_k$ for an arbitrary U(1)-covariant derivative on $T|M|$. Then the lift of $\nabla^{T|M|}$ to S is given by

$$\nabla^S s_\alpha = \frac{1}{2}\left(\omega^{LC} + A\right)\mathrm{I}\, s_\alpha.$$

Consequently, with respect to the dual spinor basis s^α the expression for the twisted Dirac operator is given by

$$\not{D}\left(s^\alpha \otimes {}_\alpha\psi\right) = s^\beta \otimes \left(\frac{1}{2}\left(\omega_k^{LC} + A_k\right)\gamma^k{}_\beta{}^\mu\, \mathrm{I}_\mu{}^\alpha\, {}_\alpha\psi - \gamma^k{}_\beta{}^\alpha \nabla_k^{\varphi^*TN}{}_\alpha\psi\right)$$

In the case that the connection on $S^\vee$ is also given by the Levi-Civita connection $(A = 0)$, we will write $\not{D}^{LC}$ for the corresponding twisted Dirac operator. Notice that the expression $\langle \not{D}\psi, \psi\rangle$ does not depend on the connection on $S^\vee$ because we use a supersymmetric bilinear form on the odd vector bundle S:

$$\begin{aligned}
&\left\langle \not{D}\psi, \psi\right\rangle_{g_S^\vee \otimes \varphi^* n} \\
&= \varepsilon^{\beta\sigma}\varphi^* n\left(\frac{1}{2}\left(\omega_k^{LC} + A_k\right)\gamma^k{}_\beta{}^\mu\, \mathrm{I}_\mu{}^\alpha\, {}_\alpha\psi - \gamma^k{}_\beta{}^\alpha \nabla_k^{\varphi^*TN}{}_\alpha\psi, {}_\sigma\psi\right) \\
&= \frac{1}{4}\left(\omega_k + A_k\right)\varepsilon^{\beta\sigma}\varepsilon_{\alpha\sigma}\gamma^k{}_\beta{}^\mu\, \mathrm{I}_\mu{}^\alpha\, \varepsilon^{\kappa\lambda}\varphi^* n\left({}_\kappa\psi, {}_\lambda\psi\right) \\
&\quad - \varepsilon^{\beta\sigma}\varphi^* n\left(\gamma^k{}_\beta{}^\alpha \nabla_k^{\varphi^*TN}{}_\alpha\psi, {}_\sigma\psi\right) \\
&= -\varepsilon^{\beta\sigma}\varphi^* n\left(\gamma^k{}_\beta{}^\alpha \nabla_k^{\varphi^*TN}{}_\alpha\psi, {}_\sigma\psi\right) = \left\langle \not{D}^{LC}\psi, \psi\right\rangle_{g_S^\vee \otimes \varphi^* n}
\end{aligned}$$

Lemma 13.6.2 *For a Wess–Zumino frame in Wess–Zumino coordinates it holds that*

$$
\begin{aligned}
& i^*\nabla_{F_\alpha} F_\beta \Phi = \Gamma^s_{\alpha\beta}\left(f_s\varphi + \chi_s^\tau{}_\tau\psi\right) - \varepsilon_{\alpha\beta}F,\\
& i^*\varepsilon^{\alpha\beta}\nabla_{F_\alpha}\nabla_{F_\beta}F_\gamma\Phi = 2{}_\gamma\left(D\!\!\!\!/^{LC}\psi\right) + 2\chi_s^\tau\gamma^t{}_\tau{}^\beta\Gamma^s_{\beta\gamma}f_t\varphi + 2\|Q\chi\|^2{}_\gamma\psi\\
& \qquad + 2\chi_s^\tau\Gamma^s_{\gamma\tau}F - \frac{1}{3}{}_\gamma SR^N(\psi),\\
& i^*\varepsilon^{\alpha\beta}\nabla_{F_\gamma}\nabla_{F_\alpha}F_\beta\Phi = -2{}_\gamma\left(D\!\!\!\!/^{LC}\psi\right) - 2\chi_s^\tau\gamma^t{}_\tau{}^\beta\Gamma^s_{\beta\gamma}f_t\varphi - 2\|Q\chi\|^2{}_\gamma\psi\\
& \qquad - 2\chi_s^\tau\Gamma^s_{\gamma\tau}F + \frac{2}{3}{}_\gamma SR^N(\psi) - \frac{1}{2}\gamma^s{}_\gamma{}^\sigma\left(\mathrm{I}_s{}^t\omega_s^{LC} + 2g_S\left(\delta_\gamma\chi, \chi_s\right)\right)_\sigma\psi.\\
& i^*\varepsilon^{\mu\nu}\varepsilon^{\alpha\beta}\nabla_{F_\mu}\nabla_{F_\nu}\nabla_{F_\alpha}F_\beta\Phi = 4\delta^{kl}\left(\nabla_{f_k}f_l\varphi + \left(f_k\chi_l{}^\lambda\right)_\lambda\psi + 2\chi_k{}^\kappa\nabla_{f_l}{}_\kappa\psi\right)\\
& \qquad + 2\left(f_m\varphi + \langle\chi_m + (P\chi)_m, \psi\rangle\right)\delta^{ms}\left(\mathrm{I}_s{}^t\omega_t^{LC} + 2g_S\left(\delta_\gamma\chi, \chi_s\right)\right)\\
& \qquad - 4\|\chi\|^2F + \varepsilon^{\mu\nu}\varepsilon^{\alpha\beta}\frac{2}{3}\left(\nabla_{\mu\psi}R^N\right)\left({}_\nu\psi, {}_\alpha\psi\right){}_\beta\psi\\
& \qquad + 2\varepsilon^{\alpha\beta}\gamma^k{}_\beta{}^\nu R^N\left({}_\nu\psi, {}_\alpha\psi\right)\left(f_k\varphi + \chi_k{}^\kappa{}_\kappa\psi\right)
\end{aligned}
$$

Proof The proof combines Lemma 13.5.1 with Lemma 13.4.9. For the first equation it holds that

$$
\begin{aligned}
2i^*\nabla_{F_\alpha}F_\beta\Phi &= i^*[F_\alpha, F_\beta]\Phi + \varepsilon_{\alpha\beta}i^*\varepsilon^{\gamma\delta}\nabla_{F_\gamma}F_\delta\Phi\\
&= 2\Gamma^s_{\alpha\beta}\left(f_s\varphi + \chi_s^\tau{}_\tau\psi\right) - 2\varepsilon_{\alpha\beta}F.
\end{aligned}
$$

The second and third equation use the following:

$$
\begin{aligned}
i^*\varepsilon^{\alpha\beta}\nabla_{[F_\gamma, F_\alpha]}F_\beta\Phi &= i^*2\Gamma^s_{\gamma\alpha}\varepsilon^{\alpha\beta}\nabla_{dif_s + \chi_s^\tau F_\tau}F_\beta\Phi\\
&= 2\gamma^s{}_\gamma{}^\beta\left(\nabla_{f_s}{}_\beta\psi + \chi_s^\tau\left(\Gamma^t_{\tau\beta}\left(f_t\varphi + \chi_t^\rho{}_\rho\psi\right) - \varepsilon_{\tau\beta}F\right)\right)\\
&= 2\left(\gamma^s{}_\gamma{}^\beta\nabla_{f_s}{}_\beta\psi - \chi_s^\tau\gamma^t{}_\tau{}^\beta\Gamma^s_{\beta\gamma}\left(f_t\varphi + \chi_t^\rho{}_\rho\psi\right) - \chi_s^\tau\Gamma^s_{\gamma\tau}F\right)
\end{aligned}
$$

Using Eq. (13.4.7) this allows to conclude:

$$
\begin{aligned}
& i^*\varepsilon^{\alpha\beta}\nabla_{F_\alpha}\nabla_{F_\beta}F_\gamma\Phi\\
&= -\varepsilon^{\alpha\beta}i^*\left(\left(\nabla_{[F_\gamma, F_\alpha]} + \frac{1}{3}R^N(F_\gamma\Phi, F_\alpha\Phi)\right)F_\beta\Phi - \frac{2}{3}[[F_\gamma, F_\alpha], F_\beta]\Phi\right)\\
&= -2\left(\gamma^s{}_\gamma{}^\beta\nabla_{f_s}{}_\beta\psi - \chi_s^\tau\gamma^t{}_\tau{}^\beta\Gamma^s_{\beta\gamma}\left(f_t\varphi + \chi_t^\rho{}_\rho\psi\right) - \chi_s^\tau\Gamma^s_{\gamma\tau}F\right)\\
&\quad - \frac{1}{3}{}_\gamma SR^N(\psi) - \gamma^s{}_\gamma{}^\sigma\left(\mathrm{I}_s{}^t\omega_t^{LC} + 2g_S\left(\delta_\gamma\chi, \chi_s\right)\right)_\sigma\psi
\end{aligned}
$$

$$= 2\,{}_{\gamma}\!\left(\not{D}^{LC}\psi\right) + 2\chi_s^{\tau}\gamma^{t}{}_{\tau}{}^{\beta}\Gamma^{s}_{\beta\gamma} f_t\varphi + 2\|Q\chi\|^2{}_{\gamma}\psi + 2\chi_s^{\tau}\Gamma^{s}_{\gamma\tau}F$$
$$- \frac{1}{3}{}_{\gamma}SR^N(\psi)$$

and

$$i^*\varepsilon^{\alpha\beta}\nabla_{F_\gamma}\nabla_{F_\alpha}F_\beta\Phi$$
$$= \varepsilon^{\alpha\beta}i^*\left(\left(\nabla_{[F_\gamma,F_\alpha]} + \frac{2}{3}R^N(F_\gamma\Phi, F_\alpha\Phi)\right)F_\beta\Phi - \frac{1}{3}[[F_\gamma, F_\alpha], F_\beta]\Phi\right)$$
$$= 2\left(\gamma^{s}_{\gamma}{}^{\beta}\nabla_{f_s\,\beta}\psi - \chi_s^{\tau}\gamma^{t}{}_{\tau}{}^{\beta}\Gamma^{s}_{\beta\gamma}\left(f_t\varphi + \chi_t^{\rho}{}_{\rho}\psi\right) - \chi_s^{\tau}\Gamma^{s}_{\gamma\tau}F\right)$$
$$+ \frac{2}{3}{}_{\gamma}SR^N(\psi) + \frac{1}{2}\gamma^{s}_{\gamma}{}^{\sigma}\left(\mathrm{I}_s{}^{t}\,\omega_t^{LC} + 2gs\left(\delta_\gamma\chi, \chi_s\right)\right)_{\sigma}\psi$$
$$= -2\,{}_{\gamma}\!\left(\not{D}^{LC}\psi\right) - 2\chi_s^{\tau}\gamma^{t}{}_{\tau}{}^{\beta}\Gamma^{s}_{\beta\gamma} f_t\varphi - 2\|Q\chi\|^2{}_{\gamma}\psi - 2\chi_s^{\tau}\Gamma^{s}_{\gamma\tau}F$$
$$+ \frac{2}{3}{}_{\gamma}SR^N(\psi) - \frac{1}{2}\gamma^{s}_{\gamma}{}^{\sigma}\left(\mathrm{I}_s{}^{t}\,\omega_s^{LC} + 2gs\left(\delta_\gamma\chi, \chi_s\right)\right)_{\sigma}\psi.$$

For the fourth order derivative we verify

$$i^*\varepsilon^{\mu\nu}\varepsilon^{\alpha\beta}\nabla_{F_\mu}\nabla_{F_\nu}\nabla_{F_\alpha}F_\beta\Phi = \varepsilon^{\mu\nu}\varepsilon^{\alpha\beta}i^*\left(\frac{1}{2}\nabla_{[F_\nu,F_\alpha]}[F_\mu, F_\beta]\Phi\right.$$
$$+ \frac{2}{3}\nabla_{[[F_\alpha,F_\mu],F_\nu,]}F_\beta\Phi - \frac{1}{6}[[F_\mu, F_\alpha], [F_\nu, F_\beta]]\Phi$$
$$\left.+ \frac{2}{3}\left(\nabla_{F_\mu}R^N\right)(F_\nu\Phi, F_\alpha\Phi)F_\beta\Phi - R^N\left(F_\mu\Phi, F_\alpha\Phi\right)[F_\nu, F_\beta]\Phi\right)$$
$$= 4\delta^{kl}\left(\nabla_{f_k}\left(f_l\varphi + \chi_l{}^{\lambda}{}_{\lambda}\psi\right) + \chi_k{}^{\kappa}i^*\nabla_{F_\kappa}F_l\Phi\right) - \frac{2}{3}\left(\mathrm{I}_\alpha{}^{\nu}\,\delta^{\kappa\lambda} + \varepsilon^{\lambda\nu}\delta^{\kappa}_{\alpha}\right)$$
$$\cdot {}_{\nu}d^{\sigma}_{\kappa\lambda}\varepsilon^{\alpha\beta}i^*\nabla_{F_\sigma}F_\beta\Phi + \varepsilon^{\mu\nu}\varepsilon^{\alpha\beta}\frac{2}{3}\left(\nabla_{\mu\psi}R^N\right)({}_{\nu}\psi, {}_{\alpha}\psi)\,{}_{\beta}\psi$$
$$+ 2\varepsilon^{\alpha\beta}\gamma^{k}_{\beta}{}^{\nu}R^N\left({}_{\nu}\psi, {}_{\alpha}\psi\right)\left(f_k\varphi + \chi_k{}^{\kappa}{}_{\kappa}\psi\right)$$
$$= 4\delta^{kl}\nabla_{f_k}\left(f_l\varphi + \chi_l{}^{\lambda}{}_{\lambda}\psi\right)$$
$$+ 4\delta^{kl}\chi_k{}^{\kappa}\left(\nabla_{f_l\,\kappa}\psi + \chi_l{}^{\lambda}i^*\nabla_{F_\lambda}F_\kappa\Phi - i^*[F_l, F_\kappa]\Phi\right)$$
$$- \frac{2}{3}\left(\mathrm{I}_\alpha{}^{\nu}\,\delta^{\kappa\lambda} + \varepsilon^{\lambda\nu}\delta^{\kappa}_{\alpha}\right){}_{\nu}d^{\sigma}_{\kappa\lambda}\left(-\gamma^{m}{}_{\sigma}{}^{\alpha}\left(f_m\varphi + \chi_m{}^{\mu}{}_{\mu}\psi\right) + \delta^{\alpha}_{\sigma}F\right)$$
$$+ \varepsilon^{\mu\nu}\varepsilon^{\alpha\beta}\frac{2}{3}\left(\nabla_{\mu\psi}R^N\right)({}_{\nu}\psi, {}_{\alpha}\psi)\,{}_{\beta}\psi$$
$$+ 2\varepsilon^{\alpha\beta}\gamma^{k}_{\beta}{}^{\nu}R^N\left({}_{\nu}\psi, {}_{\alpha}\psi\right)\left(f_k\varphi + \chi_k{}^{\kappa}{}_{\kappa}\psi\right)$$

$$= 4\delta^{kl}\left(\nabla_{f_k} f_l\varphi + (f_k\chi_l{}^\lambda)\,{}_\lambda\psi + 2\chi_k{}^\kappa\nabla_{f_l}\,{}_\kappa\psi\right) + \frac{2}{3}\left(\gamma^m{}_\sigma{}^\alpha\,\mathrm{I}_\alpha{}^\nu\,\delta^{\kappa\lambda}\right.$$

$$\left. + \gamma^m{}_\sigma{}^\kappa\varepsilon^{\lambda\nu}\right){}_\nu d^\sigma_{\kappa\lambda} f_m\varphi - \left(\frac{2}{3}\varepsilon^{\lambda\nu}\delta^\kappa_\sigma\,{}_\nu d^\sigma_{\kappa\lambda} + 4\|\chi\|^2\right)F$$

$$+ \chi_m{}^\mu\left(\frac{2}{3}\left(\gamma^m{}_\sigma{}^\alpha\,\mathrm{I}_\alpha{}^\nu\,\delta^{\kappa\lambda} + \gamma^m{}_\sigma{}^\kappa\varepsilon^{\lambda\nu}\right)\delta^\tau_\mu - \Gamma^m_{\mu\rho}\delta^{\rho\nu}\delta^{\kappa\lambda}\delta^\tau_\sigma\right){}_\nu d^\sigma_{\kappa\lambda}\,{}_\tau\psi$$

$$+ \varepsilon^{\mu\nu}\varepsilon^{\alpha\beta}\frac{2}{3}\left(\nabla_{\mu\psi} R^N\right)({}_\nu\psi, {}_\alpha\psi)\,{}_\beta\psi$$

$$+ 2\varepsilon^{\alpha\beta}\gamma^k{}_\beta{}^\nu R^N\left({}_\nu\psi, {}_\alpha\psi\right)\left(f_k\varphi + \chi_k{}^\kappa\,{}_\kappa\psi\right)$$

Let us now treat the coefficients of $f_m\varphi$, ${}_\tau\psi$ and F:

$$\frac{2}{3}\left(\gamma^m{}_\sigma{}^\alpha\,\mathrm{I}_\alpha{}^\nu\,\delta^{\kappa\lambda} + \gamma^m{}_\sigma{}^\kappa\varepsilon^{\lambda\nu}\right){}_\nu d^\sigma_{\kappa\lambda} f_m\varphi$$

$$= \frac{1}{3}\left(\gamma^m{}_\sigma{}^\alpha\,\mathrm{I}_\alpha{}^\nu\,\delta^{\kappa\lambda} + \gamma^m{}_\sigma{}^\kappa\varepsilon^{\lambda\nu}\right)\left(\varepsilon_{\kappa\nu}\delta^\rho_\lambda + \varepsilon_{\lambda\nu}\delta^\rho_\kappa - \Gamma^p_{\kappa\lambda}\gamma_{p\nu}{}^\rho\right)$$

$$\cdot\,\gamma^s{}_\rho{}^\sigma\left(\mathrm{I}_s{}^t\,\omega^{LC}_t + 2g_S(\delta_\gamma\chi, \chi_s)\right)f_m\varphi$$

$$= \gamma^m{}_\sigma{}^\rho\gamma^s{}_\rho{}^\sigma\left(\mathrm{I}_s{}^t\,\omega^{LC}_t + 2g_S(\delta_\gamma\chi, \chi_s)\right)f_m\varphi$$

$$= 2\delta^{ms}\left(\mathrm{I}_s{}^t\,\omega^{LC}_t + 2g_S(\delta_\gamma\chi, \chi_s)\right)f_m\varphi$$

$$\chi_m{}^\mu\left(\frac{2}{3}\left(\gamma^m{}_\sigma{}^\alpha\,\mathrm{I}_\alpha{}^\nu\,\delta^{\kappa\lambda} + \gamma^m{}_\sigma{}^\kappa\varepsilon^{\lambda\nu}\right)\delta^\tau_\mu - \Gamma^m_{\mu\rho}\delta^{\rho\nu}\delta^{\kappa\lambda}\delta^\tau_\sigma\right){}_\nu d^\sigma_{\kappa\lambda}\,{}_\tau\psi$$

$$= \chi_m{}^\mu\left(\frac{2}{3}\left(\gamma^m{}_\sigma{}^\alpha\,\mathrm{I}_\alpha{}^\nu\,\delta^{\kappa\lambda} + \gamma^m{}_\sigma{}^\kappa\varepsilon^{\lambda\nu}\right)\delta^\tau_\mu - \Gamma^m_{\mu\rho}\delta^{\rho\nu}\delta^{\kappa\lambda}\delta^\tau_\sigma\right)$$

$$\cdot\,\frac{1}{2}\left(\varepsilon_{\kappa\nu}\delta^\rho_\lambda + \varepsilon_{\lambda\nu}\delta^\rho_\kappa - \Gamma^p_{\kappa\lambda}\gamma_{p\nu}{}^\rho\right)\gamma^s{}_\rho{}^\sigma\left(\mathrm{I}_s{}^t\,\omega^{LC}_t + 2g_S(\delta_\gamma\chi, \chi_s)\right){}_\tau\psi$$

$$= \chi_m{}^\mu\left(\gamma^m{}_\sigma{}^\rho\delta^\tau_\mu + \gamma^m{}_\mu{}^\rho\delta^\tau_\sigma\right)\gamma^s{}_\rho{}^\sigma\left(\mathrm{I}_s{}^t\,\omega^{LC}_t + 2g_S(\delta_\gamma\chi, \chi_s)\right){}_\tau\psi$$

$$= 2\left\langle\chi_m + (P\chi)_m, \psi\right\rangle\delta^{ms}\left(\mathrm{I}_s{}^t\,\omega^{LC}_t + 2g_S(\delta_\gamma\chi, \chi_s)\right)$$

$$\left(\frac{2}{3}\varepsilon^{\lambda\nu}\delta^{\kappa}_{\sigma}{}_{\nu}d^{\sigma}_{\kappa\lambda}+4\|\chi\|^2\right)F=\left(\frac{1}{3}\varepsilon^{\lambda\nu}\delta^{\kappa}_{\sigma}\left(\varepsilon_{\kappa\nu}\delta^{\rho}_{\lambda}+\varepsilon_{\lambda\nu}\delta^{\rho}_{\kappa}-\Gamma^{p}_{\kappa\lambda}\gamma_{p\nu}{}^{\rho}\right)\right.$$

$$\left.\cdot\,\gamma^{s}{}_{\rho}{}^{\sigma}\left(\mathrm{I}_s{}^t\,\omega_t^{LC}+2gs(\delta_\gamma\chi,\chi_s)\right)+4\|\chi\|^2\right)F$$

$$=\left(\frac{1}{3}\delta^{\rho}_{\sigma}\gamma^{s}{}_{\rho}{}^{\sigma}\left(\mathrm{I}_s{}^t\,\omega_t^{LC}+2gs\left(\delta_\gamma\chi,\chi_s\right)\right)+4\|\chi\|^2\right)F$$

$$=4\|\chi\|^2F$$

This shows the claim. □

13.7 Berezinian

In this section we calculate the Berezinian as well as the divergence for a Wess–Zumino pair.

Lemma 13.7.1 *Let y^a and $X^A=(x^a,\eta^\alpha)$ be Wess–Zumino coordinates for F_A and i. The Berezinian of the matrix of frame coefficients is given by*

$$\begin{aligned}\operatorname{Ber}F=\det f&+\eta^\mu\left(f_a{}^b\left({}_\mu F_c{}^d+\chi_c{}^\gamma\,{}_\mu F_\gamma{}^d\right)\varepsilon^{ac}\varepsilon_{bd}-(\det f)\,{}_\mu F_\delta{}^\delta\right)\\&+\eta^3\eta^4\left(-\det f\left({}_{34}F_\delta{}^\delta-\frac{1}{2}\varepsilon^{\mu\nu}\,{}_\mu F_\alpha{}^\beta\,{}_\nu F_\gamma{}^\delta\varepsilon^{\alpha\gamma}\varepsilon_{\beta\delta}+\varepsilon^{\mu\nu}\,{}_\mu F_\gamma{}^\gamma\,{}_\nu F_\delta{}^\delta\right)\right.\\&+\left(\varepsilon^{\mu\nu}\,{}_\mu F_\delta{}^\delta f_a{}^b\left({}_\nu F_c{}^d+\chi_c{}^\gamma\,{}_\nu F_\gamma{}^d\right)\right.\\&+f_a{}^b\left({}_{34}F_c{}^d-\varepsilon^{\mu\sigma}\,{}_\mu F_c{}^\gamma\,{}_\sigma F_\gamma{}^d-\varepsilon^{\nu\sigma}\chi_c{}^\gamma\,{}_\nu F_\gamma{}^\delta\,{}_\sigma F_\delta{}^d-\chi_c{}^\gamma\,{}_{34}F_\gamma{}^d\right)\\&\left.\left.-\frac{1}{2}\varepsilon^{\mu\nu}\left({}_\mu F_a{}^b+\chi_a{}^\gamma\,{}_\mu F_\gamma{}^b\right)\left({}_\nu F_c{}^d+\chi_c{}^\gamma\,{}_\nu F_\gamma{}^d\right)\right)\varepsilon^{ac}\varepsilon_{bd}\right)\end{aligned}$$

Here $\det f$ *denotes the determinant of the frame coefficients of* $f_a=f_a{}^b\partial_{y^b}$.

Proof By Definition 2.3.7 the Berezinian of the frame coefficients F is given by a quotient of determinants

$$\operatorname{Ber}F=\frac{\det A}{\det B},$$

where the matrix B coincides with the lower right block of F and A is given by $A_a{}^b=F_a{}^b-F_a{}^\gamma G_\gamma{}^\delta F_\delta{}^b$. Here G is the inverse of B. Let us first calculate the determinant of $B_\alpha{}^\beta = F_\alpha{}^\beta$

$$\det B = \frac{1}{2}\left(\delta_\alpha^\beta + \eta^\mu{}_\mu F_\alpha{}^\beta + \eta^3\eta^4{}_{34}F_\alpha{}^\beta\right)\left(\delta_\gamma^\delta + \eta^\nu{}_\nu F_\gamma{}^\delta + \eta^3\eta^4{}_{34}F_\gamma{}^\delta\right)\varepsilon^{\alpha\gamma}\varepsilon_{\beta\delta}$$
$$= 1 + \eta^\mu{}_\mu F_\delta{}^\delta + \eta^3\eta^4\left({}_{34}F_\delta{}^\delta - \frac{1}{2}\varepsilon^{\mu\nu}{}_\mu F_\alpha{}^\beta{}_\nu F_\gamma{}^\delta \varepsilon^{\alpha\gamma}\varepsilon_{\beta\delta}\right)$$

With the help of Eq. (13.1.3), the inverse can be calculated as

$$(\det B)^{-1} = 1 - \eta^\mu{}_\mu F_\delta{}^\delta$$
$$- \eta^3\eta^4\left({}_{34}F_\delta{}^\delta - \frac{1}{2}\varepsilon^{\mu\nu}{}_\mu F_\alpha{}^\beta{}_\nu F_\gamma{}^\delta\varepsilon^{\alpha\gamma}\varepsilon_{\beta\delta} + \varepsilon^{\mu\nu}{}_\mu F_\gamma{}^\gamma{}_\nu F_\delta{}^\delta\right).$$

Let us now turn to A. The matrix G is determined by

$$\delta_\alpha^\gamma = \left(\delta_\alpha^\beta + \eta^\mu{}_\mu F_\alpha{}^\beta + \eta^3\eta^4{}_{34}F_\alpha{}^\beta\right)\left(\delta_\beta^\gamma + \eta^\nu{}_\nu G_\beta{}^\gamma + \eta^3\eta^4{}_{34}G_\beta{}^\gamma\right)$$
$$= \delta_\alpha^\gamma + \eta^\mu\left({}_\mu G_\alpha{}^\gamma + {}_\mu F_\alpha{}^\gamma\right) + \eta^3\eta^4\left({}_{34}G_\alpha{}^\gamma - \varepsilon^{\mu\nu}{}_\mu F_\alpha{}^\beta{}_\nu G_\beta{}^\gamma + {}_{34}F_\alpha{}^\gamma\right).$$

We conclude

$$G_\beta{}^\gamma = \delta_\beta^\gamma - \eta^\nu{}_\nu F_\beta{}^\gamma - \eta^3\eta^4\left({}_{34}F_\beta{}^\gamma + \varepsilon^{\mu\nu}{}_\mu F_\beta{}^\sigma{}_\nu F_\sigma{}^\gamma\right)$$

and hence, the components of A are given by

$$A_a{}^b = F_a{}^b - F_a{}^\gamma\left(F^{-1}\right)_\gamma{}^\delta F_\delta{}^b$$
$$= f_a{}^b + \eta^\nu{}_\nu F_a{}^b + \eta^3\eta^4{}_{34}F_a{}^b$$
$$- \left(\chi_a{}^\gamma + \eta^\mu{}_\mu F_a{}^\gamma + \ldots\right)\left(\delta_\gamma^\delta - \eta^\nu{}_\nu F_\gamma{}^\delta + \ldots\right)\left(\eta^\sigma{}_\sigma F_\delta{}^b + \eta^3\eta^4{}_{34}F_\delta{}^b\right)$$
$$= f_a{}^b + \eta^\mu\left({}_\mu F_a{}^b + \chi_a{}^\gamma{}_\mu F_\gamma{}^b\right)$$
$$+ \eta^3\eta^4\left({}_{34}F_a{}^b - \varepsilon^{\mu\sigma}{}_\mu F_a{}^\gamma{}_\sigma F_\gamma{}^b - \varepsilon^{\nu\sigma}\chi_a{}^\gamma{}_\nu F_\gamma{}^\delta{}_\sigma F_\delta{}^b - \chi_a{}^\gamma{}_{34}F_\gamma{}^b\right).$$

The determinant of A can be calculated as follows

$$\det A = \frac{1}{2}\Big(f_a{}^b + \eta^\mu\left({}_\mu F_a{}^b + \chi_a{}^\gamma{}_\mu F_\gamma{}^b\right) + \eta^3\eta^4\Big({}_{34}F_a{}^b - \varepsilon^{\mu\sigma}{}_\mu F_a{}^\gamma{}_\sigma F_\gamma{}^b$$
$$- \varepsilon^{\nu\sigma}\chi_a{}^\gamma{}_\nu F_\gamma{}^\delta{}_\sigma F_\delta{}^b - \chi_a{}^\gamma{}_{34}F_\gamma{}^b\Big)\Big)\cdot\Big(f_c{}^d + \eta^\nu\left({}_\nu F_c{}^d + \chi_c{}^\gamma{}_\nu F_\gamma{}^d\right)$$
$$+ \eta^3\eta^4\left({}_{34}F_c{}^d - \varepsilon^{\mu\sigma}{}_\mu F_c{}^\gamma{}_\sigma F_\gamma{}^d - \varepsilon^{\nu\sigma}\chi_c{}^\gamma{}_\nu F_\gamma{}^\delta{}_\sigma F_\delta{}^d - \chi_c{}^\gamma{}_{34}F_\gamma{}^d\right)\Big)$$
$$\cdot\,\varepsilon^{ac}\varepsilon_{bd}$$

$$
\begin{aligned}
&= \det f + \eta^\nu f_a{}^b \left({}_\nu F_c{}^d + \chi_c{}^\gamma\, {}_\nu F_\gamma{}^d\right) \varepsilon^{ac}\varepsilon_{bd} + \eta^3\eta^4 \\
&\cdot \Big(f_a{}^b \left({}_{34}F_c{}^d - \varepsilon^{\mu\sigma}\, {}_\mu F_c{}^\gamma\, {}_\sigma F_\gamma{}^d - \varepsilon^{\nu\sigma} \chi_c{}^\gamma\, {}_\nu F_\gamma{}^\delta\, {}_\sigma F_\delta{}^d - \chi_c{}^\gamma\, {}_{34}F_\gamma{}^d\right) \\
&- \frac{1}{2}\varepsilon^{\mu\nu} \left({}_\mu F_a{}^b + \chi_a{}^\gamma\, {}_\mu F_\gamma{}^b\right)\left({}_\nu F_c{}^d + \chi_c{}^\gamma\, {}_\nu F_\gamma{}^d\right)\Big) \varepsilon^{ac}\varepsilon_{bd}.
\end{aligned}
$$

Finally, we obtain Ber F as the product of the expressions for $\det A$ and $(\det B)^{-1}$.

$$
\begin{aligned}
\operatorname{Ber} F = &\Big(\det f + \eta^\nu f_a{}^b \left({}_\nu F_c{}^d + \chi_c{}^\gamma\, {}_\nu F_\gamma{}^d\right) \varepsilon^{ac}\varepsilon_{bd} + \eta^3\eta^4 \\
&\cdot \Big(f_a{}^b \left({}_{34}F_c{}^d - \varepsilon^{\mu\sigma}\, {}_\mu F_c{}^\gamma\, {}_\sigma F_\gamma{}^d - \varepsilon^{\nu\sigma} \chi_c{}^\gamma\, {}_\nu F_\gamma{}^\delta\, {}_\sigma F_\delta{}^d - \chi_c{}^\gamma\, {}_{34}F_\gamma{}^d\right) \\
&- \frac{1}{2}\varepsilon^{\mu\nu} \left({}_\mu F_a{}^b + \chi_a{}^\gamma\, {}_\mu F_\gamma{}^b\right)\left({}_\nu F_c{}^d + \chi_c{}^\gamma\, {}_\nu F_\gamma{}^d\right)\Big) \varepsilon^{ac}\varepsilon_{bd}\Big) \\
&\cdot \Big(1 - \eta^\mu\, {}_\mu F_\delta{}^\delta \\
&- \eta^3\eta^4 \left({}_{34}F_\delta{}^\delta - \frac{1}{2}\varepsilon^{\mu\nu}\, {}_\mu F_\alpha{}^\beta\, {}_\nu F_\gamma{}^\delta \varepsilon^{\alpha\gamma}\varepsilon_{\beta\delta} + \varepsilon^{\mu\nu}\, {}_\mu F_\gamma{}^\gamma\, {}_\nu F_\delta{}^\delta\right)\Big) \\
= &\det f + \eta^\mu \left(f_a{}^b \left({}_\mu F_c{}^d + \chi_c{}^\gamma\, {}_\mu F_\gamma{}^d\right) \varepsilon^{ac}\varepsilon_{bd} - (\det f)\, {}_\mu F_\delta{}^\delta\right) + \eta^3\eta^4 \\
&\cdot \Big(-\det f \left({}_{34}F_\delta{}^\delta - \frac{1}{2}\varepsilon^{\mu\nu}\, {}_\mu F_\alpha{}^\beta\, {}_\nu F_\gamma{}^\delta \varepsilon^{\alpha\gamma}\varepsilon_{\beta\delta} + \varepsilon^{\mu\nu}\, {}_\mu F_\gamma{}^\gamma\, {}_\nu F_\delta{}^\delta\right) \\
&+ \Big(\varepsilon^{\mu\nu}\, {}_\mu F_\delta{}^\delta f_a{}^b \left({}_\nu F_c{}^d + \chi_c{}^\gamma\, {}_\nu F_\gamma{}^d\right) \\
&+ f_a{}^b \left({}_{34}F_c{}^d - \varepsilon^{\mu\sigma}\, {}_\mu F_c{}^\gamma\, {}_\sigma F_\gamma{}^d - \varepsilon^{\nu\sigma} \chi_c{}^\gamma\, {}_\nu F_\gamma{}^\delta\, {}_\sigma F_\delta{}^d - \chi_c{}^\gamma\, {}_{34}F_\gamma{}^d\right) \\
&- \frac{1}{2}\varepsilon^{\mu\nu} \left({}_\mu F_a{}^b + \chi_a{}^\gamma\, {}_\mu F_\gamma{}^b\right)\left({}_\nu F_c{}^d + \chi_c{}^\gamma\, {}_\nu F_\gamma{}^d\right)\Big) \varepsilon^{ac}\varepsilon_{bd}\Big)
\end{aligned}
$$

This shows the claim. □

Lemma 13.7.2 *Let F_A be a Wess–Zumino frame and y^a and $X^A = (x^a, \eta^\alpha)$ be Wess–Zumino coordinates for F_A and the embedding i. Then the Berezinian of the frame coefficients is given by*

$$
\operatorname{Ber} F = \det f \left(1 + \eta^\mu \Gamma^k_{\mu\kappa} \chi_k{}^\kappa - \eta^3\eta^4 \|\chi\|^2\right).
$$

Proof We proceed by inserting the formulas from Lemma 11.3.4 in the result for the Berezinian obtained in Lemma 13.7.1. In order to keep the calculation simple we proceed order by order. The zero order term does not simplify further. For the first order it holds that

$$\begin{aligned}
{}_{\mu}(\mathrm{Ber}\,F) &= f_a{}^b\left({}_{\mu}F_c{}^d + \chi_c{}^\gamma\,{}_{\mu}F_\gamma{}^d\right)\varepsilon^{ac}\varepsilon_{bd} - (\det f)\,{}_{\mu}F_\delta{}^\delta\\
&= f_a{}^b\left(\Gamma^k_{\mu\nu}\chi_c{}^\nu f_k{}^d + \chi_c{}^\gamma\,\Gamma^k_{\mu\gamma} f_k{}^d\right)\varepsilon^{ac}\varepsilon_{bd} - (\det f)\,\Gamma^k_{\mu\delta}\chi_k{}^\delta\\
&= (\det f)\,\Gamma^k_{\mu\kappa}\chi_k{}^\kappa.
\end{aligned}$$

Here and in the following we use $2f_a{}^b f_k{}^d\varepsilon_{bd} = \varepsilon_{ak}\det f$. The second order terms are calculated as follows:

$$\begin{aligned}
{}_{34}(\mathrm{Ber}\,F) = &-\det f\left({}_{34}F_\delta{}^\delta - \frac{1}{2}\varepsilon^{\mu\nu}\,{}_{\mu}F_\alpha{}^\beta\,{}_{\nu}F_\gamma{}^\delta\varepsilon^{\alpha\gamma}\varepsilon_{\beta\delta} + \varepsilon^{\mu\nu}\,{}_{\mu}F_\gamma{}^\gamma\,{}_{\nu}F_\delta{}^\delta\right)\\
&+\Bigg(\varepsilon^{\mu\nu}\,{}_{\mu}F_\delta{}^\delta f_a{}^b\left({}_{\nu}F_c{}^d + \chi_c{}^\gamma\,{}_{\nu}F_\gamma{}^d\right)\\
&+ f_a{}^b\left({}_{34}F_c{}^d - \varepsilon^{\mu\sigma}\,{}_{\mu}F_c{}^\gamma\,{}_{\sigma}F_\gamma{}^d - \varepsilon^{\nu\sigma}\chi_c{}^\gamma\,{}_{\nu}F_\gamma{}^\delta\,{}_{\sigma}F_\delta{}^d - \chi_c{}^\gamma\,{}_{34}F_\gamma{}^d\right)\\
&- \frac{1}{2}\varepsilon^{\mu\nu}\left({}_{\mu}F_a{}^b + \chi_a{}^\gamma\,{}_{\mu}F_\gamma{}^b\right)\left({}_{\nu}F_c{}^d + \chi_c{}^\gamma\,{}_{\nu}F_\gamma{}^d\right)\Bigg)\varepsilon^{ac}\varepsilon_{bd}\\
= &-\mathrm{dct}\, f\left(-2\|Q\chi\|^2 - \frac{1}{2}\varepsilon^{\mu\nu}\Gamma^s_{\mu\alpha}\chi_s{}^\beta\,\Gamma^t_{\nu\gamma}\chi_t{}^\delta\varepsilon^{\alpha\gamma}\varepsilon_{\beta\delta}\right.\\
&\left.+\,\varepsilon^{\mu\nu}\Gamma^s_{\mu\gamma}\chi_s{}^\gamma\,\Gamma^t_{\nu\delta}\chi_t{}^\delta\right) + \Bigg(2\varepsilon^{\mu\nu}\Gamma^s_{\mu\delta}\chi_s{}^\delta f_a{}^b\chi_c{}^\gamma\,\Gamma^t_{\nu\gamma} f_t{}^d\\
&+ f_a{}^b\left(-\left(\|Q\chi\|^2\delta^s_c + 2g_S\left((P\chi)_c, (Q\chi)_t\right)\delta^{ts}\right) f_s{}^d\right.\\
&+\gamma^k_\gamma{}^\mu\left(\Gamma^s_{\mu\tau}\chi_c{}^\tau\chi_s{}^\gamma + \frac{1}{4}\gamma_{c\mu}{}^\lambda\gamma^t_\lambda{}^\gamma\left(\mathrm{I}_t{}^s\,\omega^{LC}_s + g_S\left(\delta_\gamma\chi, \chi_t\right)\right)\right) f_k{}^d\\
&\left.-\,\varepsilon^{\nu\sigma}\chi_c{}^\gamma\,\Gamma^s_{\nu\gamma}\chi_s{}^\delta\,\Gamma^t_{\sigma\delta} f_t{}^d - \chi_c{}^\gamma\gamma^t_\gamma{}^\lambda\,\Gamma^s_{\lambda\mu} f_s{}^d\chi_t{}^\mu\right)\\
&- 2\varepsilon^{\mu\nu}\chi_a{}^\gamma\,\Gamma^s_{\mu\gamma} f_s{}^b\chi_c{}^\delta\,\Gamma^t_{\nu\delta} f_t{}^d\Bigg)\varepsilon^{ac}\varepsilon_{bd}\\
= &\det f\left(\left(2\|Q\chi\|^2 + \|\chi\|^2 - 2\|P\chi\|^2\right) + \left(4\|P\chi\|^2 - 2\|Q\chi\|^2\right.\right.\\
&\left.\left.-\,2\|Q\chi\|^2 - 2\|Q\chi\|^2 - 2\|Q\chi\|^2\right) - 4\left(\|P\chi\|^2 - \|Q\chi\|^2\right)\right)\\
= &-\det f\,\|\chi\|^2
\end{aligned}$$

□

We apply Eq. (13.1.3) to the result of the preceding Lemma 13.7.2 to obtain the Berezinian of the inverse E of the matrix of frame coefficients of a Wess–Zumino pair:

$$\begin{aligned}\operatorname{Ber} E &= (\det f)^{-1}\left(1 - \eta^\mu \Gamma^k_{\mu\delta}\chi_k{}^\delta - \eta^3\eta^4\left(\varepsilon^{\mu\nu}\Gamma^k_{\mu\delta}\chi_k{}^\delta \Gamma^l_{\nu\gamma}\chi_l{}^\gamma - \|\chi\|^2\right)\right)\\ &= (\det f)^{-1}\left(1 - \eta^\mu \Gamma^k_{\mu\delta}\chi_k{}^\delta + \eta^3\eta^4\left(\|Q\chi\|^2 - \|P\chi\|^2\right)\right)\end{aligned} \tag{13.7.3}$$

Lemma 13.7.4 *Let $X^A = (x^a, \eta^\alpha)$ be Wess–Zumino coordinates for a Wess–Zumino frame F_A and the embedding $i\colon |M| \to M$. Then*

$$\begin{aligned}\operatorname{div}_{[F^\bullet]} F_\mu &= \frac{1}{2}\eta^\nu \Gamma^s_{\nu\mu}\left(\mathrm{I}_s{}^t\,\omega^{LC}_t + 2g_S\left(\delta_\gamma\chi, \chi_s\right)\right)\\ &\quad + \eta^3\eta^4\left(\mathrm{I}_s{}^t\,\omega^{LC}_t \delta^{sk}\left(2\left(Q\chi\right)_k{}^\kappa + \left(P\chi\right)_k{}^\kappa\right)\varepsilon_{\kappa\mu} + 2\varepsilon^{kl}\delta_{\mu\kappa}\left(f_l\chi_k{}^\kappa\right)\right).\end{aligned}$$

Proof Note that even though the result looks similar we cannot use Lemma 10.2.13 directly. The frame F_μ here might differ from a superconformal frame by a rescaling and change of splitting, as in Lemma 10.2.13, but also by a U(1)-transformation. Furthermore, the Wess–Zumino coordinates used here differ from superconformal coordinates.

Using the rules given in Proposition 8.1.2, we obtain as in Lemma 10.2.13:

$$\left(\operatorname{div}_{[F^\bullet]} F_\mu\right)[F^\bullet] = L_{F_\mu}[F^\bullet] = \left(\sum_B \partial_{X^B} F_\mu{}^B - \frac{F_\mu \operatorname{Ber} F}{\operatorname{Ber} F}\right)[F^\bullet]$$

We now develop this expression in orders of the odd coordinate η using Lemmas 11.3.4 and 13.7.2:

$${}_0\left(\operatorname{div}_{[F^\bullet]} F_\mu\right) = i^*\left(\sum_B \partial_{X^B} F_\mu{}^B - \frac{F_\mu \operatorname{Ber} F}{\operatorname{Ber} F}\right) = {}_\beta F_\mu{}^\beta - \Gamma^k_{\mu\delta}\chi_k{}^\delta = 0$$

The first order term is given by

$$\begin{aligned}{}_\nu\left(\operatorname{div}_{[F^\bullet]} F_\mu\right) &= i^* F_\nu\left(\sum_B \partial_{X^B} F_\mu{}^B - \frac{F_\mu \operatorname{Ber} F}{\operatorname{Ber} F}\right)\\ &= \partial_{x^b}\left({}_\nu F_\mu{}^b\right) - \varepsilon_{\nu\beta}\,{}_{34}F_\mu{}^\beta - \frac{i^* F_\nu F_\mu \operatorname{Ber} F}{{}_0(\operatorname{Ber} F)} - \frac{{}_\mu(\operatorname{Ber} F)\,{}_\nu(\operatorname{Ber} F)}{{}_0(\operatorname{Ber} F)^2}\\ &= \partial_{x^b}\left({}_\nu F_\mu{}^b\right) - \varepsilon_{\nu\beta}\,{}_{34}F_\mu{}^\beta - \frac{\Gamma^s_{\nu\mu}\left(f_s\,{}_0(\operatorname{Ber} F) + \chi_s{}^\sigma\,{}_\sigma(\operatorname{Ber} F)\right)}{{}_0(\operatorname{Ber} F)}\\ &\quad + \varepsilon_{\nu\mu}\frac{{}_{34}(\operatorname{Ber} F)}{{}_0(\operatorname{Ber} F)} - \frac{{}_\mu(\operatorname{Ber} F)\,{}_\nu(\operatorname{Ber} F)}{{}_0(\operatorname{Ber} F)^2}\end{aligned}$$

$$= \Gamma^s_{\nu\mu}\left(\partial_{x^b} {f_s}^{b} - \frac{f_s \det f}{\det f}\right) + \varepsilon_{\nu\beta}\left(\|Q\chi\|^2 \delta^\beta_\mu + \frac{1}{2}{\gamma^t}_\mu{}^\sigma\, {\mathrm{I}_\sigma}^\beta\, \omega^{LC}_t\right)$$
$$- \Gamma^s_{\mu\nu}{\chi_s}^\sigma \Gamma^t_{\sigma\tau}{\chi_t}^\tau - \varepsilon_{\nu\mu}\|\chi\|^2 - \Gamma^t_{\mu\tau}{\chi_t}^\tau \Gamma^s_{\nu\sigma}{\chi_s}^\sigma$$
$$= \frac{1}{2}\Gamma^s_{\mu\nu}\left({\mathrm{I}_s}^t\, \omega^{LC}_t + 2gs\left(\delta_\gamma\chi, \chi_s\right)\right)$$

In the last step we used the fact that the divergence of a f_s can be calculated with the help of the Levi-Civita covariant derivative:

$$\partial_{x^b} {f_s}^b - \frac{f_s \det f}{\det f} = \operatorname{div} f_s = \left(\nabla_{f_b} f_s\right)^b = \omega^{LC}_b\, {\mathrm{I}_s}^b,$$

as well as Eq. (13.4.7).

The term of order two is determined as follows:

$$_{34}\left(\operatorname{div}_{[F\bullet]} F_\mu\right) = -\frac{1}{2} i^* \varepsilon^{\alpha\beta} F_\alpha F_\beta \left(\sum_B \partial_{X^B} {F_\mu}^B - \frac{F_\mu \operatorname{Ber} F}{\operatorname{Ber} F}\right)$$
$$= \left(\partial_{x^b}\, {}_{34}{F_\mu}^b\right) + \varepsilon^{\alpha\beta} \frac{1}{2} i^* F_\alpha \left(\frac{F_\beta F_\mu \operatorname{Ber} F}{\operatorname{Ber} F} + \frac{\left(F_\mu \operatorname{Ber} F\right)\left(F_\beta \operatorname{Ber} F\right)}{(\operatorname{Ber} F)^2}\right)$$
$$= \left(\partial_{x^b}\, {}_{34}{F_\mu}^b\right) + \varepsilon^{\alpha\beta} \frac{1}{2} i^* \left(\frac{F_\alpha F_\beta F_\mu \operatorname{Ber} F}{\operatorname{Ber} F} + \frac{2\left(F_\alpha F_\mu \operatorname{Ber} F\right)\left(F_\beta \operatorname{Ber} F\right)}{(\operatorname{Ber} F)^2}\right.$$
$$\left. - \frac{\left(F_\mu \operatorname{Ber} F\right)\left(F_\alpha F_\beta \operatorname{Ber} F\right)}{(\operatorname{Ber} F)^2} + 2\frac{\left(F_\mu \operatorname{Ber} F\right)\left(F_\alpha \operatorname{Ber} F\right)\left(F_\beta \operatorname{Ber} F\right)}{(\operatorname{Ber} F)^3}\right)$$
$$= \left(\partial_{x^b}\, {}_{34}{F_\mu}^b\right) + \varepsilon^{\alpha\beta} \frac{1}{2} i^* \left(-\frac{\left([F_\mu, F_\alpha] F_\beta - \frac{2}{3}[[F_\mu, F_\alpha], F_\beta]\right) \operatorname{Ber} F}{\operatorname{Ber} F}\right.$$
$$\left. + \frac{\left([F_\alpha, F_\mu] \operatorname{Ber} F\right)\left(F_\beta \operatorname{Ber} F\right)}{(\operatorname{Ber} F)^2} + 2\frac{\left(F_\mu \operatorname{Ber} F\right)\left(F_\alpha \operatorname{Ber} F\right)\left(F_\beta \operatorname{Ber} F\right)}{(\operatorname{Ber} F)^3}\right)$$
$$= \left(\partial_{x^b}\, {}_{34}{F_\mu}^b\right) - (\det f)^{-1}\left({\gamma^k}_\mu{}^\beta\, f_{k\ \beta}(\operatorname{Ber} F)\right.$$
$$+ {\chi_k}^\kappa {\gamma^k}_\mu{}^\beta \left(\Gamma^l_{\kappa\beta}\left(\left(f_l \det f\right) + {\chi_l}^\lambda{}_\lambda(\operatorname{Ber} F)\right) - \varepsilon_{\kappa\beta}\ {}_{34}(\operatorname{Ber} F)\right)$$
$$\left. - \frac{1}{2}{\gamma^s}_\mu{}^\lambda\left({\mathrm{I}_s}^t\, \omega^{LC}_t + 2gs\left(\delta_\gamma\chi, \chi_s\right)\right)_\lambda(\operatorname{Ber} F)\right)$$
$$+ {\gamma^k}_\mu{}^\beta (\det f)^{-2}\left(\left(f_k \det f\right) + \left({\chi_k}^\kappa{}_\kappa \operatorname{Ber} F\right)\right)_\beta(\operatorname{Ber} F)$$
$$+ (\det f)^{-3}\ {}_\mu(\operatorname{Ber} F)\varepsilon^{\alpha\beta}\ {}_\alpha(\operatorname{Ber} F)\ {}_\beta(\operatorname{Ber} F)$$

$$
\begin{aligned}
&= \gamma^{k}_{\mu}{}^{\lambda}\Gamma^{l}_{\lambda\kappa}\partial_{x^b}\left(f_l{}^b\chi_k{}^\kappa\right) - \gamma^{k}_{\mu}{}^{\beta}(\det f)^{-1}f_k\left(\det f\,\Gamma^{l}_{\beta\lambda}\chi_l{}^\lambda\right)\\
&\quad - \chi_k{}^\kappa\left(\gamma^{k}_{\mu}{}^{\beta}\Gamma^{l}_{\kappa\beta}\left(\frac{f_l(\det f)}{(\det f)} + \chi_l{}^\lambda\Gamma^{s}_{\lambda\sigma}\chi_s{}^\sigma\right) + \Gamma^{k}_{\mu\kappa}\|\chi\|^2\right)\\
&\quad + \frac{1}{2}\left(\mathrm{I}_s{}^t\,\omega^{LC}_t + 2gs\left(\delta_\gamma\chi,\chi_s\right)\right)\gamma^{s}_{\mu}{}^{\lambda}\Gamma^{k}_{\lambda\kappa}\chi_k{}^\kappa\\
&\quad + \gamma^{k}_{\mu}{}^{\beta}\left((\det f)^{-1}\left(f_k\det f\right) + \left(\chi_k{}^\kappa\Gamma^{l}_{\kappa\lambda}\chi_l{}^\lambda\right)\right)\Gamma^{s}_{\beta\sigma}\chi_s{}^\sigma\\
&\quad + \Gamma^{r}_{\mu\rho}\chi_r{}^\rho\chi_s{}^\sigma\Gamma^{s}_{\sigma\alpha}\varepsilon^{\alpha\beta}\Gamma^{t}_{\beta\tau}\chi_t{}^\tau\\
&= \left(\left(\partial_{x^b}f_l{}^b\right) - \frac{f_l(\det f)}{(\det f)} + gs\left(\delta_\gamma,\chi_l\right)\right)\chi_n{}^\nu\gamma^{l}_{\nu}{}^{\beta}\gamma^{n}_{\beta}{}^{\kappa}\varepsilon_{\kappa\mu}\\
&\quad + \frac{1}{2}\mathrm{I}_s{}^t\,\omega^{LC}_t\chi_n{}^\nu\gamma^{n}_{\nu}{}^{\beta}\gamma^{s}_{\beta}{}^{\kappa}\varepsilon_{\kappa\mu} + \left(\gamma^{k}_{\mu}{}^{\beta}\Gamma^{l}_{\lambda\kappa} - \gamma^{l}_{\mu}{}^{\beta}\Gamma^{k}_{\lambda\kappa}\right)\left(f_l\chi_k{}^\kappa\right)\\
&\quad + \left(\|\chi\|^2 - 2\|P\chi\|^2\right)\chi_s{}^\sigma\gamma^{s}_{\sigma}{}^{\kappa}\varepsilon_{\kappa\mu}\\
&= \mathrm{I}_s{}^t\,\omega^{LC}_t\delta^{sk}\left(2\left(Q\chi\right)_k{}^\kappa + \left(P\chi\right)_k{}^\kappa\right)\varepsilon_{\kappa\mu} + 2\varepsilon^{kl}\delta_{\mu\kappa}\left(f_l\chi_k{}^\kappa\right)
\end{aligned}
$$

Here in the last step we need the following two identities for terms of third order in the gravitino:

$$
\begin{aligned}
2gs\left(\delta_\gamma\chi,\chi_s\right)\delta^{sn}\left(Q\chi\right)_n{}^\nu &= -\|Q\chi\|^2\left(\delta_\gamma\chi\right)^\nu,\\
2gs\left(\delta_\gamma\chi,\chi_s\right)\delta^{sn}\left(P\chi\right)_n{}^\nu &= -\|P\chi\|^2\left(\delta_\gamma\chi\right)^\nu = 0.
\end{aligned}
\tag{13.7.5}
$$

The following is a consequence of Eq. (13.4.7):

$$
\begin{aligned}
2gs\left(\delta_\gamma\chi,\chi_s\right)\delta^{sn}\left(Q\chi\right)_n{}^\nu\varepsilon_{\nu\mu} &= -\gamma^{t}_{\mu}{}^{\sigma}\Gamma^{l}_{\sigma\tau}\chi_t{}^\tau\chi_l{}^\lambda\Gamma^{k}_{\lambda\kappa}\chi_k{}^\kappa\\
&= \|Q\chi\|^2\Gamma^{k}_{\mu\kappa}\chi_k{}^\kappa + gs\left(\delta_\gamma\chi,\chi_s\right)\gamma^{s}_{\mu}{}^{\lambda}\Gamma^{k}_{\lambda\kappa}\chi_k{}^\kappa\\
&= -\|Q\chi\|^2\left(\delta_\gamma\chi\right)^\nu\varepsilon_{\nu\mu} + 2gs\left(\delta_\gamma\chi,\chi_s\right)\delta^{sn}\left(P\chi\right)_n{}^\nu\varepsilon_{\nu\mu}.
\end{aligned}
$$

Similarly to Eq. (13.4.7) we have

$$
\gamma^{l}_{\mu}{}^{\sigma}\Gamma^{t}_{\sigma\tau}\chi_t{}^\tau\chi_l{}^\lambda = -\|P\chi\|^2\delta^\lambda_\mu + gs\left(\delta_\gamma\chi,\chi_s\right)\gamma^{s}_{\mu}{}^{\lambda}
$$

and

$$
\begin{aligned}
2gs\left(\delta_\gamma\chi,\chi_s\right)\delta^{sn}\left(P\chi\right)_n{}^\nu\varepsilon_{\nu\mu} &= -\gamma^{l}_{\mu}{}^{\sigma}\Gamma^{t}_{\sigma\tau}\chi_t{}^\tau\chi_l{}^\lambda\Gamma^{k}_{\lambda\kappa}\chi_k{}^\kappa\\
&= -\|P\chi\|^2\left(\delta_\gamma\chi\right)^\nu\varepsilon_{\nu\mu} - 2gs\left(\delta_\gamma\chi,\chi_s\right)\delta^{sn}\left(P\chi\right)_n{}^\nu\varepsilon_{\nu\mu}.
\end{aligned}
$$

As the coefficients $\chi_s{}^\sigma$ of the gravitino are odd, one obtains $\|P\chi\|^2\delta_\gamma\chi = 0$ by expanding. Equation (13.7.5) follow. This completes the proof. □

13.8 Supersymmetry

In this section we will prove the Lemma 11.4.14:

Lemma 11.4.14 *Let F_A be the Wess–Zumino frame determined by f_a, s_α and χ and m the superconformal metric determined by F_A. There exists a $s \in \mathcal{N}_{M/|M|}{}^{\vee} \otimes S$ such that*

$$\delta_q g[m] = \operatorname{susy}_q g,$$
$$\left(\delta_q \chi[m]\right)(X) = \left(\operatorname{susy}_q \chi\right)(X) - \gamma(X)\langle q, s\rangle .$$

Proof The proof proceeds by calculating the standard connection ∇ for the metric m induced by the Wess–Zumino frame F_A and its torsion. Since for Wess–Zumino pairs all arising quantities can be expressed in terms of the frames f_a, s_α and the gravitino χ, the formulas for the variation of metric and gravitino obtained in Proposition 11.4.9 can be made explicit.

First, recall that the standard connection ∇ can be written in the following form

$$\nabla_{F_A} F_b = 2\omega_A \operatorname{I} F_b, \qquad \nabla_{F_A} F_\beta = \omega_A \operatorname{I} F_\beta .$$

By Proposition 10.2.3, we have $\omega_\alpha = -\frac{1}{3} d_{\alpha\beta}^{\gamma} \operatorname{I}_\gamma{}^\beta$ and hence $i^*\omega_\alpha = 0$. Furthermore, write $q = q^\mu s_\mu$ and $X = X^a f_a$ For thc variation of the metric, we have to calculate

$$\begin{aligned} i^* p_{\mathcal{D}^\perp} i^* T(q, di X) &= q^\mu X^a i^* p_{\mathcal{D}^\perp} i^* T(i^* F_\mu, i^* F_a - \chi_a{}^\alpha i^* F_\alpha) \\ &= q^\mu X^a \left(i^* T_{\mu a}{}^c + \chi_a{}^\alpha i^* T_{\mu\alpha}{}^c\right) i^* F_c . \end{aligned}$$

For the torsion tensor of the standard connection to m we know that $i^* T_{\mu\alpha}{}^c = -2\Gamma_{\mu\alpha}^c$ and that $i^* T_{\mu a}{}^c = t_\mu \operatorname{I}_a{}^c$ for some t_μ, see Propositions 10.1.5 and 10.2.2 using $i^* d_{\mu a}^c = 0$. Consequently,

$$\begin{aligned} i^* p_{\mathcal{D}^\perp} i^* T(q, di X) &= q^\mu t_\mu X^a \operatorname{I}_a{}^c i^* F_c - 2 q^\mu X^a \chi_a{}^\alpha \Gamma_{\alpha\mu}^c i^* F_c \\ &= \langle q, t\rangle X^a \operatorname{I}_a{}^c i^* F_c + 2 g_S(q, \gamma^c \chi(X)) i^* F_c \end{aligned}$$

and

$$\begin{aligned} \left(\delta_q g[m]\right)(X, Y) &= i^* m\left(i^* p_{\mathcal{D}^\perp} i^* T(q, di X), i^* p_{\mathcal{D}^\perp} di Y\right) \\ &\quad + i^* m\left(i^* p_{\mathcal{D}^\perp} di X, i^* p_{\mathcal{D}^\perp} i^* T(q, di Y)\right) \\ &= 2\left(g(g_S(q, \gamma^c \chi(X)) f_c, Y) + g(X, g_S(q, \gamma^c \chi(Y) f_c))\right) \\ &= 2 g_S(q, \gamma(X)\chi(Y) + \gamma(Y)\chi(X)). \end{aligned}$$

For the time-derivative of b_t it follows:

$$\left.\frac{d}{dt}\right|_{t=0} b_t f_a = -g_S(q, \gamma_a \chi(f_b) + \gamma_b \chi(f_a)) \delta^{bc} f_c .$$

In order to calculate $\chi\left(\frac{d}{dt}b_t X\right)$ we need the following Fierz-type equality:

$$\begin{aligned}
\chi_k{}^\kappa \chi_l{}^\lambda &= \frac{1}{2}\chi_k{}^\sigma \chi_l{}^\tau \left(\delta^{\kappa\lambda}\delta_{\sigma\tau} + \varepsilon^{\kappa\pi}\gamma_{l\pi}{}^\lambda \Gamma^l_{\sigma\tau} + \varepsilon^{\kappa\lambda}\varepsilon_{\sigma\tau}\right) \\
&= \frac{1}{4}\chi_s{}^\sigma \chi_t{}^\tau \left(\varepsilon_{kl}\delta^{\kappa\lambda}\varepsilon^{st}\delta_{\sigma\tau} + \varepsilon_{kl}\varepsilon^{\kappa\pi}\gamma_{m\pi}{}^\lambda \varepsilon^{st}\Gamma^m_{\sigma\tau} + 2\delta^s_k\delta^t_l\varepsilon^{\kappa\lambda}\varepsilon_{\sigma\tau}\right) \\
&= \frac{1}{4}\varepsilon_{kl}\delta^{\kappa\lambda}\left(\|P\chi\|^2 - \|Q\chi\|^2\right) + \frac{1}{2}\varepsilon_{kl}\varepsilon^{\kappa\pi}\gamma^m{}_\pi{}^\lambda \mathrm{I}_m{}^n g_S\left(\delta_\gamma\chi, \chi_n\right) \\
&\quad - \frac{1}{2}\varepsilon^{\kappa\lambda} g_S\left(\chi_k, \chi_l\right).
\end{aligned} \tag{13.8.1}$$

Thus,

$$\begin{aligned}
\chi\left(\left.\frac{d}{dt}\right|_{t=0} b_t X\right) &= -X^a g_S(q, \gamma_a\chi(f_b) + \gamma_b\chi(f_a))\delta^{bc}\chi(f_c) \\
&= X^a q^\mu \left(\Gamma^d_{\mu\kappa}\delta_{ad}\delta^{kl} + \Gamma^l_{\mu\kappa}\delta^k_a\right)\chi_k{}^\kappa\chi_l{}^\lambda s_\lambda \\
&= X^a q^\mu \left(\frac{1}{4}\varepsilon_{al}\Gamma^l_{\mu\kappa}\delta^{\kappa\lambda}\left(\|P\chi\|^2 - \|Q\chi\|^2\right)\right. \\
&\quad + \frac{1}{2}\varepsilon_{al}\gamma^l_\mu{}^\pi \gamma^m{}_\pi{}^\lambda \mathrm{I}_m{}^n g_S\left(\delta_\gamma\chi, \chi_n\right) \\
&\quad \left. - \frac{1}{2}\left(\gamma_{a\mu}{}^\lambda\delta^{kl} + \gamma^l_\mu{}^\lambda\delta^k_a\right) g_S\left(\chi_k, \chi_l\right)\right) s_\lambda \\
&= \frac{1}{4}\left(\|P\chi\|^2 - \|Q\chi\|^2\right)\gamma(X)q - \frac{1}{2}g_S\left(\delta_\gamma\chi, \chi_m\right)\gamma^m\gamma(X)q \\
&\quad - \frac{1}{2}\|\chi\|^2\gamma(X)q - \frac{1}{2}g_S\left(\chi(X), \chi_l\right)\gamma^l q \\
&= g_S\left(\delta_\gamma\chi, \chi(\mathrm{I}X)\right)\mathrm{I}q - \frac{1}{2}g_S\left(\chi(X), \chi_l\right)\gamma^l q \\
&\quad - \gamma(X)\left(\frac{1}{4}\left(\|P\chi\|^2 + 3\|Q\chi\|^2\right)q + \frac{1}{2}g_S\left(\delta_\gamma\chi, \chi_m\right)\gamma^m q\right).
\end{aligned}$$

For the variation of the gravitino, note that $i^*T_{\alpha\beta}{}^\gamma = 0$, since $i^*d^\gamma_{\alpha\beta} = 0$ and calculate

$$\begin{aligned}
&i^*p_{\mathcal{D}}i^*T(q, diX) + \nabla^{i^*TM}_X q \\
&\quad = q^\mu X^a\left(i^*T_{\mu a}{}^\gamma + \chi_a{}^\alpha T_{\mu\alpha}{}^\gamma\right)i^*F_\gamma + X(q^\mu)s_\mu + q^\mu X^a\left\langle dif_a, i^*\nabla F_\mu\right\rangle \\
&\quad = q^\mu X^a\left(-i^*\omega_a \mathrm{I}_\mu{}^\gamma - i^*d^\gamma_{\mu a}\right)i^*F_\gamma + X(q^\mu)s_\mu + q^\mu X^a\omega_a \mathrm{I}_\mu{}^\gamma\, i^*F_\gamma \\
&\quad = X(q^\mu)s_\mu - q^\mu X^a i^*d^\gamma_{\mu a}s_\gamma
\end{aligned}$$

$$= X(q^\mu)s_\mu - \frac{1}{4}\left(\mathrm{I}_s{}^t\,\omega_t^{LC} + 2g_S\left(\delta_\gamma\chi, \chi_s\right)\right)\gamma^s\gamma(X)q$$

$$= \nabla_X^{LC}q + g_S\left(\delta_\gamma\chi, \chi(\mathrm{I}X)\right)\mathrm{I}q - \frac{1}{4}\left(\mathrm{I}_s{}^t\,\omega_t^{LC} + 2g_S\left(\delta_\gamma\chi, \chi_s\right)\right)\gamma(X)\gamma^s q$$

where the last but one step is justified by Lemma 13.4.9. Hence for the variation of the gravitino, we obtain:

$$\left(\delta_q\chi[m]\right)(X) = \chi\left(\left.\frac{d}{dt}\right|_{t=0} b_t X\right) - i^*p_{\mathcal{D}}i^*T(q, di X) - \nabla_X^{i^*TM}q$$

$$= -\nabla_X^{LC}q - \frac{1}{2}g_S\left(\chi(X), \chi(f_l)\right)\gamma^l q$$

$$+ \frac{1}{4}\gamma(X)\left(\mathrm{I}_s{}^t\,\omega_t^{LC}\gamma^s q - \left(\|P\chi\|^2 + 3\|Q\chi\|^2\right)q\right)$$

This shows the claim. □

13.9 Action

In this section we present a detailed calculation which leads to the proof of Theorem 12.3.1, restated here for the convenience of the reader:

Theorem 12.3.1 *Let M be a fiberwise compact family of super Riemann surfaces and $i\colon |M| \to M$ an underlying even manifold. We denote by g, χ, and g_S respectively the metric, gravitino and spinor metric on $|M|$. Let $\Phi\colon M \to N$ be a morphism to a Riemannian supermanifold (N, n) and φ, ψ, and F its component fields. The action functional $A(\varphi, g, \psi, \chi, F)$ defined by*

$$A(\varphi, g, \psi, \chi, F) = \int_{|M|/B}\left(\|d\varphi\|^2_{g^\vee\otimes\varphi^*n} + g_S^\vee\otimes\varphi^*n\left(\not{D}\psi, \psi\right) - \|F\|^2_{\varphi^*n}\right.$$
$$+ 4g^\vee\otimes\varphi^*n\left(d\varphi, \langle Q\chi, \psi\rangle\right) + \|Q\chi\|^2_{g^\vee\otimes g_S}\|\psi\|^2_{g_S^\vee\otimes\varphi^*n}$$
$$\left.- \frac{1}{6}g_S^\vee\otimes\varphi^*n\left(SR^N(\psi), \psi\right)\right)dvol_g \tag{12.3.2}$$

equals $A(\Phi, m)$.

Here, by slight abuse of notation we denote by $\langle Q\chi, \psi\rangle \in \Gamma\left(T^\vee|M| \otimes \varphi^*TN\right)$ the contraction of $Q\chi$ with ψ along the spinor factor. Furthermore,

$$SR^N(\psi) = s^\alpha \otimes \left({g_S}^{\mu\nu}R^N\left({}_\alpha\psi, {}_\mu\psi\right){}_\nu\psi\right)$$

is the contraction of the pullback of the curvature tensor R^N of the Levi-Civita connection on N along φ with $\psi = s^{\alpha}{}_{\alpha}\psi$.

Proof As has already been outlined in Sect. 12.3, the proof proceeds by calculating local expressions for a Wess–Zumino frame F_A and corresponding Wess–Zumino coordinates $X^A = (x^a, \eta^\alpha)$ on M and y^a on $|M|$. This allows in particular to use Lemma 11.3.4 and its consequences derived in this chapter. The action $A(m, \Phi)$ is given in this Wess–Zumino pair by

$$\begin{aligned} A(\Phi, m) &= \frac{1}{2}\int_{M/B} \| d\Phi|_{\mathcal{D}} \|^2_{m^\vee|_{\mathcal{D}^\vee}\otimes\Phi^* n}[dvol_m] \\ &= \frac{1}{2}\int_{M/B} \varepsilon^{\alpha\beta}\Phi^* n\left(F_\alpha\Phi, F_\beta\Phi\right)\operatorname{Ber} E[dX^\bullet]. \end{aligned}$$

Here Ber E denotes the Berezinian of the inverse of the frame coefficients $F_A{}^B$, see Eq. (13.7.3). By Proposition 8.4.4, there is a top form $|\mathcal{L}|$ such that

$$A(\Phi, m) = \int_{|M|/B} |\mathcal{L}|,$$

where the volume form $|\mathcal{L}|$ can be determined as follows:

$$\begin{aligned} |\mathcal{L}| &= \frac{1}{2}i^*\partial_{\eta^3}\partial_{\eta^4}\left(\varepsilon^{\alpha\beta}\Phi^* n\left(F_\alpha\Phi, F_\beta\Phi\right)\operatorname{Ber} E\right)dy^1\,dy^2 \\ &= \frac{1}{4}i^*\varepsilon^{\alpha\beta}\varepsilon^{\mu\nu}F_\mu F_\nu\left(\Phi^* n(F_\alpha\Phi, F_\beta\Phi)\operatorname{Ber} E\right)dy^1\,dy^2 \\ &= \frac{1}{4}i^*\varepsilon^{\alpha\beta}\varepsilon^{\mu\nu}F_\mu\left(\left(\Phi^* n(\nabla_{F_\nu}F_\alpha\Phi, F_\beta\Phi) - \Phi^* n(F_\alpha\Phi, \nabla_{F_\nu}F_\beta\Phi)\right)\operatorname{Ber} E\right. \\ &\quad \left.+\ \Phi^* n(F_\alpha\Phi, F_\beta\Phi)F_\nu \operatorname{Ber} E\right)dy^1\,dy^2 \\ &= \frac{1}{4}i^*\varepsilon^{\alpha\beta}\varepsilon^{\mu\nu}F_\mu\left(2\Phi^* n(\nabla_{F_\nu}F_\alpha\Phi, F_\beta\Phi)\operatorname{Ber} E\right. \\ &\quad \left.+\ \Phi^* n(F_\alpha\Phi, F_\beta\Phi)F_\nu \operatorname{Ber} E\right)dy^1\,dy^2 \\ &= \frac{1}{4}i^*\varepsilon^{\alpha\beta}\varepsilon^{\mu\nu}\left(2\Phi^* n(\nabla_{F_\mu}\nabla_{F_\nu}F_\alpha\Phi, F_\beta\Phi)\operatorname{Ber} E\right. \\ &\quad +2\Phi^* n(\nabla_{F_\nu}F_\alpha\Phi, \nabla_{F_\mu}F_\beta\Phi)\operatorname{Ber} E - 2\Phi^* n(\nabla_{F_\nu}F_\alpha\Phi, F_\beta\Phi)F_\mu \operatorname{Ber} E \\ &\quad \left.+\ 2\Phi^* n(\nabla_{F_\mu}F_\alpha\Phi, F_\beta\Phi)F_\nu \operatorname{Ber} E + \Phi^* n(F_\alpha\Phi, F_\beta\Phi)F_\mu F_\nu \operatorname{Ber} E\right) \\ &\quad \cdot dy^1\,dy^2 \end{aligned}$$

$$= i^* \varepsilon^{\alpha\beta} \varepsilon^{\mu\nu} \left(\frac{1}{2} \Phi^* n(\nabla_{F_\mu} \nabla_{F_\nu} F_\alpha \Phi, F_\beta \Phi) \operatorname{Ber} E \right.$$

$$+ \frac{1}{2} \Phi^* n(\nabla_{F_\nu} F_\alpha \Phi, \nabla_{F_\mu} F_\beta \Phi) \operatorname{Ber} E + \Phi^* n(\nabla_{F_\mu} F_\alpha \Phi, F_\beta \Phi) F_\nu \operatorname{Ber} E$$

$$\left. + \frac{1}{4} \Phi^* n(F_\alpha \Phi, F_\beta \Phi) F_\mu F_\nu \operatorname{Ber} E \right) dy^1 \, dy^2$$

We now treat the four summands separately using Lemma 13.6.2:

$$\frac{1}{2} i^* \varepsilon^{\alpha\beta} \varepsilon^{\mu\nu} \Phi^* n(\nabla_{F_\mu} \nabla_{F_\nu} F_\alpha \Phi, F_\beta \Phi) \operatorname{Ber} E$$

$$= \varepsilon^{\alpha\beta} \varphi^* n \left({}_\alpha \left(\not{D}^{LC} \psi \right) + \chi_s^\tau \gamma^t{}_\tau{}^\beta \Gamma_{\beta\alpha}^s f_t \varphi + \| Q\chi \|^2 {}_\alpha \psi + \chi_s^\tau \Gamma_{\alpha\tau}^s F \right.$$

$$\left. - \frac{1}{6} {}_\alpha SR^N(\psi), {}_\beta \psi \right) (\det f)^{-1}$$

$$= \left(g_S^\vee \otimes \varphi^* n \left(\not{D}\psi, \psi \right) + 2 g^\vee \otimes \varphi^* n \left(d\varphi, \langle Q\chi, \psi \rangle \right) + \| Q\chi \|^2 \| \psi \|^2 \right.$$

$$\left. + \varphi^* n \left(F, \langle \gamma^t \chi(f_t), \psi \rangle \right) - \frac{1}{6} g_S^\vee \otimes \varphi^* n \left(SR^N(\psi), \psi \right) \right) (\det f)^{-1}$$

$$\frac{1}{2} i^* \varepsilon^{\alpha\beta} \varepsilon^{\mu\nu} \Phi^* n(\nabla_{F_\nu} F_\alpha \Phi, \nabla_{F_\mu} F_\beta \Phi) \operatorname{Ber} E$$

$$= \frac{1}{2} \varepsilon^{\alpha\beta} \varepsilon^{\mu\nu} \varphi^* n \left(\Gamma_{\nu\alpha}^s \left(f_s \varphi + \chi_s^\sigma {}_\sigma \psi \right) - \varepsilon_{\nu\alpha} F, \Gamma_{\mu\beta}^t \left(f_t \varphi + \chi_t^\tau {}_\tau \psi \right) - \varepsilon_{\mu\beta} F \right)$$

$$\cdot (\det f)^{-1}$$

$$= \left(\| d\varphi \|^2 + 2 g^\vee \otimes \varphi^* n \left(d\varphi, \langle \chi, \psi \rangle \right) + \frac{1}{2} \| \chi \|^2 \| \psi \|^2 - \| F \|^2 \right) (\det f)^{-1}$$

$$i^* \varepsilon^{\alpha\beta} \varepsilon^{\mu\nu} \Phi^* n(\nabla_{F_\mu} F_\alpha \Phi, F_\beta \Phi) F_\nu \operatorname{Ber} E$$

$$= -\varepsilon^{\alpha\beta} \varepsilon^{\mu\nu} \varphi^* n \left(\Gamma_{\mu\alpha}^s \left(f_s \varphi + \chi_s^\sigma {}_\sigma \psi \right) - \varepsilon_{\mu\alpha} F, {}_\beta \psi \right) \Gamma_{\nu\tau}^t \chi_t^\tau (\det f)^{-1}$$

$$= \left(-2 g^\vee \otimes \varphi^* n \left(d\varphi, \langle P\chi, \psi \rangle \right) - \| P\chi \|^2 \| \psi \|^2 - \varphi^* n \left(F, \langle \gamma^t \chi(f_t), \psi \rangle \right) \right)$$

$$\cdot (\det f)^{-1}$$

$$\frac{1}{4} i^* \varepsilon^{\alpha\beta} \varepsilon^{\mu\nu} \Phi^* n(F_\alpha \Phi, F_\beta \Phi) F_\mu F_\nu \operatorname{Ber} E = \frac{1}{2} \left(\| P\chi \|^2 - \| Q\chi \|^2 \right) \| \psi \|^2 (\det f)^{-1}$$

Now we can add up the four terms and obtain:

$$
\begin{aligned}
|\mathcal{L}| = i^* \varepsilon^{\alpha\beta} \varepsilon^{\mu\nu} \Big(& \frac{1}{2} \Phi^* n(\nabla_{F_\mu} \nabla_{F_\nu} F_\alpha \Phi, F_\beta \Phi) \operatorname{Ber} E \\
& + \frac{1}{2} \Phi^* n(\nabla_{F_\nu} F_\alpha \Phi, \nabla_{F_\mu} F_\beta \Phi) \operatorname{Ber} E + \Phi^* n(\nabla_{F_\mu} F_\alpha \Phi, F_\beta \Phi) F_\nu \operatorname{Ber} E \\
& + \frac{1}{4} \Phi^* n(F_\alpha \Phi, F_\beta \Phi) F_\mu F_\nu \operatorname{Ber} E \Big) dy^1 dy^2 \\
= \Big(& g_S^\vee \otimes \varphi^* n \left(\not{D} \psi, \psi \right) + 2 g^\vee \otimes \varphi^* n \left(d\varphi, \langle Q\chi, \psi \rangle \right) + \| Q\chi \|^2 \| \psi \|^2 \\
& + \varphi^* n \left(F, \langle \gamma^t \chi(f_t), \psi \rangle \right) - \frac{1}{6} g_S^\vee \otimes \varphi^* n \left(SR^N(\psi), \psi \right) + \| d\varphi \|^2 \\
& + 2 g^\vee \otimes \varphi^* n \left(d\varphi, \langle \chi, \psi \rangle \right) + \frac{1}{2} \| \chi \|^2 \| \psi \|^2 - \| F \|^2 \\
& - 2 g^\vee \otimes \varphi^* n \left(d\varphi, \langle P\chi, \psi \rangle \right) - \| P\chi \|^2 \| \psi \|^2 - \varphi^* n \left(F, \langle \gamma^t \chi(f_t), \psi \rangle \right) \\
& + \frac{1}{2} \left(\| P\chi \|^2 - \| Q\chi \|^2 \right) \| \psi \|^2 \Big) (\det f)^{-1} dy^1 dy^2 \\
= \Big(& \| d\varphi \|^2 + g_S^\vee \otimes \varphi^* n \left(\not{D} \psi, \psi \right) - \| F \|^2 + 4 g^\vee \otimes \varphi^* n \left(d\varphi, \langle Q\chi, \psi \rangle \right) \\
& + \| Q\chi \|^2 \| \psi \|^2 - \frac{1}{6} g_S^\vee \otimes \varphi^* n \left(SR^N(\psi), \psi \right) \Big) \, dvol_g
\end{aligned}
$$

This finishes the proof. □

13.10 Components of $\Delta^{\mathcal{D}} \Phi$

Let $E \to M$ be a vector bundle over a super Riemann surface with connection ∇^E. For any section X of E and embedding $i \colon |M| \to M$, we call the fields

$$
\begin{aligned}
i^* X &\in \Gamma \left(i^* E \right), \\
s^\mu \otimes i^* \nabla^E_{F_\mu} X &\in \Gamma \left(S^\vee \otimes i^* E \right), \\
-\frac{1}{2} i^* \Delta^{\mathcal{D}} X = -\frac{1}{2} i^* m^{\mu\nu} \left(\nabla^E_{F_\mu} \nabla^E_{F_\nu} X - \left(\operatorname{div} F_\mu \right) \nabla^E_{F_\nu} X \right) &\in \Gamma \left(i^* E \right),
\end{aligned}
$$

component fields of X. It is an easy variant of Corollary 11.2.9 and Proposition 12.2.2 that X is completely determined by its component fields. In this section we compute explicitly the component fields of $\Delta^{\mathcal{D}} \Phi^* TN$.

Proposition 13.10.1 *The component fields of $\Delta^{\mathcal{D}}\,\Phi$ are given by*

$$\begin{aligned}
i^*\,\Delta^{\mathcal{D}}\,\Phi &= -2F,\\
s^\mu\otimes\nabla_{F_\mu}\,\Delta^{\mathcal{D}}\,\Phi &= -2\not{D}^{LC}\psi + 4\,d\varphi\left(\vee_{g^\vee\otimes gs}\,Q\chi\right) - 2\|Q\chi\|^2\psi\\
&\quad - \left(2\vee_{gs}\,\delta_\gamma\chi\right)\otimes F + \frac{2}{3}SR^N(\psi),\\
-\frac{1}{2}i^*\,\Delta^{\mathcal{D}}\,\Delta^{\mathcal{D}}\,\Phi &= -2\,\mathrm{Tr}_g\,(\nabla\,d\varphi) - 4\,\mathrm{Tr}_g\,\nabla\,\langle Q\chi,\psi\rangle\\
&\quad + 2\left\langle\delta_\gamma\chi,\not{D}\psi - 2\,d\varphi\left(\vee_{g^\vee\otimes gs}\,Q\chi\right) + \|Q\chi\|^2\psi\right\rangle\\
&\quad + 2\|\chi\|^2F - \frac{1}{3}gs^{\mu\nu}gs^{\alpha\beta}\left(\nabla_{\mu\psi}R^N\right)\left({}_\nu\psi,{}_\alpha\psi\right){}_\beta\psi\\
&\quad - gs^{\alpha\beta}\gamma^k{}_\beta{}^\nu R^N\left({}_\nu\psi,{}_\alpha\psi\right)\left(f_k\varphi + \langle\chi_k,\psi\rangle\right).
\end{aligned}$$

Here, by slight abuse of notation, we denote by $d\varphi\left(\vee_{g^\vee\otimes gs}\,Q\chi\right)\in\Gamma\left(S^\vee\otimes\varphi^*TN\right)$ the field arising from $\vee Q\chi$ by application of $d\varphi$ to the vector part and the identification $\varphi^*TN\otimes S^\vee\simeq S^\vee\otimes\varphi^*TN$.

Proof We have $i^*\,\Delta^{\mathcal{D}}\,\Phi = -2F$ by Definition 12.2.1. The first order term can be calculated with the help of Lemmas 13.6.2 and 13.7.4:

$$\begin{aligned}
i^*\nabla_{F_\mu}\,\Delta^{\mathcal{D}}\,\Phi &= i^*\varepsilon^{\alpha\beta}\nabla_{F_\mu}\nabla_{F_\alpha}F_\beta\Phi + i^*\varepsilon^{\alpha\beta}\left(F_\mu\,\mathrm{div}\,F_\alpha\right)F_\beta\Phi\\
&= -2\,{}_\mu\!\left(\not{D}^{LC}\psi\right) - 2\chi_s^\tau\gamma^t{}_\tau{}^\beta\Gamma^s_{\beta\mu}f_t\varphi - 2\|Q\chi\|^2\,{}_\mu\psi - 2\chi_s^\tau\Gamma^s_{\mu\tau}F\\
&\quad + \frac{2}{3}\,{}_\mu SR^N(\psi) - \frac{1}{2}\gamma^s{}_\mu{}^\sigma\left(\mathrm{I}_s{}^t\,\omega^{LC}_s + 2gs\left(\delta_\mu\chi,\chi_s\right)\right){}_\sigma\psi\\
&\quad + \frac{1}{2}\varepsilon^{\alpha\beta}\Gamma^s_{\mu\alpha}\left(\mathrm{I}_s{}^t\,\omega^{LC}_s + 2gs\left(\delta_\gamma\chi,\chi_s\right)\right){}_\beta\psi\\
&= -2\,{}_\mu\!\left(\not{D}^{LC}\psi\right) - 2\chi_s^\tau\gamma^t{}_\tau{}^\beta\Gamma^s_{\beta\mu}f_t\varphi - 2\|Q\chi\|^2\,{}_\mu\psi - 2\chi_s^\tau\Gamma^s_{\mu\tau}F\\
&\quad + \frac{2}{3}\,{}_\mu SR^N(\psi)
\end{aligned}$$

This completes the calculation of the first-order component field. We now turn to the term of order two using again Lemmas 13.6.2 and 13.7.4:

$$\begin{aligned}
i^*\,\Delta^{\mathcal{D}}\,\Delta^{\mathcal{D}}\,\Phi &= i^*\varepsilon^{\mu\nu}\varepsilon^{\alpha\beta}\left(\nabla_{F_\mu}\nabla_{F_\nu} + \left(\mathrm{div}\,F_\mu\right)\nabla_{F_\nu}\right)\left(\nabla_{F_\alpha}F_\beta\Phi + \left(\mathrm{div}\,F_\alpha\right)F_\beta\Phi\right)\\
&= i^*\varepsilon^{\mu\nu}\varepsilon^{\alpha\beta}\left(\nabla_{F_\mu}\nabla_{F_\nu}\nabla_{F_\alpha}F_\beta\Phi - 2\left(F_\mu\,\mathrm{div}\,F_\alpha\right)\nabla_{F_\nu}F_\beta\Phi\right.\\
&\quad \left.+ \left(F_\mu F_\nu\,\mathrm{div}\,F_\alpha\right)F_\beta\Phi\right)
\end{aligned}$$

$$\begin{aligned}
&= 4\delta^{kl} \left(\nabla_{f_k} f_l \varphi + \left(f_k \chi_l{}^{\lambda}\right)_{\lambda} \psi + 2\chi_k{}^{\kappa} \nabla_{f_l}{}_{\kappa} \psi\right) \\
&\quad + 2\left(f_m \varphi + \langle \chi_m + (P\chi)_m, \psi\rangle\right) \delta^{ms} \left(\mathrm{I}_s{}^t\, \omega_t^{LC} + 2 g_S\left(\delta_\gamma \chi, \chi_s\right)\right) \\
&\quad - 4\|\chi\|^2 F + \varepsilon^{\mu\nu} \varepsilon^{\alpha\beta} \frac{2}{3} \left(\nabla_{\mu\psi} R^N\right) \left({}_\nu \psi, {}_\alpha \psi\right) {}_\beta \psi \\
&\quad + 2\varepsilon^{\alpha\beta} \gamma^k{}_\beta{}^\nu R^N \left({}_\nu \psi, {}_\alpha \psi\right) \left(f_k \varphi + \chi_k{}^\kappa{}_\kappa \psi\right) \\
&\quad - \varepsilon^{\mu\nu} \varepsilon^{\alpha\beta} \Gamma^s_{\mu\alpha} \left(\mathrm{I}_s{}^t\, \omega_t^{LC} + 2 g_S\left(\delta_\gamma \chi, \chi_s\right)\right) \\
&\quad \cdot \left(\Gamma^n_{\nu\beta} \left(f_n \varphi + \chi_n{}^\nu{}_\nu \psi\right) + \varepsilon_{\nu\beta} F\right) - 2\varepsilon^{\alpha\beta}{}_{34}(\operatorname{div} F_\alpha)\, {}_\beta \psi \\
&= 4\delta^{ms} \left(\nabla_{f_m} f_s \varphi + f_m \varphi \left(\mathrm{I}_s{}^t\, \omega_t^{LC} + 2 g_S\left(\delta_\gamma \chi, \chi_s\right)\right)\right) \\
&\quad + 4\delta^{ms} \left(\left(f_m \chi_s{}^\lambda\right)_\lambda \psi + 2\chi_m{}^\kappa \nabla_{f_s}{}_\kappa \psi\right) - 2\varepsilon^{\alpha\beta}{}_{34}(\operatorname{div} F_\alpha)\, {}_\beta \psi \\
&\quad + \langle 4(Q\chi)_m + 6(P\chi)_m, \psi\rangle \delta^{ms} \left(\mathrm{I}_s{}^t\, \omega_t^{LC} + 2 g_S\left(\delta_\gamma \chi, \chi_s\right)\right) \\
&\quad - 4\|\chi\|^2 F + \varepsilon^{\mu\nu} \varepsilon^{\alpha\beta} \frac{2}{3} \left(\nabla_{\mu\psi} R^N\right) \left({}_\nu \psi, {}_\alpha \psi\right) {}_\beta \psi \\
&\quad + 2\varepsilon^{\alpha\beta} \gamma^k{}_\beta{}^\nu R^N \left({}_\nu \psi, {}_\alpha \psi\right) \left(f_k \varphi + \chi_k{}^\kappa{}_\kappa \psi\right)
\end{aligned}$$

Let us now treat the terms involving $d\varphi$ and ψ separately. For the terms of $i^* \Delta^{\mathcal{D}} \Delta^{\mathcal{D}} \Phi$ involving derivatives of φ we have:

$$\begin{aligned}
&4\delta^{ms} \left(\nabla_{f_m} f_s \varphi + f_m \varphi \left(\mathrm{I}_s{}^t\, \omega_t^{LC} + 2 g_S\left(\delta_\gamma \chi, \chi_s\right)\right)\right) \\
&\qquad = 4\,\mathrm{Tr}_g\,(\nabla\, d\varphi) + 8\, d\varphi \left(\langle \delta_\gamma \chi, \vee_{g^\vee \otimes g_S} Q\chi\rangle\right)
\end{aligned}$$

Now we gather all terms involving ψ:

$$\begin{aligned}
&4\delta^{ms} \left(\left(f_m \chi_s{}^\lambda\right)_\lambda \psi + 2\chi_m{}^\kappa \nabla_{f_s}{}_\kappa \psi\right) - 2\varepsilon^{\alpha\beta}{}_{34}(\operatorname{div} F_\alpha)\, {}_\beta \psi \\
&\qquad + \langle 4(Q\chi)_m + 6(P\chi)_m, \psi\rangle \delta^{ms} \left(\mathrm{I}_s{}^t\, \omega_t^{LC} + 2 g_S\left(\delta_\gamma \chi, \chi_s\right)\right) \\
&\quad = 4\delta^{ms} \left(\left(f_m \chi_s{}^\lambda\right)_\lambda \psi + 2\chi_m{}^\kappa \nabla_{f_s}{}_\kappa \psi\right) \\
&\qquad - 2\varepsilon^{\alpha\beta} \left(\mathrm{I}_s{}^t\, \omega_t^{LC} \delta^{sk} \left(2\left(Q\chi\right)_k{}^\kappa + \left(P\chi\right)_k{}^\kappa\right) \varepsilon_{\kappa\alpha}\right. \\
&\qquad \left.+\, 2\varepsilon^{kl} \delta_{\alpha\kappa} \left(f_l \chi_k{}^\kappa\right)\right) {}_\beta \psi \\
&\qquad + \langle 4(Q\chi)_m + 6(P\chi)_m, \psi\rangle \delta^{ms} \left(\mathrm{I}_s{}^t\, \omega_t^{LC} + 2 g_S\left(\delta_\gamma \chi, \chi_s\right)\right)
\end{aligned}$$

$$
\begin{aligned}
&= 4\delta^{ms}\Big(\big(f_m\chi_k{}^{\kappa}\big)\gamma_{s\kappa}{}^{\tau}\gamma^{k}{}_{\tau}{}^{\lambda}{}_{\lambda}\psi + \chi_k{}^{\kappa}\gamma_{s\kappa}{}^{\tau}\gamma^{k}{}_{\tau}{}^{\lambda}\nabla_{f_m\,\lambda}\psi \\
&\quad + \chi_k{}^{\kappa}\gamma^{k}{}_{\kappa}{}^{\tau}\gamma_{s\tau}{}^{\lambda}\nabla_{f_m\,\lambda}\psi\Big) + 8\big\langle(Q\chi)_m + (P\chi)_m, \psi\big\rangle\,\delta^{ms}\,\mathrm{I}_s{}^{t}\,\omega_t^{LC} \\
&\quad + 2\big\langle 4(Q\chi)_m + 6(P\chi)_m, \psi\big\rangle\,\delta^{ms}\,g_S\left(\delta_\gamma\chi, \chi_s\right) \\
&= 8\,\mathrm{Tr}_g\,\nabla\,\langle Q\chi, \psi\rangle - 4\Big\langle\delta_\gamma\chi, \not{D}\psi + \|Q\chi\|^2\psi\Big\rangle
\end{aligned}
$$

The last but one step uses Eq. (13.7.5). This completes the calculation of the second order component fields of $\Delta^{\mathcal{D}}\Phi$. □

Appendix A
Spinors on Riemann Surfaces

In this chapter we gather some formulas and well-known results on Clifford algebras, spinors and spin structures in two dimensions and fix notation and sign conventions. As a byproduct, we remark that the algebraic structures of spinors on Riemann surfaces extend to families of Riemann surfaces over a supermanifold.

The literature on spinors is vast. Details can, for example, be found in Lawson and Michelsohn (1989). Notice, however, that they use slightly different sign conventions.

A.1 Linear Algebra

Let V be a two dimensional real, oriented vector space with a symmetric, non-degenerate, positive bilinear form g and metric volume form $dvol_g$. The vector space V possesses also a canonical almost complex structure I determined by

$$g(\mathrm{I}\,v, w) = dvol_g(v, w).$$

for all $v, w \in V$. Following Deligne and Freed (1999a), we use the convention that the Clifford algebra $\mathrm{Cl}(V, g)$ is the quotient of the tensor algebra of V by the ideal generated by

$$v \otimes w + w \otimes v - 2g(v, w).$$

As the ideal is of even degree in the tensor algebra, the Clifford algebra decomposes in an even and an odd part

$$\mathrm{Cl}(V, g) = \mathrm{Cl}^0(V, g) \oplus \mathrm{Cl}^1(V, g)$$

E. Keßler, *Supergeometry, Super Riemann Surfaces and the Superconformal Action Functional*, Lecture Notes in Mathematics 2230,
https://doi.org/10.1007/978-3-030-13758-8

where the odd part is given by $\mathrm{Cl}^1(V, g) = V$. The even part can be identified with the complex numbers by setting $\omega = -\mathrm{i}$. Here $\omega \in \mathrm{Cl}^0(V, g)$ is the volume form that is given with respect to any orthonormal basis e_1, e_2 of V by $\omega = e_1 e_2$. The sign is chosen so that the multiplication of $i \in \mathrm{Cl}(V, g)$ from the left coincides with I on V. As an algebra $\mathrm{Cl}(V, g)$ is non-canonically isomorphic to $\mathrm{Aut}(\mathbb{R}^2)$ and hence the only irreducible Clifford module is isomorphic to $\mathbb{R}^2$.

The spin group Spin(2) is the multiplicative subgroup of the even part of the Clifford algebra generated by products of vectors of norm one. For any $A \in \mathrm{Spin}(2)$ the map $v \mapsto AvA^{-1}$ from V to $V \subset \mathrm{Cl}^1(V, g)$ is a special orthogonal transformation and establishes Spin(2) as a double cover of SO(2). It is particular to the two dimensional case that $AvA^{-1} = A^2 v$ and hence

$$\begin{aligned} \lambda\colon \mathrm{Spin}(2) &\to \mathrm{SO}(2) \\ A &\mapsto A^2 \end{aligned}$$

is an isomorphism. Furthermore, we will use that also the unitary group U(1) and the symplectic group Sp(2) are isomorphic to SO(2).

Since Spin(2) is isomorphic to SO(2) all irreducible representations of Spin(2) are isomorphic to

$$\begin{aligned} \rho_n\colon \mathrm{Spin}(2) &\to \mathrm{GL}(2) \\ A &\mapsto A^n \end{aligned}$$

We will follow the convention from physics to say that ρ_n is the representation of type $\frac{n}{2}$. In particular the representation of Spin(2) on V given by AvA^{-1} is of type 1. Any representation respects the standard metric, volume form and almost complex structure on $\mathbb{R}^2$.

Let now S be an irreducible module over $\mathrm{Cl}(V, g)$ with Clifford map γ and a Clifford-invariant scalar product g_S. As the spinor module S is considered to be purely odd in this text, we will work with an anti-symmetric (supersymmetric) scalar product g_S on S, that is, $g_S(s, s') = -g_S(s', s)$. Left-multiplication with $i = -\omega \in \mathrm{Cl}^0(V, g)$ induces a complex structure on S.

Let us define the bilinear form Γ by

$$g(\Gamma(t, t'), v) = -g_S(\gamma(v)t, t'). \tag{A.1.1}$$

for all $t, t' \in S$ and $v \in V$. Using the compatibility of the Clifford map γ with the almost complex structures and metrics, one can check that Γ is symmetric, $\Gamma(t, t') = \Gamma(t', t)$ and that Γ is complex linear, that is $\Gamma(\mathrm{I}t, t') = \Gamma(t, \mathrm{I}t') = \mathrm{I}\Gamma(t, t')$. Hence $\Gamma\colon S \otimes_{\mathbb{C}} S \to V$ is a complex linear isomorphism. It follows that for any $u \in \mathrm{U}(1) = \mathrm{SO}(2)$ we have that $\Gamma(us \otimes us) = u^2 \Gamma(s \otimes s)$ and hence Γ is also an isomorphism of U(1)-representations.

In the remainder of this section we present the above algebraic facts in an explicit basis, as is most convenient for the calculations. In any orthonormal basis e_1, e_2

of $\mathbb{R}^2$, the standard Euclidean metric and its volume form $dvol_g$ are given by

$$g(e_a, e_b) = \delta_{ab}, \qquad dvol_g(e_a, e_b) = \varepsilon_{ab},$$

where δ is the Kronecker-delta and ε is the completely anti-symmetric tensor with $\varepsilon_{12} = 1$. Consequently the matrix of the standard almost complex structure I is given by

$$\mathrm{I}\, e_a = \mathrm{I}_a^{\ c}\, e_c = \varepsilon_{ab}\delta^{bc} e_c.$$

The almost complex structure I gives $\mathbb{R}^2$ the structure of a complex vector space that corresponds to the standard identification $\mathbb{C} = \mathbb{R}^2$. Similarly for any orthonormal basis s_α of S,

$$g_S(s_\alpha, s_\beta) = \varepsilon_{\alpha\beta}, \qquad \mathrm{I}\, s_\alpha = \mathrm{I}_\alpha^{\ \tau}\, s_\tau = \varepsilon_{\alpha\beta}\delta^{\beta\tau} s_\tau.$$

The corresponding complex bases are given by

$$e = \frac{1}{2}\left(e_1 - \mathrm{i}e_2\right), \qquad s = \frac{1}{2}\left(s_1 - \mathrm{i}s_2\right).$$

We may assume without loss of generality that $\Gamma(s \otimes s) = e$. Hence in the real basis, the map Γ is given by

$$\Gamma^1 = \begin{pmatrix} 1 & 0 \\ 0 & -1 \end{pmatrix}, \qquad \Gamma^2 = \begin{pmatrix} 0 & 1 \\ 1 & 0 \end{pmatrix}.$$

By Eq. (A.1.1), we have that

$$\gamma_{k\alpha}^{\ \ \ \beta} = \delta_{kl}\Gamma^l_{\alpha\sigma}\varepsilon^{\sigma\beta},$$

and hence the Clifford multiplication is given by

$$\gamma(e_1) = \gamma_1 = \begin{pmatrix} 0 & 1 \\ 1 & 0 \end{pmatrix}, \qquad \gamma(e_2) = \gamma_2 = \begin{pmatrix} -1 & 0 \\ 0 & 1 \end{pmatrix}. \tag{A.1.2}$$

Using the convention $\gamma^l = \gamma_k\delta^{kl}$, the following are consequences of the Clifford relations

$$\begin{aligned}
\gamma^{k\ \beta}_{\ \alpha}\gamma^{l\ \delta}_{\ \beta} &= \delta^{kl}\delta^\delta_\alpha + \varepsilon^{kl}\,\mathrm{I}_\alpha^{\ \delta},\\
\gamma^{k\ \mu}_{\ \alpha}\gamma^{l\ \beta}_{\ \mu}\delta_{kl} &= \Gamma^k_{\alpha\nu}\varepsilon^{\nu\mu}\Gamma^l_{\mu\tau}\varepsilon^{\tau\beta}\delta_{kl} = 2\delta^\beta_\alpha,\\
\gamma^{k\ \mu}_{\ \alpha}\gamma^{l\ \alpha}_{\ \mu} &= \Gamma^k_{\alpha\nu}\varepsilon^{\nu\mu}\Gamma^l_{\mu\tau}\varepsilon^{\tau\alpha} = 2\delta^{kl},
\end{aligned}$$

whereas the following express the compatibility of γ and Γ with the complex structure:

$$\mathrm{I}_a^{\ b}\,\gamma_{b\alpha}^{\ \ \ \delta} s_\delta = \gamma(\mathrm{I}\,e_a)s_\alpha = \gamma(e_a)\,\mathrm{I}\,s_\alpha = \mathrm{I}_\alpha^{\ \beta}\,\gamma_{a\beta}^{\ \ \ \delta} s_\delta,$$
$$\gamma_{a\alpha}^{\ \ \ \beta}\,\mathrm{I}_\beta^{\ \delta}\, s_\delta = \mathrm{I}\,\gamma(e_a)s_\alpha = -\gamma(e_a)\,\mathrm{I}\,s_\alpha = -\,\mathrm{I}_\alpha^{\ \beta}\,\gamma_{a\beta}^{\ \ \ \delta} s_\delta,$$
$$\mathrm{I}_\alpha^{\ \delta}\,\Gamma_{\delta\beta}^k e_k = \Gamma(\mathrm{I}\,s_\alpha, s_\beta) = \mathrm{I}\,\Gamma(s_\alpha, s_\beta) = \Gamma_{\alpha\beta}^l\,\mathrm{I}_l^{\ k}\, e_k,$$
$$\mathrm{I}_\alpha^{\ \delta}\,\Gamma_{\delta\beta}^k e_k = \Gamma(\mathrm{I}\,s_\alpha, s_\beta) = \Gamma(s_\alpha, \mathrm{I}\,s_\beta) = \mathrm{I}_\beta^{\ \delta}\,\Gamma\alpha\delta^k e_k.$$

We will also use the following Fierz identities

$$\gamma_a\gamma^b\gamma^a = \gamma^c\delta_{ac}\gamma^b\gamma^a = 0,$$
$$\gamma^{a\ \ \nu}_{\ \mu}\gamma^{b\ \ \sigma}_{\ \nu}\gamma^{c\ \ \mu}_{\ \sigma} = 0.$$

The first identity is proven as follows

$$\gamma_a\gamma^b\gamma^a = \gamma^c\delta_{ac}\gamma^b\gamma^a = 2\delta^{cb}\delta_{ac}\gamma^a - \gamma^b\gamma^c\gamma^a\delta_{ca} = 2\gamma^b - 2\gamma^b = 0.$$

For the second identity, one can without loss of generality assume that $a = b$. Then

$$\gamma^{a\ \ \nu}_{\ \mu}\gamma^{b\ \ \sigma}_{\ \nu}\gamma^{c\ \ \mu}_{\ \sigma} = \gamma^{c\ \ \mu}_{\ \mu} = 0.$$

A.2 Riemann Surfaces

Let $|M|$ be a Riemann surface, that is, an oriented manifold of dimension $2|0$ together with a metric g on the tangent bundle. The oriented Riemannian volume form $dvol_g$ is a two form on $|M|$ and there exists an almost complex structure I such that

$$g(\mathrm{I}\,X, Y) = dvol_g(X, Y) \tag{A.2.1}$$

for all vector fields X and Y. By the Newlander–Nirenberg-Theorem (see Theorem 7.2.5), the almost complex structure I is actually integrable, turning $|M|$ into a complex manifold of dimension $1|0$. As rescaling the metric in Eq. (A.2.1) leaves the almost complex structure I invariant, one obtains bijections between the sets of conformal classes of metrics, almost complex structures and complex structures on $|M|$.

We will now turn to spinor bundles on $|M|$. Recall that the spin-group in two dimensions Spin(2) is canonically isomorphic to SO(2) and the double cover

$\lambda\colon \mathrm{Spin}(2) \to \mathrm{SO}(2)$ is given by squaring. A spin structure on $|M|$ is the choice of a λ-morphism $P_{\mathrm{Spin}(2)}(g) \to P_{\mathrm{SO}(2)}(g)$, where $P_{\mathrm{Spin}(2)}(g)$ is a Spin(2)-principal bundle over $|M|$ and $P_{\mathrm{SO}(2)}(g)$ is the SO(2)-principal bundle of g-orthonormal frames on $|M|$. The spinor bundle is the associated vector bundle $S = P_{\mathrm{Spin}(2)}(g) \times_\rho \mathbb{R}^2$, where $\rho = \rho_1$ is the defining representation of $\mathrm{Spin}(2) \simeq \mathrm{SO}(2)$ on $\mathbb{R}^2$. As an associated bundle to an SO(2)-principal bundle, the spinor bundle is canonically equipped with an almost complex structure I and a supersymmetric metric g_S. The map Γ (see Eq. (A.1.1)) defines a complex linear isomorphism $S \otimes_{\mathbb{C}} S \to T|M|$, showing that S is actually a holomorphic vector bundle. As a side remark, note that the opposite choice of almost complex structure on S, that is, $i = \omega$, would lead to the identification $S \otimes S \simeq T^\vee|M|$ to be found elsewhere, for example in Atiyah (1971) and Jost et al. (2018a)

A square root of the canonical bundle, that is, a holomorphic line bundle $S^\vee$ such that $S^\vee \otimes_{\mathbb{C}} S^\vee \simeq T^\vee|M|$ determines a spin structure, see, for example, Atiyah (1971, Proposition 3.2). Notice that such a holomorphic square root of the canonical bundle does a priori only depend on the almost complex structure or the conformal class of metrics. However, for any metric in the conformal class, the square root S determines a spin structure to this metric. The spin structure is, for example, determined by choosing for every normed complex frame e of $T|M|$ a frame s of S such that $s \otimes s = e$. This reduces the structure group of the frame bundle of S to Spin(2) and defines via Γ a Clifford module structure on S.

Note that, while the above is best known in the case of classical manifolds, the constructions presented here also hold in the case that $|M|$ is a non-trivial family of supermanifolds of dimension $0|2$ over an arbitrary base B.

A.3 Gravitinos

The gravitino χ is a section of $T^\vee|M| \otimes S$ or, by isometric identification, $T^\vee|M| \otimes S^\vee$. In this section, we will give the decompositions of $T|M| \otimes S$, $T^\vee|M| \otimes S$ and $T^\vee|M| \otimes S^\vee$ in a sum of two irreducible representations of SO(2).

The problem of reducing to irreducible representations is a local one, so it is sufficient to look at the linear algebra case. The quantization map

$$\begin{aligned}\delta_\gamma\colon V \otimes S &\to S\\ v \otimes s &\to \gamma(v)s\end{aligned}$$

is a surjective linear map. All elements $v \otimes s$ such that $\mathrm{I}\,v \otimes s = v \otimes \mathrm{I}\,s$ are in the kernel of δ_γ. Consequently it holds that

$$\ker \delta_\gamma = V \otimes_{\mathbb{C}} S = S \otimes_{\mathbb{C}} S \otimes_{\mathbb{C}} S.$$

One can check that the following definition is independent of the orthonormal basis e_a of V

$$\gamma : S \to V \otimes S$$
$$s \mapsto \delta^{ab} e_a \otimes \gamma(e_b)s$$

and gives an orthogonal splitting of the following short exact sequence

$$0 \longrightarrow \ker \delta_\gamma \longrightarrow V \otimes S \xrightarrow{\delta_\gamma} S \longrightarrow 0$$

Indeed, the projection operators $P = \gamma \circ \delta_\gamma$ and $Q = \mathrm{id} - P$ are self-adjoint because the Clifford map γ is compatible with the metric. Hence the direct sum $V \otimes S = \ker \delta_\gamma \oplus S$ is the decomposition in irreducible representations of Spin(2), where P projects on a representation of type $\frac{1}{2}$ and Q on one of type $\frac{3}{2}$.

Using the conventions of Sect. A.1 the projection operators P and Q are given by

$$P(e_a \otimes s) = \frac{1}{2}\delta^{bc} e_b \otimes \gamma_c \gamma_a s \qquad Q(e_a \otimes s) = \frac{1}{2}\delta^{bc} e_b \otimes \gamma_a \gamma_b s.$$

With the help of the metrics g and g_S the above decomposition can be transferred to $T^\vee|M| \otimes S$ and $T^\vee|M| \otimes S^\vee$. The bundle $T^\vee|M| \otimes S^\vee$ decomposes into

$$T^\vee|M| \otimes S^\vee = S^\vee \otimes_{\mathbb{C}} S^\vee \otimes_{\mathbb{C}} S^\vee \oplus S^\vee.$$

In the case $T^\vee|M| \otimes S$ the kernel of δ_γ is given by all elements $\alpha \otimes s$ such that $\mathrm{I}\alpha \otimes s = -\alpha \otimes \mathrm{I}s$ and hence

$$T^\vee|M| \otimes S = T^\vee|M| \otimes_{\mathbb{C}} \overline{S} \oplus S,$$

where $\overline{S}$ denotes the conjugate almost complex structure on S.

In Chap. 13, we need several component expressions for the norm of χ, $Q\chi$ and $P\chi$ with respect to the metric $g^\vee \otimes g_S$:

$$\begin{aligned}
\|\chi\|^2 &= \|P\chi\|^2 + \|Q\chi\|^2 = g^\vee \otimes g_S(\chi, \chi) \\
&= -\chi_a{}^\alpha \chi_b{}^\beta g^\vee \otimes g_S\left(f^a \otimes s_\alpha, f^b \otimes s_\beta\right) = -\chi_a{}^\alpha \chi_b{}^\beta \delta^{ab} \varepsilon_{\alpha\beta} \\
\|Q\chi\|^2 &= g^\vee \otimes g_S(Q\chi, Q\chi) = g^\vee \otimes g_S(Q\chi, \chi) \\
&= -\frac{1}{2}\chi_a{}^\alpha \gamma_{n\alpha}{}^\mu \gamma^a{}_\mu{}^\nu \chi_b{}^\beta g^\vee \otimes g_s\left(f^n \otimes s_\nu, f^b \otimes s_\beta\right) \\
&= \frac{1}{2}\chi_a{}^\alpha \gamma^b{}_\alpha{}^\mu \Gamma^a_{\mu\beta} \chi_b{}^\beta
\end{aligned}$$

$$\begin{aligned}
\|P\chi\|^2 &= g^\vee \otimes g_S(P\chi, P\chi) = g^\vee \otimes g_S(P\chi, \chi) \\
&= -\frac{1}{2}\chi_a{}^\alpha \gamma^a_\alpha{}^\mu \gamma_{n\mu}{}^\nu \chi_b{}^\beta g^\vee \otimes g_s\left(f^n \otimes s_\nu, f^b \otimes s_\beta\right) \\
&= \frac{1}{2}\chi_a{}^\alpha \gamma^a_\alpha{}^\mu \Gamma^b_{\mu\beta}\chi_b{}^\beta \\
\|P\chi\|^2 - \|Q\chi\|^2 &= \frac{1}{2}\chi_a{}^\alpha \chi_b{}^\beta \left(\gamma^a_\alpha{}^\mu \Gamma^b_{\mu\beta} - \gamma^b_\alpha{}^\mu \Gamma^a_{\mu\beta}\right) = \chi_a{}^\alpha \chi_b{}^\beta \varepsilon^{ab}\delta_{\alpha\beta}
\end{aligned}$$

A.4 Spinors and Change of the Metric

As explained in Sect. A.2, the construction of spinor bundles depends on the Riemannian metric on the surface. In this section we explain how to compare spinors for different metrics, mainly based on the work in Bourguignon and Gauduchon (1992).

Let g and $\tilde{g}$ be two different Riemannian metrics on $T|M|$. There exists a unique, self-adjoint endomorphism $H \in \operatorname{End} T|M|$ such that $\tilde{g}(X, Y) = g(HX, Y)$ for all vector fields X and Y. Setting $b = H^{-\frac{1}{2}}$ yields an isometry of Riemannian vector bundles

$$b\colon (T|M|, g) \to (T|M|, g').$$

Since b is SO-equivariant, it gives an isomorphism of the corresponding principal bundles of orthonormal frames $P_{\mathrm{SO}}(g) \to P_{\mathrm{SO}}(g')$.

If the spin structures $\xi\colon P_{\mathrm{Spin}(2)}(g) \to P_{\mathrm{SO}(2)}(g)$ and $\xi'\colon P_{\mathrm{Spin}(2)}(g') \to P_{\mathrm{SO}(2)}(g')$ represent the same topological spin structure the map b lifts to an equivariant isomorphism of principal bundles $\tilde{b}\colon P_{\mathrm{Spin}(2)}(g) \to P_{\mathrm{Spin}(2)}(g')$ such that $b \circ \xi = \xi' \circ \tilde{b}$. In turn, for ρ the defining representation of Spin(2) on $\mathbb{R}^2$ (as for any other representation) the map $\tilde{b}$ induces a vector bundle isomorphism

$$\beta\colon S_g = P_{\mathrm{Spin}(2)}(g) \times_\rho \mathbb{R}^2 \to S_{g'} = P_{\mathrm{Spin}(2)}(g') \times_\rho \mathbb{R}^2.$$

This construction of β is compatible with Clifford multiplication, that is, $\beta(\gamma(v)s) = \gamma'(b(v))\beta(s)$ where γ' denotes Clifford multiplication on $S_{g'}$. In addition, the map β is an isometry between (S_g, g_S) and $(S_{g'}, g'_S)$.

The map β allows to compare spinors with respect to different metrics. As an example, suppose that g_t is a time-indexed family of Riemannian metrics such that $g_0 = g$ and $\frac{d}{dt}\big|_{t=0} g_t = h$. We obtain a time-indexed family of spinor bundles S_{g_t} and the corresponding Dirac-operators $\partial\!\!\!/_{g_t}\colon \Gamma\left(S_{g_t}\right) \to \Gamma\left(S_{g_t}\right)$. As the operators $\partial\!\!\!/_t$ are defined on different bundles, they cannot be compared directly. Instead, we use the map $\beta_t\colon S_g \to S_{g_t}$ and form the operator $\beta_t{}^{-1}\partial\!\!\!/_{g_t}\beta_t\colon \Gamma\left(S_g\right) \to \Gamma\left(S_g\right)$.

Then, one can calculate for all $s \in \Gamma\left(S_g\right)$, see Bourguignon and Gauduchon (1992, Théorème 21),

$$\left.\frac{d}{dt}\right|_{t=0} \beta_t{}^{-1} \partial\!\!\!/_{g_t} \beta_t s = -\frac{1}{2}\gamma(e_a)\nabla^{LC}_{h_a{}^b e_b} s + \frac{1}{4}\gamma\left(d\mathrm{Tr}_g\, h - \mathrm{div}_g(h)\right).$$

Here $h_a{}^b = h_{ac}g^{cb}$ is the endomorphism associated to h and the indices are taken with respect to the orthonormal basis e_a.

Example A.4.1 (Conformal Rescalings of the Metric) An example that is particularly important for this book is the case of the conformally rescaled metric $g_t = e^{4t\sigma}g$. In this case $H_t X = e^{4t\sigma}X$ and $b_t X = e^{-2t\sigma}X$. The spinor bundles S_g and S_{g_t} are isomorphic as vector bundles and we can assume $g_{S,t} = e^{2t\sigma}g_S$. Consequently, $\beta_t = e^{-t\sigma}\,\mathrm{id}_{S_g}$ and the map Γ constructed in Eq. (A.1.1) is conformally invariant, that is, $\Gamma(\beta_t s, \beta_t \tilde{s}) = b_t \Gamma(s, \tilde{s})$. The Dirac operator $\partial\!\!\!/_{g_t}$ satisfies

$$\partial\!\!\!/_{g_t}\beta s = e^{-2\sigma}\beta\left(\partial\!\!\!/_g s + \gamma(d\sigma)s\right),$$

that is, the Dirac operator is homogeneous under the additional rescaling $\partial\!\!\!/_{g_t}\beta(e^{-\sigma}s) = e^{-3\sigma}\beta(\partial\!\!\!/_g s)$, see Ginoux (2009, Prop. 1.3.10).

Another important case is the one of pull-back metric along a diffeomorphism. Let $\xi\colon |M| \to |M|$ be a diffeomorphism and g_ξ be the corresponding pull-back metric on $T|M|$. The differential $d\xi$ is an isometry between $(T|M|, g_\xi)$ and $(\xi^*T|M|, \xi^*g)$ and hence yields an SO(2)-equivariant bundle isomorphism $\xi_{\mathrm{SO}(2)}\colon P_{\mathrm{SO}(2)}(T|M|, g_\xi) \to \xi^* P_{\mathrm{SO}(2)}(g)$. Let $P_{\mathrm{Spin}(2)}(g_\xi) \to P_{\mathrm{SO}(2)}(g_\xi)$ be the spin structure which is topologically equivalent to $\xi^* P_{\mathrm{Spin}(2)}(g) \to \xi^* P_{\mathrm{SO}(2)}(g)$. Then, the map $\xi_{\mathrm{SO}(2)}$ can be lifted to an equivariant bundle map $\xi_{\mathrm{Spin}(2)}\colon P_{\mathrm{Spin}(2)}(g_\xi) \to \xi^* P_{\mathrm{Spin}(2)}(g)$. For any representation $\mu\colon \mathrm{Spin}(2) \to \mathrm{GL}(V)$ on a vector space V, we obtain an associated map of vector bundles $P_{\mathrm{Spin}(2)}(g_\xi) \times_\mu V \to \xi^*\left(P_{\mathrm{Spin}(2)}(g) \times_\mu V\right)$. In particular, we obtain an isometry $\xi_S\colon S_{g_\xi} \to \xi^* S_g$. For any section $s \in \Gamma\left(S_g\right)$, let us denote by $s_\xi = (\xi_S)^{-1}\xi^* s \in \Gamma\left(S_{g_\xi}\right)$.

Assume now that ξ_t is the flow generated by the vector field $X \in \Gamma\left(T|M|\right)$ and let $g_t = g_{\xi_t}$. Then, $(\beta_t)^{-1}s_{\xi_t}$ is a section of S_g for all times t. Its time derivative at zero is called the Bourguignon–Gauduchon Lie derivative

$$\mathcal{L}_X s = \left.\frac{d}{dt}\right|_{t=0} (\beta_t)^{-1} s_{\xi_t}.$$

As was shown in Bourguignon and Gauduchon (1992), the Bourguignon–Gauduchon Lie derivative of a spinor field can be computed with the help of the Levi-Civita covariant derivative,

$$\mathcal{L}_X s = \nabla^{LC}_X s - \frac{1}{4}\gamma\left(d(\vee X)\right)s. \tag{A.4.2}$$

Furthermore, the Bourguignon–Gauduchon Lie derivative is tensorial and extends to all representations of Spin(2). In particular, for vector fields $X, Y, Z \in \Gamma(T|M|)$, the Bourguignon–Gauduchon Lie derivative compares to the ordinary Lie derivative as follows

$$g(\mathcal{L}_X Y, Z) = g(L_X Y, Z) + \frac{1}{2}(L_X g)(Y, Z). \tag{A.4.3}$$

Recall that for any metric g on a compact $|M|$, a symmetric bilinear form h and any vector field X it holds

$$\int_{|M|} g^\vee \otimes g^\vee (h, L_X g)\, dvol_g = -\int_{|M|} \langle X, \operatorname{div}_g h\rangle\, dvol_g.$$

Here $\operatorname{div}_g h = \operatorname{Tr}_g \nabla^{LC} h \in \Gamma\left(T^\vee|M|\right)$. We are aiming at the following analogous result:

Proposition A.4.4 *For* $\chi, \rho \in \Gamma\left(T^\vee|M| \otimes S^\vee\right)$ *define* $\operatorname{div}_\chi \rho \in \Gamma\left(T^\vee|M|\right)$ *by*

$$\langle X, \operatorname{div}_\chi \rho\rangle = g^\vee \otimes g_S^\vee\left(\nabla_X^{LC}\chi, \rho\right) + \mathrm{I}X\left(g^\vee \otimes g_S^\vee\left(\frac{1}{2}\chi\circ\mathrm{I} - \frac{1}{4}\mathrm{I}\circ\chi, \rho\right)\right),$$

for $X \in \Gamma(T|M|)$ *and where* $\chi\circ\mathrm{I}$, $\mathrm{I}\circ\chi \in \Gamma\left(T^\vee|M| \otimes S^\vee\right)$ *are given by* $(\chi\circ\mathrm{I})(X) = \chi(\mathrm{I}X)$ *and* $(\mathrm{I}\circ\chi)(X) = \mathrm{I}\chi(X)$ *respectively. Then,*

$$\int_{|M|} g^\vee \otimes g_S^\vee(\rho, \mathcal{L}_X\chi)\, dvol_g = \int_{|M|} \langle X, \operatorname{div}_\chi \rho\rangle\, dvol_g.$$

Proof Since the Bourguignon–Gauduchon Lie-derivative respects the metric structure, we have

$$g^\vee \otimes g_S^\vee(\rho, \mathcal{L}_X\chi) = g \otimes g_S(\overline{\rho}, \mathcal{L}_X\overline{\chi}),$$

where $\overline{\rho} = \vee_{g\otimes g_S}$ and $\overline{\chi} = \vee_{g\otimes g_S}\chi$ are the metric duals of ρ and χ respectively. Choose a local orthonormal frame f_a of $T|M|$ and s_α of S. For $\overline{\chi} = \overline{\chi}^{a\beta} f_a \otimes s^\beta$,

$$\mathcal{L}_X\overline{\chi} = X(\overline{\chi}^{a\beta}) f_a \otimes s_\beta + \overline{\chi}^{a\beta}(\mathcal{L}_X f_a)\otimes s_\beta + \overline{\chi}^{a\beta} f_a \otimes \left(\mathcal{L}_X s_\beta\right)$$

Using Eq. (A.4.3) and the fact that $|M|$ has dimension two, we obtain

$$\begin{aligned}
g(\mathcal{L}_X f_a, f_b) &= g(L_X f_a, f_b) + \frac{1}{2}(L_X g)(f_a, f_b) \\
&= g\left(\nabla_X^{LC} f_a, f_b\right) + \frac{1}{2}\left(g\left(f_a, \nabla_{f_b}^{LC} X\right) - g\left(\nabla_{f_a}^{LC} X, f_b\right)\right) \\
&= g\left(\nabla_X^{LC} f_a, f_b\right) + \frac{1}{2}(\operatorname{div}_g \mathrm{I}X)\, g(\mathrm{I} f_a, f_b).
\end{aligned}$$

Similarly, for the spinors we use the Eq. (A.4.2) to obtain

$$g_S\left(\mathcal{L}_X s_\alpha, s_\beta\right) = g_S\left(\nabla_X^{LC} s_\alpha - \frac{1}{4}\gamma\left(d\left(\vee X\right)\right) s_\alpha, s_\beta\right)$$
$$= g_S\left(\nabla_X^{LC} s_\alpha + \frac{1}{4}\left(\mathrm{div}_g\, \mathrm{I}\, X\right) \mathrm{I}\, s_\alpha, s_\beta\right).$$

Consequently,

$$\int_{|M|} g^\vee \otimes g_S^\vee\left(\mathcal{L}_X \chi, \rho\right) dvol_g = \int_{|M|} g \otimes g_S\left(\mathcal{L}_X \overline{\chi}, \overline{\rho}\right) dvol_g$$
$$= \int_{|M|} g \otimes g_S\left(\nabla_X^{LC} \overline{\chi}, \overline{\rho}\right) + \left(\mathrm{div}_g\, \mathrm{I}\, X\right) \overline{\chi}^{a\alpha} \overline{\rho}^{b\beta}$$
$$\cdot\left(\frac{1}{4} g\left(f_a, f_b\right) g_S\left(\mathrm{I}\, s_\alpha, s_\beta\right) - \frac{1}{2} g_S\left(\mathrm{I}\, f_a, f_b\right) g_S\left(s_\alpha, s_\beta\right)\right) dvol_g$$
$$= \int_{|M|} g^\vee \otimes g_S^\vee\left(\nabla_X^{LC} \chi, \rho\right)$$
$$+ \mathrm{I}\, X\left(g^\vee \otimes g_S^\vee\left(\frac{1}{2}\chi \circ \mathrm{I} - \frac{1}{4}\mathrm{I} \circ \chi, \rho\right)\right) dvol_g.$$

□

Appendix B
Supersymmetry in Components

The goal of this chapter is to give a direct verification of the invariance of the action functional $A(\varphi, g, \psi, \chi, F)$ (compare Theorem 12.3.1) under the supersymmetry transformations. For simplicity of the calculations, we will constrain ourselves to the case of a flat target ($R = 0$) and $F = 0$.

This direct verification has two purposes: Even though the calculations are straightforward, they are error prone and usually not presented in all details in the literature. The presentation here may thus be seen as a service to the reader of, for example Deser and Zumino (1976), Brink et al. (1976), D'Hoker and Phong (1988), Jost (2009, Chapter 2.4.7) to quickly verify the claimed supersymmetry.

The other purpose lies in this work itself. The conclusion of Chap. 12 that $A(\varphi, g, \psi, \chi, F)$ is invariant under supersymmetry relies on the long calculations in Chap. 13. The direct verification of the supersymmetry of $A(\varphi, g, \psi, \chi, F)$ yields a cross check of the calculations in Chap. 13. Indeed, the calculations presented here have helped to detect several errors while calculating.

The verification of supersymmetry is done locally on $|M|$, in local coordinates x^a, a local orthonormal frame f_a and a local spin frame s_α that covers f_a. With respect to those frames we denote the coefficients of ψ and χ as follows:

$$\psi = s^\mu{}_\mu\psi \qquad\qquad \chi(f_a) = \chi_a{}^\alpha s_\alpha$$

Note that with respect to the frames f_a and s_α the following tensors are constant:

$$g_{ab} = \delta_{ab} \qquad\qquad gs_{\alpha\beta} = \varepsilon_{\alpha\beta} \qquad\qquad \gamma^k{}_\alpha{}^\beta$$

The supersymmetry variations are parametrized by a spinor $q = q^\mu s_\mu$ and according to Definitions 11.4.5 and 12.2.4 given by

$$\begin{aligned}
\mathrm{susy}_q\,\varphi &= q^\mu{}_\mu\psi,\\
\mathrm{susy}_q\,\psi &= s^\alpha \otimes q^\mu \Gamma^k_{\mu\alpha}\left(f_k\varphi + \chi_k{}^\nu{}_\nu\psi\right),
\end{aligned}$$

E. Keßler, *Supergeometry, Super Riemann Surfaces and the Superconformal Action Functional*, Lecture Notes in Mathematics 2230,
https://doi.org/10.1007/978-3-030-13758-8

$$\mathrm{susy}_q\, f_a = q^\mu \Gamma^k_{\mu\nu} \chi_n{}^\nu \left(\delta_{ka}\delta^{nb} + \delta^n_a \delta^b_k\right) f_b + q^\mu{}_\mu U\, \mathrm{I}_a{}^b\, f_b,$$

$$\left(\mathrm{susy}_q\, \chi\right)(f_a) = -\left(f_a q^\sigma + \frac{1}{2}\omega^{LC}_a q^\mu\, \mathrm{I}_\mu{}^\sigma\right) s_\sigma - \frac{1}{2} g_S\left(\chi_a, \chi_s\right) q^\mu \gamma^s{}_\mu{}^\sigma s_\sigma.$$

Notice that we express the variation of the metric g in terms of the variation of the frame f_a. The supersymmetry transformation of f_a is defined up to an even multiple of $\mathrm{I}_a{}^b$. For $q^\mu{}_\mu U = q^\mu \Gamma^d_{\mu\nu} \mathrm{I}_d{}^n \chi_n{}^\nu$ we obtain the same expression for $\mathrm{susy}_q\, f_a$ as, for example, in Deser and Zumino (1976). For further comment on the differences between $\mathrm{susy}_q\, f_a$ and $\mathrm{susy}_q\, g$ as well as $\left(\mathrm{susy}_q\, \chi\right)(f_a)$ and $\mathrm{susy}_q\, \chi(f_a)$, see Remark 11.4.15.

Recall that the action functional for $F = 0$ and $R = 0$ is given by

$$A(\varphi, g, \psi, \chi) = \int_{|M|/B} \Big(\|d\varphi\|^2_{g^\vee\otimes\varphi^* n} + g_S^\vee \otimes \varphi^* n\left(D\!\!\!/\,\psi, \psi\right)$$
$$+ 4g^\vee \otimes \varphi^* n\left(d\varphi, \langle Q\chi, \psi\rangle\right) + \|Q\chi\|^2_{g^\vee\otimes g_S} \|\psi\|^2_{g_S^\vee\otimes\varphi^* n}\Big)\, dvol_g.$$

We now vary the summands of $A(\varphi, g, \psi, \chi)$ separately:

$$\mathrm{susy}_q \|d\varphi\|^2_{g^\vee\otimes\varphi^* n} = \mathrm{susy}_q\left(\delta^{ab}\varphi^* n\left(f_a\varphi, f_b\varphi\right)\right)$$
$$= 2\delta^{ab}\varphi^* n\left(q^\mu \Gamma^k_{\mu\nu}\chi_n{}^\nu \left(\delta_{ak}\delta^{nc} + \delta^n_a\delta^c_k\right) f_c\varphi + q^\mu{}_\mu U\, \mathrm{I}_a{}^c\, f_c\varphi, f_b\varphi\right)$$
$$+ 2\delta^{ab}\varphi^* n\left(\nabla_{f_a}\left(q^\mu{}_\mu\psi\right), f_b\varphi\right)$$

Here and in the following, ∇ denotes the Levi-Civita covariant derivative on TN with respect to the metric n.

$$\mathrm{susy}_q\, g_S^\vee \otimes \varphi^* n\left(D\!\!\!/\,\psi, \psi\right) = \mathrm{susy}_q\left(-\delta^{kl}\varepsilon^{\alpha\beta}\gamma(f_k)_\alpha{}^\delta \varphi^* n\left(\nabla_{f_l\,\delta}\psi, {}_\beta\psi\right)\right)$$
$$= -\left(q^\mu \Gamma^d_{\mu\nu}\chi_n{}^\nu\left(\delta_{kd}\delta^{nc} + \delta^n_k\delta^c_d\right) + q^\mu{}_\mu U\, \mathrm{I}_k{}^c\right)\delta^{kl}\varepsilon^{\alpha\beta}\gamma_{c\alpha}{}^\delta \varphi^* n\left(\nabla_{f_l\,\delta}\psi, {}_\beta\psi\right)$$
$$- \varepsilon^{\alpha\beta}\gamma^k{}_\alpha{}^\delta \varphi^* n\left(\left(\left(q^\mu \Gamma^d_{\mu\nu}\chi_n{}^\nu\left(\delta_{ld}\delta^{nc} + \delta^n_l\delta^c_d\right) + q^\mu{}_\mu U\, \mathrm{I}_l{}^c\right)\delta^{kl}\nabla_{f_c\,\delta}\psi, {}_\beta\psi\right)\right.$$
$$- \varepsilon^{\alpha\beta}\gamma^k{}_\alpha{}^\delta \varphi^* n\left(\nabla_{f_k}\left(q^\mu \Gamma^m_{\mu\delta}\left(f_m\varphi + \chi_m{}^\nu{}_\nu\psi\right)\right), {}_\beta\psi\right)$$
$$- \varepsilon^{\alpha\beta}\gamma^k{}_\alpha{}^\delta \varphi^* n\left(\nabla_{f_k\,\delta}\psi, q^\mu \Gamma^m_{\mu\beta}\left(f_m\varphi + \chi_m{}^\nu{}_\nu\psi\right)\right)$$

We have obtained in Sect. 11.4 that supersymmetry can be obtained as a time derivative of a time-indexed family of metrics and gravitinos under the canonical isometries. In particular, $Q\,\mathrm{susy}_q\,\chi = \mathrm{susy}_q\, Q\chi$ and $\mathrm{susy}_q\, Q\chi(f_a) = \left(\mathrm{susy}_q\, Q\chi\right)(f_a) + Q\chi(q^\mu{}_\mu U\, \mathrm{I}_a{}^c\, f_c)$. Consequently,

$$\begin{aligned}
\operatorname{susy}_q 4g^\vee \otimes \varphi^* n\left(d\varphi, \langle Q\chi, \psi\rangle\right) &= \operatorname{susy}_q \left(2\chi_p{}^\pi \gamma(f_a)_\pi{}^\tau \gamma^p{}_\tau{}^\beta \delta^{ab} \varphi^* n\left(f_b\varphi, {}_\beta\psi\right)\right) \\
&= 2\Big(-\Big(\left(f_p q^\pi\right) + \frac{1}{2}\omega_p^{LC} q^\mu \, \mathrm{I}_\mu{}^\pi + \frac{1}{2} g_S\left(\chi_p, \chi_s\right) q^\mu \gamma^s{}_\mu{}^\pi\Big) \gamma_{a\pi}{}^\tau \gamma^p{}_\tau{}^\beta \\
&\qquad + q^\mu{}_\mu U \, \mathrm{I}_a{}^c \chi_p{}^\pi \gamma_{c\pi}{}^\tau \gamma^p{}_\tau{}^\beta\Big) \delta^{ab} \varphi^* n\left(f_b\varphi, {}_\beta\psi\right) \\
&\quad + 2\chi_p{}^\pi \gamma^b{}_\pi{}^\gamma \gamma^p{}_\gamma{}^\beta \varphi^* \left(\left(q^\mu \Gamma_{\mu\nu}^r \left(\delta_{rb}\delta^{ns} + \delta_b^n \delta_r^s\right) \chi_n{}^\nu + q^\mu{}_\mu U \, \mathrm{I}_b{}^s\right) f_s\varphi, {}_\beta\psi\right) \\
&\quad + 2\chi_p{}^\pi \gamma^b{}_\pi{}^\gamma \gamma^p{}_\gamma{}^\beta \varphi^* \left(\nabla_{f_b}\left(q^\mu{}_\mu\psi\right), {}_\beta\psi\right) \\
&\quad + 2\chi_p{}^\pi \gamma^b{}_\pi{}^\gamma \gamma^p{}_\gamma{}^\beta \varphi^* n\left(f_b\varphi, q^\mu \Gamma_{\mu\beta}^k \left(f_k\varphi + \chi_k{}^\nu{}_\nu\psi\right)\right).
\end{aligned}$$

Similarly, we have $\operatorname{susy}_q \|Q\chi\|^2 = 2g^\vee \otimes g_S\left(\operatorname{susy}_q \chi, Q\chi\right)$. Hence,

$$\begin{aligned}
\operatorname{susy}_q \|Q\chi\|^2 \|\psi\|^2 &= \operatorname{susy}_q \left(\frac{1}{2}\chi_p{}^\pi \gamma^r{}_\pi{}^\tau \Gamma_{\tau\rho}^p \chi_r{}^\rho \varepsilon^{\alpha\beta} \varphi^* n\left({}_\alpha\psi, {}_\beta\psi\right)\right) \\
&= -\left(\left(f_p q^\pi\right) + \frac{1}{2}\omega_p^{LC} q^\mu \, \mathrm{I}_\mu{}^\pi + \frac{1}{2} g_S\left(\chi_p, \chi_s\right) q^\mu \gamma^s{}_\mu{}^\pi\right) \gamma^r{}_\pi{}^\tau \Gamma_{\tau\rho}^p \chi_r{}^\rho \|\psi\|^2 \\
&\quad + 2\|Q\chi\|^2 \varepsilon^{\alpha\beta} \varphi^* n\left(q^\mu \Gamma_{\mu\alpha}^k \left(f_k\varphi + \chi_k{}^\nu{}_\nu\psi\right), {}_\beta\psi\right)
\end{aligned}$$

The variation of the volume form is given by

$$\begin{aligned}
\operatorname{susy}_q dvol_g &= \operatorname{susy}_q \left(\frac{1}{\det f} dx^1 dx^2\right) \\
&= -2q^\mu \Gamma_{\mu\nu}^k \chi_k{}^\nu \frac{1}{\det f} dx^1 dx^2 = -2q^\mu \Gamma_{\mu\nu}^k \chi_k{}^\nu \, dvol_g.
\end{aligned}$$

We obtain the following additional terms for $\operatorname{susy}_q A(\varphi, g, \psi, \chi)$:

$$\begin{aligned}
&\Big(\|d\varphi\|^2_{g^\vee \otimes \varphi^* n} + g_S^\vee \otimes \varphi^* n\left(\not{D}\psi, \psi\right) + 4g^\vee \otimes \varphi^* n\left(d\varphi, \langle Q\chi, \psi\rangle\right) \\
&\qquad + \|Q\chi\|^2_{g^\vee \otimes g_S} \|\psi\|^2_{g_S^\vee \otimes \varphi^* n}\Big) \left(-2q^\mu \Gamma_{\mu\nu}^k \chi_k{}^\nu\right).
\end{aligned}$$

The variation of the functional $A(\varphi, g, \psi, \chi)$ is given by the integral with respect to $dvol_g$ over the five expressions above. In each summand the terms proportional to $q^\mu{}_\mu U$ vanish, as they should. We will reorder the remaining terms in the integrand of the variation in five different summands that cancel independently: BLUE, VIOLET, GREEN, RED and ORANGE. The summands differ in particular in the order of the gravitino. The blue terms are of order zero in the gravitino:

$$
\begin{aligned}
\text{BLUE} = & \, 2\delta^{ab}\varphi^* n \left(\nabla_{f_a}\left(q^\mu{}_\mu\psi\right), f_b\varphi\right) \\
& - \varepsilon^{\alpha\beta}\gamma^k{}_\alpha{}^\delta \varphi^* n \left(\nabla_{f_k}\left(q^\mu \Gamma^m_{\mu\delta} f_m\varphi\right), {}_\beta\psi\right) \\
& - \varepsilon^{\alpha\beta}\gamma^k{}_\alpha{}^\delta \varphi^* n \left(\nabla_{f_k}{}_\delta\psi, q^\mu \Gamma^m_{\mu\beta} f_m\varphi\right) \\
& - \left(2\left(f_p q^\pi\right) + q^\mu \omega_p^{LC}\, \mathrm{I}_\mu{}^\pi\right)\gamma^b{}_\pi{}^\gamma \gamma^p{}_\gamma{}^\beta \varphi^* n\left(f_b\varphi, {}_\beta\psi\right) \\
= & \, 2\delta^{km}\varphi^* n\left(\left(f_k q^\mu\right){}_\mu\psi + q^\mu \nabla_{f_k}{}_\mu\psi, f_m\varphi\right) \\
& - \gamma^m{}_\mu{}^\alpha \gamma^k{}_\alpha{}^\beta \varphi^* n\left({}_\beta\psi, \left(f_k q^\mu\right) f_m\varphi + q^\mu \nabla_{f_k} f_m\varphi\right) \\
& - q^\mu \gamma^m{}_\mu{}^\alpha \gamma^k{}_\alpha{}^\beta \varphi^* n\left(\nabla_{f_k}{}_\beta\psi, f_m\varphi\right) \\
& - \left(2\left(f_k q^\mu\right) + \omega_k^{LC} q^\nu\, \mathrm{I}_\nu{}^\mu\right)\gamma^m{}_\mu{}^\alpha \gamma^k{}_\alpha{}^\beta \varphi^* n\left({}_\beta\psi, f_m\varphi\right) \\
= & \left(f_k q^\mu\right)\left(2\delta^{km}\delta^\beta_\mu - \gamma^m{}_\mu{}^\alpha\gamma^k{}_\alpha{}^\beta\right)\varphi^* n\left({}_\beta\psi, f_m\varphi\right) \\
& + q^\mu\left(2\delta^{km}\delta^\beta_\mu - \gamma^m{}_\mu{}^\alpha\gamma^k{}_\alpha{}^\beta\right)\varphi^* n\left(\nabla_{f_k}{}_\beta\psi, f_m\varphi\right) \\
& + q^\mu \gamma^m{}_\mu{}^\alpha\gamma^k{}_\alpha{}^\beta \varphi^* n\left({}_\beta\psi, \nabla_{f_k} f_m\varphi\right) \\
& - \omega_k^{LC} q^\mu\, \mathrm{I}_\mu{}^\sigma\left(2\delta^{mk}\delta^\beta_\sigma - \gamma^k{}_\sigma{}^\alpha\gamma^m{}_\alpha{}^\beta\right)\varphi^* n\left({}_\beta\psi, f_m\varphi\right) \\
= & \, f_k\left(q^\mu \gamma^k{}_\mu{}^\alpha \gamma^m{}_\alpha{}^\beta \varphi^* n\left({}_\beta\psi, f_m\varphi\right)\right) \\
& + (\operatorname{div} f_k)\, q^\mu \gamma^k{}_\mu{}^\alpha \gamma^m{}_\alpha{}^\beta \varphi^* n\left({}_\beta\psi, f_m\varphi\right) \\
& + q^\mu \gamma^m{}_\mu{}^\alpha \gamma^k{}_\alpha{}^\beta \varphi^* n\left({}_\beta\psi, \nabla_{f_k} f_m\varphi - \nabla_{f_m} f_k\varphi - [f_m, f_k]\varphi\right) \\
= & \, f_k\left(q^\mu \gamma^k{}_\mu{}^\alpha \gamma^m{}_\alpha{}^\beta \varphi^* n\left({}_\beta\psi, f_m\varphi\right)\right) \\
& + (\operatorname{div} f_k)\, q^\mu \gamma^k{}_\mu{}^\alpha \gamma^m{}_\alpha{}^\beta \varphi^* n\left({}_\beta\psi, f_m\varphi\right)
\end{aligned}
$$

Here in the last but one step we have used that div $f_k = \omega_l^{LC}\, \mathrm{I}_k{}^l$ and $\omega_k^{LC} = -d^a_{12}\delta_{ak}$. Consequently the integral over the blue terms with respect to the volume form $dvol_g$ vanishes.

We now turn to the terms of first order in the gravitino. The violet terms are those that are linear in the gravitino and quadratic in the derivatives of φ:

$$\begin{aligned}
\text{VIOLET} &= 2\delta^{ab}\varphi^*n\left(q^\mu \Gamma^k_{\mu\nu}\chi_n{}^\nu\left(\delta_{ak}\delta^{nc}+\delta^n_a\delta^c_k\right)f_c\varphi,\, f_b\varphi\right)\\
&\quad + 2\chi_p{}^\pi\gamma^b{}_\pi{}^\gamma\gamma^p{}_\gamma{}^\beta\varphi^*n\left(f_b\varphi,\, q^\mu\Gamma^k_{\mu\beta}f_k\varphi\right) - \|d\varphi\|^2_{g^\vee\otimes\varphi^*n}2q^\mu\Gamma^k_{\mu\nu}\chi_k{}^\nu\\
&= 2q^\mu\left(2\delta^{pn}\Gamma^m_{\mu\pi} - \gamma^m{}_\pi{}^\gamma\gamma^p{}_\gamma{}^\beta\Gamma^n_{\beta\mu} - \delta^{mn}\Gamma^p_{\mu\pi}\right)\chi_p{}^\pi\varphi^*n\left(f_m\varphi,\, f_n\varphi\right)\\
&= 2q^\mu\left(\gamma^m{}_\pi{}^\gamma\gamma^n{}_\gamma{}^\beta\Gamma^p_{\beta\mu} - \delta^{mn}\Gamma^p_{\mu\pi}\right)\chi_p{}^\pi\varphi^*n\left(f_m\varphi,\, f_n\varphi\right)\\
&= 0.
\end{aligned}$$

The summand GREEN consists of all terms that are linear in the gravitino and depend on ψ in second order:

$$\begin{aligned}
\text{GREEN} &= -q^\mu\Gamma^d_{\mu\nu}\chi_n{}^\nu\left(\delta_{kd}\delta^{nc}+\delta^n_k\delta^c_d\right)\delta^{kl}\varepsilon^{\alpha\beta}\gamma_{c\alpha}{}^\delta\varphi^*n\left(\nabla_{f_l}\,{}_\delta\psi,\,{}_\beta\psi\right)\\
&\quad - \varepsilon^{\alpha\beta}\gamma_{k\alpha}{}^\delta\varphi^*n\left(q^\mu\Gamma^d_{\mu\nu}\chi_n{}^\nu\left(\delta_{ld}\delta^{nc}+\delta^n_l\delta^c_d\right)\delta^{kl}\nabla_{f_c}\,{}_\delta\psi,\,{}_\beta\psi\right)\\
&\quad - \varepsilon^{\alpha\beta}\gamma^k{}_\alpha{}^\delta\varphi^*n\left(\nabla_{f_k}\left(q^\mu\Gamma^m_{\mu\delta}\chi_m{}^\nu\,{}_\nu\psi\right),\,{}_\beta\psi\right)\\
&\quad - \varepsilon^{\alpha\beta}\gamma^k{}_\alpha{}^\delta\varphi^*n\left(\nabla_{f_k}\,{}_\delta\psi,\, q^\mu\Gamma^m_{\mu\beta}\chi_m{}^\nu\,{}_\nu\psi\right)\\
&\quad + 2\chi_p{}^\pi\gamma^b{}_\pi{}^\gamma\gamma^p{}_\gamma{}^\beta\varphi^*\left(\nabla_{f_b}\left(q^\mu\,{}_\mu\psi\right),\,{}_\beta\psi\right)\\
&\quad - \left(\left(f_p q^\pi\right) + \frac{1}{2}\omega^{LC}_p q^\mu \mathrm{I}_\mu{}^\pi\right)\gamma^r{}_\pi{}^\tau\Gamma^p_{\tau\rho}\chi_r{}^\rho\|\psi\|^2\\
&\quad + g^\vee_S\otimes\varphi^*n\left(\not{D}\psi,\psi\right)\left(-2q^\mu\Gamma^k_{\mu\nu}\chi_k{}^\nu\right).\\
&= q^\mu\left(\Gamma^m_{\mu\nu}\varepsilon^{\sigma\alpha}\gamma^n{}_\alpha{}^\tau + \Gamma^c_{\mu\nu}\varepsilon^{\sigma\alpha}\gamma_{c\alpha}{}^\tau\delta^{mn} + \Gamma^c_{\mu\nu}\varepsilon^{\sigma\alpha}\gamma_{c\alpha}{}^\tau\delta^{mn}\right.\\
&\qquad + \Gamma^m_{\mu\nu}\varepsilon^{\sigma\alpha}\gamma^n{}_\alpha{}^\tau + \gamma^n{}_\mu{}^\alpha\gamma^m{}_\alpha{}^\tau\delta^\sigma_\nu + \gamma^n{}_\mu{}^\alpha\gamma^m{}_\alpha{}^\sigma\delta^\tau_\nu\\
&\qquad \left. - 2\delta^\sigma_\mu\gamma^m{}_\nu{}^\alpha\gamma^n{}_\alpha{}^\tau - 2\Gamma^n_{\mu\nu}\varepsilon^{\sigma\alpha}\gamma^m{}_\alpha{}^\tau\right)\chi_n{}^\nu\varphi^*n\left(\nabla_{f_m}\,{}_\sigma\psi,\,{}_\tau\psi\right)\\
&\quad + \left(\left(f_m q^\mu\right)\left(\frac{1}{2}\gamma^k{}_\mu{}^\delta\Gamma^m_{\delta\nu} + \gamma^k{}_\mu{}^\delta\Gamma^m_{\delta\nu} - \gamma^k{}_\mu{}^\delta\Gamma^m_{\delta\nu}\right)\chi_k{}^\nu\right.\\
&\qquad \left. + \frac{1}{2}q^\mu\gamma^k{}_\mu{}^\delta\Gamma^m_{\delta\nu}\left(f_m\chi_k{}^\nu\right) + \frac{1}{2}q^\mu\gamma^k{}_\mu{}^\delta\Gamma^m_{\delta\nu}\chi_k{}^\nu\mathrm{I}_m{}^l\omega^{LC}_l\right)\|\psi\|^2
\end{aligned}$$

$$
\begin{aligned}
&= q^\mu \big(2\Gamma^m_{\mu\nu}\varepsilon^{\sigma\alpha}\gamma^{n}{}_{\alpha}{}^{\tau} + 2\Gamma^c_{\mu\nu}\varepsilon^{\sigma\alpha}\gamma_{c\alpha}{}^{\tau}\delta^{mn} + \gamma^{n}{}_{\mu}{}^{\alpha}\gamma^{m}{}_{\alpha}{}^{\tau}\delta^\sigma_\nu \\
&\quad + \gamma^{n}{}_{\mu}{}^{\alpha}\gamma^{m}{}_{\alpha}{}^{\sigma}\delta^\tau_\nu - 2\delta^\sigma_\mu\gamma^{m}{}_{\nu}{}^{\alpha}\gamma^{n}{}_{\alpha}{}^{\tau} - 2\Gamma^n_{\mu\nu}\varepsilon^{\sigma\alpha}\gamma^{m}{}_{\alpha}{}^{\tau} \\
&\quad - \gamma^{n}{}_{\mu}{}^{\alpha}\Gamma^m_{\alpha\nu}\varepsilon^{\sigma\tau}\big)\,\chi_n{}^\nu\varphi^* n\left(\nabla_{f_m\,\sigma}\psi, {}_\tau\psi\right) \\
&\quad + f_m\left(\delta^{mn} g_S\left(q, Q\chi_n\right)\|\psi\|^2\right) + (\operatorname{div} f_m)\left(\delta^{mn} g_S(q, Q\chi_n)\|\psi\|^2\right)
\end{aligned}
$$

The coefficient of $\chi_n{}^\nu\varphi^* n\left(\nabla_{f_m\,\sigma}\psi, {}_\tau\psi\right)$ vanishes by a Fierz-type identity. Consequently also the integral over the green term vanishes as it forms a total derivative.

The red terms consist of the terms quadratic in the gravitino.

$$
\begin{aligned}
\text{RED} = &- g_S\left(\chi_p, \chi_s\right) q^\mu\gamma^{s}{}_{\mu}{}^{\pi}\gamma^{b}{}_{\pi}{}^{\gamma}\gamma^{p}{}_{\gamma}{}^{\beta}\varphi^* n\left(f_b\varphi, {}_\beta\psi\right) \\
&+ 2\chi_p{}^\pi\gamma^{b}{}_{\pi}{}^{\gamma}\gamma^{p}{}_{\gamma}{}^{\beta}\varphi^*\left(q^\mu\Gamma^r_{\mu\nu}\left(\delta_{rb}\delta^{ns} + \delta^n_b\delta^s_r\right)\chi_n{}^\nu f_s\varphi, {}_\beta\psi\right) \\
&+ 2\chi_p{}^\pi\gamma^{b}{}_{\pi}{}^{\gamma}\gamma^{p}{}_{\gamma}{}^{\beta}\varphi^* n\left(f_b\varphi, q^\mu\Gamma^k_{\mu\beta}\chi_k{}^\nu{}_\nu\psi\right) \\
&+ 2\|Q\chi\|^2\varepsilon^{\alpha\beta}\varphi^* n\left(q^\mu\Gamma^k_{\mu\alpha} f_k\varphi, {}_\beta\psi\right) \\
&- 8g^\vee\otimes\varphi^* n\left(d\varphi, \langle Q\chi, \psi\rangle\right) q^\mu\Gamma^k_{\mu\nu}\chi_k{}^\nu \\
= &\left(\|\chi\|^2\delta^b_k - 2\delta^{pb} g_S\left(\chi_p, \chi_k\right) + 2\|Q\chi\|^2\delta^b_k\right)\varphi^* n\left(f_b\varphi, \left\langle\gamma^k q, \psi\right\rangle\right) \\
&+ q^\mu\chi_k{}^\kappa\chi_l{}^\lambda\left(2\Gamma^r_{\mu\kappa}\gamma_{r\lambda}{}^{\sigma}\gamma^{l}{}_{\sigma}{}^{\beta}\delta^{kb} + 2\Gamma^b_{\mu\kappa}\gamma^{k}{}_{\lambda}{}^{\sigma}\gamma^{l}{}_{\sigma}{}^{\beta} + 2\delta^\beta_\kappa\gamma^{k}{}_{\mu}{}^{\sigma}\gamma^{l}{}_{\sigma}{}^{\tau}\Gamma^b_{\tau\lambda}\right. \\
&\left.- 4\Gamma^k_{\mu\kappa}\gamma^{b}{}_{\lambda}{}^{\sigma}\gamma^{l}{}_{\sigma}{}^{\beta}\right)\varphi^* n\left(f_b\varphi, {}_\beta\psi\right)
\end{aligned}
$$

Now we use Eq. (13.8.1) to obtain

$$
\begin{aligned}
\text{RED} = &\left(\|\chi\|^2\delta^b_k - 2\delta^{pb} g_S\left(\chi_p, \chi_k\right) + 2\|Q\chi\|^2\delta^b_k\right)\varphi^* n\left(f_b\varphi, \left\langle\gamma^k q, \psi\right\rangle\right) \\
&+ q^\mu\left(\left(\|P\chi\|^2 - \|Q\chi\|^2\right)\left(-\mathrm{I}_\mu{}^\sigma\gamma^{l}{}_{\sigma}{}^{\beta}\mathrm{I}_l{}^b + \gamma^{b}{}_{\mu}{}^{\beta} - \gamma^{b}{}_{\mu}{}^{\beta}\right)\right. \\
&+ g_S\left(\delta_\gamma\chi, \chi_s\right)\left(-\gamma^{r}{}_{\mu}{}^{\pi}\gamma^{s}{}_{\pi}{}^{\lambda}\gamma_{r\lambda}{}^{\sigma}\gamma^{l}{}_{\sigma}{}^{\beta}\mathrm{I}_l{}^b + \gamma^{b}{}_{\mu}{}^{\pi}\gamma^{s}{}_{\pi}{}^{\lambda}\mathrm{I}_\lambda{}^\beta\right. \\
&\left.+ \mathrm{I}_\mu{}^\tau\gamma^{b}{}_{\tau}{}^{\pi}\gamma^{s}{}_{\pi}{}^{\beta} - \varepsilon_{kl}\gamma^{k}{}_{\mu}{}^{\pi}\gamma^{s}{}_{\pi}{}^{\lambda}\gamma^{b}{}_{\lambda}{}^{\sigma}\gamma^{l}{}_{\sigma}{}^{\beta}\right) \\
&\left.- g_S\left(\chi_k, \chi_l\right)\left(2\gamma^{l}{}_{\mu}{}^{\beta}\delta^{kb} + \gamma^{b}{}_{\mu}{}^{\beta}\delta^{kl} - \gamma^{b}{}_{\mu}{}^{\beta}\delta^{kl} - 2\gamma^{k}{}_{\mu}{}^{\lambda}\gamma^{s}{}_{\lambda}{}^{\sigma}\gamma^{l}{}_{\sigma}{}^{\beta}\right)\right) \\
&\cdot\varphi^* n\left(f_b\varphi, {}_\beta\psi\right) \\
= &\; 0.
\end{aligned}
$$

The orange terms are all terms of the variation that contain the gravitino to the third order:

$$\begin{aligned}
\text{ORANGE} &= -\frac{1}{2}gs\left(\chi_p, \chi_s\right) q^\mu \gamma^s{}_\mu{}^\pi \gamma^r{}_\pi{}^\tau \Gamma^p_{\tau\rho} \chi_r{}^\rho \|\psi\|^2 \\
&\quad + 2\|Q\chi\|^2 \varepsilon^{\alpha\beta} \varphi^* n \left(q^\mu \Gamma^k_{\mu\alpha} \chi_k{}^\nu{}_\nu\psi, {}_\beta\psi\right) - 2\|Q\chi\|^2\|\psi\|^2 \left(q^\mu \Gamma^k_{\mu\nu} \chi_k{}^\nu\right) \\
&= q^\mu \left(\frac{1}{2}\|\chi\|^2 \Gamma^r_{\mu\rho} - gs\left(\chi_p, \chi_s\right) \delta^{pr} \Gamma^s_{\mu\rho}\right) \chi_r{}^\rho \|\psi\|^2 \\
&\quad + q^\mu \Gamma^k_{\mu\nu} \chi_k{}^\nu \|Q\chi\|^2 \|\psi\|^2 - 2q^\mu \Gamma^k_{\mu\nu} \chi_k{}^\nu \|Q\chi\|^2 \|\psi\|^2 \\
&= 0
\end{aligned}$$

In the last step we have used $\|P\chi\|^2 \Gamma^s_{\mu\sigma} \chi_s{}^\sigma = 0$, see Eq. (13.7.5) and

$$\begin{aligned}
-gs\left(\chi_p, \chi_s\right) \delta^{pr} \Gamma^s_{\mu\rho} \chi_r{}^\rho &= -\chi_s{}^\sigma \chi_p{}^\pi \varepsilon_{\pi\sigma} \Gamma^s_{\mu\rho} \chi_r{}^\rho \delta^{pr} \\
&= -\frac{1}{2} \varepsilon_{\alpha\beta} \varepsilon^{\pi\rho} \chi_s{}^\sigma \chi_p{}^\alpha \varepsilon_{\pi\sigma} \Gamma^s_{\mu\rho} \chi_r{}^\beta \delta^{pr} = \frac{1}{2}\|\chi\|^2 \Gamma^s_{\mu\sigma} \chi_s{}^\sigma
\end{aligned}$$

The integral over the sum of the terms BLUE, VIOLET, GREEN, RED and ORANGE with respect to $dvol_g$ vanishes as all summands are in the form of a divergence. This completes the proof of the supersymmetry of $A(\varphi, \psi, g, \chi)$.

References

Alldridge, Alexander, Joachim Hilgert, and Wolfgang Palzer. 2012. Berezin integration on non-compact supermanifolds. *Journal of Geometry and Physics* 62 (2): 427–448. https://doi.org/10.1016/j.geomphys.2011.11.005.

Alldridge, Alexander, Joachim Hilgert, and Tillmann Wurzbacher. 2014. Singular superspaces. *Mathematische Zeitschrift* 278 (1–2): 441–492. https://doi.org/10.1007/s00209-014-1323-5.

Alldridge, Alexander, Joachim Hilgert, and Tillmann Wurzbacher. 2016. Superorbits. *Journal of the Institute of Mathematics of Jussieu*. https://doi.org/10.1017/S147474801600030X.

Atiyah, Michael F. 1971. Riemann surfaces and spin structures. *Annales scientifiques de l'École normale supérieure* 4 (1): 47–62. http://www.numdam.org/item?id=ASENS_1971_4_4_1_47_0. Accessed 24 April 2019.

Balduzzi, Luigi, Claudio Carmeli, and Gianni Cassinelli. 2009. Super G-spaces. In *Symmetry in mathematics and physics*. ed. by Donald Babbitt, Vyjayanthi Chari, and Rita Fioresi. Contemporary mathematics, vol. 490, 159–176. Providence: American Mathematical Society. https://doi.org/10.1090/conm/490/09594. arXiv: 0809.3870 [math-ph].

Balduzzi, Luigi, Claudio Carmeli, and Gianni Cassinelli. 2011. Super vector bundles. *Journal of Physics: Conference Series* 284. https://doi.org/10.1088/1742-6596/284/1/012010.

Baranov, M. A., I. V. Frolov, and A. S. Shvarts. 1987. Geometry of twodimensional superconformal field theories. *Theoretical and Mathematical Physics* 70 (1): 64–72. https://doi.org/10.1007/BF01017011.

Bartocci, Claudio, Ugo Bruzzo, and Daniel Hernández-Ruipérez. 1991. *The geometry of supermanifolds*. Mathematics and its applications, vol. 71. Dordrecht: Kluwer Academic Publishers.

Batchelor, Marjorie. 1979. The structure of supermanifolds. *Transactions of the American Mathematical Society* 253: 329–338. https://doi.org/10.2307/1998201.

Batchelor, Marjorie. 1980. Two approaches to supermanifolds. *Transactions of the American Mathematical Society* 258: 257–270. https://doi.org/10.1090/S0002-9947-1980-0554332-9.

Baum, Helga. 2009. *Eichfeldtheorie. Eine Einführung in die Differentialgeometrie auf Faserbündeln*. Berlin: Springer. https://doi.org/10.1007/978-3-540-38293-5.

Berezin, Felix Alexandrovich. 1987. *Introduction to superanalysis*, ed. Alexandre Kirillov. Rev. by D. Leites. Mathematical physics and applied mathematics, vol. 9. Dordrecht: D. Reidel Publishing Company. https://doi.org/10.1007/978-94-017-1963-6.

Bourguignon, Jean-Pierre and Paul Gauduchon. 1992. Spineurs, opérateurs de Dirac et variations de métriques. *Communications in Mathematical Physics* 144 (3): 581–599. https://doi.org/10.1007/BF02099184.

E. Keßler, *Supergeometry, Super Riemann Surfaces and the Superconformal Action Functional*, Lecture Notes in Mathematics 2230,
https://doi.org/10.1007/978-3-030-13758-8

Brink, L., P. Di Vecchia, and P. Howe. 1976. A locally supersymmetric and reparametrization invariant action for the spinning string. *Physics Letters B* 65 (5): 471–474. https://doi.org/10.1016/0370-2693(76)90445-7

Carmeli, Claudio, Lauren Caston, and Rita Fioresi. 2011. *Mathematical foundations of supersymmetry*. Zürich: European Mathematical Society. https://doi.org/10.4171/097.

Chen, Qun et al. 2006. Dirac-harmonic maps. *Mathematische Zeitschrift* 254 (2): 409–432. https://doi.org/10.1007/s00209-006-0961-7.

Crane, Louis and Jeffrey M. Rabin. 1988. Super Riemann surfaces: Uniformization and Teichmüller theory. *Communications in Mathematical Physics* 113 (4): 601–623. https://doi.org/10.1007/BF01223239.

Deligne, Pierre, and Daniel S. Freed 1999a. Sign manifesto. In *Quantum fields and strings: A course for mathematicians*, ed. Pierre Deligne et al. 2 vols, vol. 1, 357–363. Providence: American Mathematical Society.

Deligne, Pierre, and Daniel S. Freed. 1999b. Supersolutions. In *Quantum fields and strings: A course for mathematicians*, ed. Pierre Deligne et al. 2 vols, vol. 1, 227–356. Providence: American Mathematical Society. arXiv: hep-th/9901094.

Deligne, Pierre, and John W. Morgan. 1999. Notes on supersymmetry. (following Joseph Bernstein). In *Quantum fields and strings: A course for mathematicians*, ed. Pierre Deligne et al. 2 vols, vol. 1. Providence: American Mathematical Society.

Deligne, Pierre, Pavel Etingof, et al., eds. 1999. *Quantum fields and strings: A course for mathematicians*. 2 vols. Providence: American Mathematical Society. http://www.math.ias.edu/qft. Accessed 24 April 2019.

Deser, S., and B. Zumino. 1976. A complete action for the spinning string. *Physics Letters B* 65 (4): 369–373. https://doi.org/10.1016/0370-2693(76)90245-8.

DeWitt, Bryce Seligman. 1992. *Supermanifolds*. 2nd ed. Cambridge: Cambridge University Press.

D'Hoker, Eric and D. H. Phong. 1988. The geometry of string perturbation theory. *Reviews of Modern Physics* 60 (4): 917–1065. https://doi.org/10.1103/RevModPhys.60.917.

Donagi, Ron, and Edward Witten. 2015. Supermoduli space is not projected. In *String-math 2012*, ed. Ron Donagi et al., Proceedings of symposia in pure mathematics, vol. 90, 19–71. Providence: American Mathematical Society. https://doi.org/10.1090/pspum/090/01525. arXiv: 1304.7798[hep-th].

Eisenbud, David, and Joe Harris. 2000. *The geometry of schemes*. Graduate texts in mathematics, vol. 197. New York: Springer. https://doi.org/10.1007/b97680.

Fioresi, Rita, and Fabio Gavarini. 2012. Chevalley supergroups. *Memoirs of the American Mathematical Society* 215 (1014). https://doi.org/10.1090/S0065-9266-2011-00633-7.

Fioresi, Rita, and F. Zanchetta. 2017. Representability in supergeometry. *Expositiones Mathematicae* 35 (3), 315–325. https://doi.org/10.1016/j.exmath.2016.10.001.

Friedan, Daniel. 1986. Notes on string theory and two dimensional conformal field theory. In *Unified string theories*. Proceedings, ed. M. B. Green and D. J. Gross, 162–213. Singapore: World Scientific.

Giddings, Steven B., and Philip Nelson. 1987. Torsion constraints and super Riemann surfaces. *Physical Review Letters* 59 (23), 2619–2622. https://doi.org/10.1103/PhysRevLett.59.2619.

Giddings, Steven B., and Philip Nelson. 1988. The geometry of super Riemann surfaces. *Communications in Mathematical Physics* 116 (4), 607–634. https://doi.org/10.1007/BF01224903.

Ginoux, Nicolas. 2009. *The Dirac spectrum*. Lecture notes in mathematics, vol. 1976. Berlin: Springer. https://doi.org/10.1007/978-3-642-01570-0.

Goertsches, Oliver. 2008. Riemannian supergeometry. *Mathematische Zeitschrift* 260 (3): 557–593. https://doi.org/10.1007/s00209-007-0288-z. arXiv:math/0604143.

Green, Paul. 1982. On holomorphic graded manifolds. *Proceedings of the American Mathematical Society* 85 (4): 587–590. https://doi.org/10.1090/S0002-9939-1982-0660609-6.

Grothendieck, Alexander, and Jean Dieudonné. 1960. Éléments de géometrie algébrique. I. Le langage des schémas. *Publications Mathématiques de l'I.H.É.S.* 4: 5–228 (in French). http://www.numdam.org/item?id=PMIHES_1960__4__5_0. Accessed 24 April 2019.

Hanisch, Florian. 2009. Variational problems on supermanifolds. Ph.D. Thesis. Universität Potsdam. http://opus.kobv.de/ubp/volltexte/2012/5975/. Accessed 24 April 2019.

Hartshorne, Robin. 1977. *Algebraic geometry*. Graduate texts in mathematics, vol. 52. New York: Springer. https://doi.org/10.1007/978-1-4757-3849-0.

Hodgkin, Luke. 1987a. A direct calculation of super-Teichmüller space. *Letters in Mathematical Phyiscs* 14 (1): 47–53. https://doi.org/10.1007/BF00403469.

Hodgkin, Luke. 1987b. On metrics and super-Riemann surfaces. *Letters in Mathematical Phyiscs* 14 (2): 177–184. https://doi.org/10.1007/BF00420309.

Howe, P. 1979. Super Weyl transformations in two dimensions. *Journal of Physics A: Mathematical and General* 12 (3): 393–402. https://doi.org/10.1088/0305-4470/12/3/015.

Huybrechts, Daniel. 2005. *Complex geometry. An introduction*. Universitext. Heidelberg: Springer. https://doi.org/10.1007/b137952.

Jost, Jürgen. 2001. *Bosonic strings*. A mathematical treatment. Studies in advanced mathematics, vol. 21. Cambridge: American Mathematical Society and International Press.

Jost, Jürgen. 2006. *Compact Riemann surfaces*. An introduction to contemporary mathematics, 3rd ed. Universitext. Berlin: Springer. https://doi.org/10.1007/978-3-540-33067-7.

Jost, Jürgen. 2009. *Geometry and physics*. Berlin: Springer. https://doi.org/10.1007/978-3-642-00541-1.

Jost, Jürgen. 2011. *Riemannian geometry and geometric analysis*. 6th ed. Universitext. Berlin: Springer. https://doi.org/10.1007/978-3-642-21298-7.

Jost, Jürgen and Shing-Tung Yau. 2010. Harmonic mappings and moduli spaces of Riemann surfaces. *Geometry of Riemann surfaces and their moduli spaces*, ed. Lizhen Ji, Scott A. Wolpert, and Shing-Tung Yau. Surveys in differential geometry, vol. 14, 171–196. Somerville, MA: International Press.

Jost, Jürgen, Enno Keßler, and Jürgen Tolksdorf. 2017a. Super Riemann surfaces, metrics and gravitinos. *Advances in Theoretical and Mathematical Physics* 21 (5): 1161–1187. https://doi.org/10.4310/ATMP.2017.v21.n5.a2. arXiv: 1412.5146 [math-ph].

Jost, Jürgen, Ruijun Wu, and Miaomiao Zhu. 2017b. Coarse regularity of solutions to a nonlinear sigma-model with L^p gravitino. *Calculus of Variations and Partial Differential Equations* 56 (6): 154. https://doi.org/10.1007/s00526-017-1241-6.

Jost, Jürgen, Enno Keßler, Jürgen Tolksdorf, Ruijun Wu, and Miaomiao Zhu. 2018a. Regularity of solutions of the nonlinear sigma model with gravitino. *Communications in Mathematical Physics* 358 (1): 171–197. https://doi.org/10.1007/s00220-017-3001-z.

Jost, Jürgen, Enno Keßler, Jürgen Tolksdorf, Ruijun Wu, and Miaomiao Zhu. 2018b. Symmetries and conservation laws of a nonlinear sigma model with gravitino. *Journal of Geometry and Physics* 128: 185–198. https://doi.org/10.1016/j.geomphys.2018.01.019.

Kähler differential. 2015. nLab. http://ncatlab.org/nlab/revision/K%C3%A4hler+differential/53. Accessed 24 April 2019.

Keßler, Enno. 2016. Super Riemann surfaces and the super conformal action functional. *Quantum mathematical physics. A bridge between mathematics and physics*, ed. Felix Finster et al., 401–419. Basel: Birkhäuser. https://doi.org/10.1007/978-3-319-26902-3_17. arXiv: 1511.05001[math.DG].

Keßler, Enno, and Jürgen Tolksdorf. 2016. The functional of super Riemann surfaces – A "semi-classical" survey. *Vietnam Journal of Mathematics* 44 (1): 215–229. https://doi.org/10.1007/s10013-016-0183-1.

Kobayashi, Shoshichi, and Katsumi Nomizu. 1996. *Foundations of differential geometry*. 2 vols. Wiley Classics Library. New York: Wiley.

Kostant, Bertram. 1977. Graded manifolds, graded Lie theory, and prequantization. In *Differential geometrical methods in mathematical physics*, ed. Konrad Bleuler and Axel Reetz. Lecture notes in mathematics, vol. 570, 177–306. Berlin: Springer. https://doi.org/10.1007/BFb0087788.

Lawson, Herbert Blaine, and Marie-Louise Michelsohn. 1989. *Spin geometry*. Princeton mathematical series, vol. 38. Princeton, NJ: Princeton University Press.

LeBrun, Claude, and Mitchell Rothstein. 1988. Moduli of super Riemann surfaces. *Communications in Mathematical Physics* 117 (1): 159–176. https://doi.org/10.1007/BF01228415.

Leites, D. A. 1980. Introduction to the theory of supermanifolds. *Russian Mathematical Surveys* 35 (1): 1–64. https://doi.org/10.1070/RM1980v035n01ABEH001545.

Lott, John. 1990. Torsion constraints in supergeometry. *Communications in Mathematical Physics* 133 (3): 563–615. https://doi.org/10.1007/BF02097010.

Lück, Wolfgang. 2005. *Algebraische Topologie. Homologie und Mannigfaltigkeiten* (in German). Berlin: Vieweg+Teubner Verlag. https://doi.org/10.1007/978-3-322-80241-5.

Mac Lane, Saunders. 1998. *Categories for the working mathematician*. Graduate texts in mathematics, vol. 5, 2nd ed. New York: Springer. https://doi.org/10.1007/978-1-4757-4721-8.

Manin, Yuri I. 1988. *Gauge field theory and complex geometry*. Grundlehren der mathematischen Wissenschaften, vol. 289. Berlin: Springer.

Manin, Yuri I. 1991. *Topics in noncommutative geometry*. Porter lectures. Princeton: Princeton University Press.

Matsumura, Hideyuki. 1989. *Commutative ring theory*. Paperback ed. Cambridge studies in advanced mathematics, vol. 8. Cambridge: Cambridge University Press.

McHugh, Andrew. 1989. A Newlander–Nirenberg theorem for supermanifolds. *Journal of Mathematical Physics* 30 (5): 1039–1042. https://doi.org/10.1063/1.528373.

Molotkov, Vladimir. 2010. Infinite dimensional and colored supermanifolds. *Journal of Nonlinear Mathematical Physics* 17: 375–446. https://doi.org/10.1142/S140292511000088X.

Munkres, James R. 2000. *Topology*, 2nd ed. Upper Saddle River, NJ: Prentice Hall.

Natanzon, S. M. 2004. *Moduli of Riemann surfaces, real algebraic curves, and their superanalogs*. Translations of mathematical monographs, vol. 225. Providence: American Mathematical Society.

Penner, R. C., and Anton M. Zeitlin. 2015. *Decorated super-Teichmüller Space*. arXiv: 1509.06302 [math.GT]. (pre-published).

Rabin, Jeffrey M. 1995. Super elliptic curves. *Journal of Geometry and Physics* 15 (3): 252–280. https://doi.org/10.1016/0393-0440(94)00012-S.

Rogers, Alice. 2007. *Supermanifolds. Theory and applications*. Singapore: World Scientific Publishing.

Rothstein, Mitchell. 1987. Integration on noncompact supermanifolds. *Transactions of the American Mathematical Society* 299 (1): 387–396. https://doi.org/10.1090/S0002-9947-1987-0869418-5.

Sachse, Christoph. 2009. Global analytic approach to super Teichmüller spaces. Ph.D. Thesis. Universität Leipzig. arXiv: 0902.3289 [math.AG].

Shander, V. N. 1988. Orientations of supermanifolds. *Functional Analysis and Its Applications* 22 (1): 80–82. https://doi.org/10.1007/BF01077738.

Speyer, David. 2009. *Kahler differentials and ordinary differentials*. (answer). MathOverflow. http://mathoverflow.net/q/9723. Accessed 24 April 2019.

Tolksdorf, Jürgen. *Dirac (Type) operators, connections and gauge theories* (in preparation).

Tromba, Anthony J. 1992. *Teichmüller theory in Riemannian geometry*. Lectures in mathematics ETH Zürich. Basel: Birkhäuser Verlag.

Tuynman, Gijs M. 2004. *Supermanifolds and aupergroups*. Basic theory. Mathematics and its applications, vol. 570. Dordrecht: Kluwer Academic Publishers.

Vaintrob, A. Yu. 1988. Almost complex structures on supermanifolds. In *Reports of the Department of Mathematics, University of Stockholm*. Seminar on supermanifolds No 246, ed. D. Leites, 140–144.

Wess, J., and B. Zumino. 1974. Supergauge transformations in four dimensions. *Nuclear Physics B* 70 (1): 39–50. https://doi.org/10.1016/0550-3213(74)90355-1.

Witten, Edward. 2012. *Notes on supermanifolds and integration*. arXiv: 1209.2199 [hep-th]. (pre-published).

Wolf, Michael (1989). The Teichmüller theory of harmonic maps. *Journal of Differential Geometry* 29 (2): 449–479. https://doi.org/10.4310/jdg/1214442885.

Index

E. Keßler, *Supergeometry, Super Riemann Surfaces and the Superconformal Action Functional*, Lecture Notes in Mathematics 2230,
https://doi.org/10.1007/978-3-030-13758-8

LECTURE NOTES IN MATHEMATICS

Springer

Editorial Policy

1. Lecture Notes aim to report new developments in all areas of mathematics and their applications – quickly, informally and at a high level. Mathematical texts analysing new developments in modelling and numerical simulation are welcome.

   Manuscripts should be reasonably self-contained and rounded off. Thus they may, and often will, present not only results of the author but also related work by other people. They may be based on specialised lecture courses. Furthermore, the manuscripts should provide sufficient motivation, examples and applications. This clearly distinguishes Lecture Notes from journal articles or technical reports which normally are very concise. Articles intended for a journal but too long to be accepted by most journals, usually do not have this "lecture notes" character. For similar reasons it is unusual for doctoral theses to be accepted for the Lecture Notes series, though habilitation theses may be appropriate.

2. Besides monographs, multi-author manuscripts resulting from SUMMER SCHOOLS or similar INTENSIVE COURSES are welcome, provided their objective was held to present an active mathematical topic to an audience at the beginning or intermediate graduate level (a list of participants should be provided).

   The resulting manuscript should not be just a collection of course notes, but should require advance planning and coordination among the main lecturers. The subject matter should dictate the structure of the book. This structure should be motivated and explained in a scientific introduction, and the notation, references, index and formulation of results should be, if possible, unified by the editors. Each contribution should have an abstract and an introduction referring to the other contributions. In other words, more preparatory work must go into a multi-authored volume than simply assembling a disparate collection of papers, communicated at the event.

3. Manuscripts should be submitted either online at www.editorialmanager.com/lnm to Springer's mathematics editorial in Heidelberg, or electronically to one of the series editors. Authors should be aware that incomplete or insufficiently close-to-final manuscripts almost always result in longer refereeing times and nevertheless unclear referees' recommendations, making further refereeing of a final draft necessary. The strict minimum amount of material that will be considered should include a detailed outline describing the planned contents of each chapter, a bibliography and several sample chapters. Parallel submission of a manuscript to another publisher while under consideration for LNM is not acceptable and can lead to rejection.

4. In general, **monographs** will be sent out to at least 2 external referees for evaluation.

   A final decision to publish can be made only on the basis of the complete manuscript, however a refereeing process leading to a preliminary decision can be based on a pre-final or incomplete manuscript.

   Volume Editors of **multi-author works** are expected to arrange for the refereeing, to the usual scientific standards, of the individual contributions. If the resulting reports can be forwarded to the LNM Editorial Board, this is very helpful. If no reports are forwarded or if other questions remain unclear in respect of homogeneity etc, the series editors may wish to consult external referees for an overall evaluation of the volume.

5. Manuscripts should in general be submitted in English. Final manuscripts should contain at least 100 pages of mathematical text and should always include

   - a table of contents;
   - an informative introduction, with adequate motivation and perhaps some historical remarks: it should be accessible to a reader not intimately familiar with the topic treated;
   - a subject index: as a rule this is genuinely helpful for the reader.
   - For evaluation purposes, manuscripts should be submitted as pdf files.

6. Careful preparation of the manuscripts will help keep production time short besides ensuring satisfactory appearance of the finished book in print and online. After acceptance of the manuscript authors will be asked to prepare the final LaTeX source files (see LaTeX templates online: https://www.springer.com/gb/authors-editors/book-authors-editors/manuscriptpreparation/5636) plus the corresponding pdf- or zipped ps-file. The LaTeX source files are essential for producing the full-text online version of the book, see http://link.springer.com/bookseries/304 for the existing online volumes of LNM). The technical production of a Lecture Notes volume takes approximately 12 weeks. Additional instructions, if necessary, are available on request from lnm@springer.com.

7. Authors receive a total of 30 free copies of their volume and free access to their book on SpringerLink, but no royalties. They are entitled to a discount of 33.3 % on the price of Springer books purchased for their personal use, if ordering directly from Springer.

8. Commitment to publish is made by a *Publishing Agreement*; contributing authors of multiauthor books are requested to sign a *Consent to Publish form*. Springer-Verlag registers the copyright for each volume. Authors are free to reuse material contained in their LNM volumes in later publications: a brief written (or e-mail) request for formal permission is sufficient.

Addresses:

Professor Jean-Michel Morel, CMLA, École Normale Supérieure de Cachan, France
E-mail: moreljeanmichel@gmail.com

Professor Bernard Teissier, Equipe Géométrie et Dynamique,
Institut de Mathématiques de Jussieu – Paris Rive Gauche, Paris, France
E-mail: bernard.teissier@imj-prg.fr

Springer: Ute McCrory, Mathematics, Heidelberg, Germany,
E-mail: lnm@springer.com

Printed in the United States
By Bookmasters